Use of Modern Materials in Technological Processes Accompanied by Frictional Heating (Volume II)

Use of Modern Materials in Technological Processes Accompanied by Frictional Heating (Volume II)

Guest Editors

Aleksander Yevtushenko
Michal Kuciej
Katarzyna Topczewska

MDPI

Basel • Beijing • Wuhan • Barcelona • Belgrade • Novi Sad • Cluj • Manchester

Guest Editors
Aleksander Yevtushenko
Department of Mechanics and Applied Computer Science
Bialystok University of Technology (BUT)
Bialystok
Poland

Michal Kuciej
Department of Mechanics and Applied Computer Science
Bialystok University of Technology (BUT)
Bialystok
Poland

Katarzyna Topczewska
Department of Mechanics and Applied Computer Science
Bialystok University of Technology (BUT)
Bialystok
Poland

Editorial Office
MDPI AG
Grosspeteranlage 5
4052 Basel, Switzerland

This is a reprint of the Special Issue, published open access by the journal *Materials* (ISSN 1996-1944), freely accessible at: https://www.mdpi.com/journal/materials/special_issues/T39Q16YFV3.

For citation purposes, cite each article independently as indicated on the article page online and as indicated below:

Lastname, A.A.; Lastname, B.B. Article Title. *Journal Name* **Year**, *Volume Number*, Page Range.

ISBN 978-3-7258-5297-0 (Hbk)
ISBN 978-3-7258-5298-7 (PDF)
https://doi.org/10.3390/books978-3-7258-5298-7

Contents

Preface . . . . **vii**

Aleksander Yevtushenko, Katarzyna Topczewska and Przemysław Zamojski
Use of Functionally Graded Material to Decrease Maximum Temperature of a Coating–Substrate System
Reprinted from: *Materials* **2023**, *16*, 2265, https://doi.org/10.3390/ma16062265 . . . . **1**

Aleksander Yevtushenko, Katarzyna Topczewska and Przemysław Zamojski
Influence of Functionally Graded Protective Coating on the Temperature in a Braking System
Reprinted from: *Materials* **2023**, *16*, 4308, https://doi.org/10.3390/ma16124308 . . . . **20**

Aleksander Yevtushenko, Michał Kuciej, Katarzyna Topczewska and Przemysław Zamojski
Effect of Convective Cooling on the Temperature in a Friction System with Functionally Graded Strip
Reprinted from: *Materials* **2023**, *16*, 5228, https://doi.org/10.3390/ma16155228 . . . . **40**

Aleksander Yevtushenko, Katarzyna Topczewska and Przemysław Zamojski
The Mutual Influence of Thermal Contact Conductivity and Convective Cooling on the Temperature Field in a Tribosystem with a Functionally Graded Strip
Reprinted from: *Materials* **2023**, *16*, 7126, https://doi.org/10.3390/ma16227126 . . . . **58**

Krzysztof L. Molski and Piotr Tarasiuk
Stress Concentration Factors for Non-Load-Carrying Welded Cruciform Joints Subjected to Tension, Bending, and Shear
Reprinted from: *Materials* **2024**, *17*, 609, https://doi.org/10.3390/ma17030609 . . . . **83**

Yanning Guo and Peiyao Li
Effect of Residual Stress and Microstructure on the Fatigue Crack Growth Behavior of Aluminum Friction Stir Welded Joints
Reprinted from: *Materials* **2024**, *17*, 385, https://doi.org/10.3390/ma17020385 . . . . **113**

Asri Peni Wulandari, Erlin Karlina, Eric Tanudjaja, Abdul Rohmat, Joko Kusmoro, Muhammad Fadhlillah, et al.
Characteristics and Quality of Flame-Retarded Ramie Fabrics for the Development of Functional Textiles
Reprinted from: *Materials* **2024**, *17*, 1416, https://doi.org/10.3390/ma17061416 . . . . **129**

Mohammad Faseeulla Khan, Ramachandran Damodaram, Hussain Altammar and Gangaraju Manogna Karthik
Metallurgical and Mechanical Properties of Stellite 6 Deposition Developed Through Friction Surfacing Technique
Reprinted from: *Materials* **2025**, *18*, 1003, https://doi.org/10.3390/ma18051003 . . . . **141**

Nenghui Wang, Chuanfeng Wang, Wenhu Xu, Weiping Cheng, Haihong Wu and Hongsheng Li
Friction and Wear Performances and Mechanisms of Graphite/Copper Composites Under Electrical Contact in Marine Environments
Reprinted from: *Materials* **2025**, *18*, 1516, https://doi.org/10.3390/ma18071516 . . . . **156**

Preface

The second volume of the Special Issue "Use of Modern Materials in Technological Processes Accompanied by Frictional Heating" continues the exploration of advanced theoretical models, experimental studies, and technological applications in which friction generates significant thermal and mechanical effects. This reprint includes nine articles that reflect both fundamental investigations—such as the modeling of heat transfer in layered systems containing functionally graded materials (FGMs)—and applied research, including welding technologies, flame-retardant fabrics, advanced coatings, and the tribological behavior of composites under demanding operating conditions.

The primary aim of this collection is to deepen the scientific understanding of frictional heating phenomena while demonstrating how the integration of modern materials can improve the safety, reliability, and performance of engineering systems. The works gathered within are motivated by real-world challenges: the need for efficient and durable braking systems, lightweight and fatigue-resistant welded structures in aerospace, enhanced fire safety in textiles, and improved corrosion and wear resistance in marine and industrial environments.

This reprint is intended for a wide audience, including researchers in materials science, mechanical engineering, and applied physics, as well as engineers and designers working on the practical implementation of advanced materials in friction-intensive processes.

Aleksander Yevtushenko, Michal Kuciej, and Katarzyna Topczewska
Guest Editors

materials

MDPI

Article

Use of Functionally Graded Material to Decrease Maximum Temperature of a Coating–Substrate System

Aleksander Yevtushenko, Katarzyna Topczewska * and Przemysław Zamojski

Faculty of Mechanical Engineering, Bialystok University of Technology (BUT), 45C Wiejska Street, 15-351 Bialystok, Poland

* Correspondence: k.topczewska@pb.edu.pl

Abstract: A mathematical model for determining the temperature distribution in the system consisting of a coating deposited on the surface of substrate was proposed. The foundation material is homogeneous, while the coating is made of a functionally gradient material (FGM) with thermal conductivity increasing exponentially along the thickness. Heating processes of the outer surface of the coating were considered with a constant and linearly decreasing in time intensity of the heat flux. Such thermal loads are common in thermal problems of friction, particularly regarding frictional heating during braking. An exact (in quadrature) solution of the corresponding boundary-value problems of parabolic heat conduction was obtained. Asymptotic solutions to these problems were also found for small and large values of the Fourier number. Calculations were performed for a coating made of two-component FGM ZrO_2—Ti-6Al-4V, applied on a cast iron substrate. In order to explain the effect of FGM on temperature, corresponding analysis was carried out for the coating made of a homogeneous (ZrO_2) material.

Keywords: functionally graded materials; thermal barrier coating; frictional heating; temperature

1. Introduction

Development of protective coatings on the frictional components has revolutionized tribological systems in many industrial applications [1]. Properly selected and produced coatings allow to improve reliability and performance of the systems, by enhancing the tribological properties on the contact surfaces, which is essential in the case of friction cooperative elements. The protective layers called thermal barrier coatings (TBCs) are deposited on the outer surfaces of frictional elements of the heavy-loaded tribosystems, which operate under elevated temperature, to increase the components' durability and alleviate the damage that arises from contact in such extreme conditions. Besides the thermal resistance, the TBCs can protect them from wear and significantly increase the service life of the friction pair elements. In order to get these features, the hard coatings are applied on the relatively soft substrates [2]. Metals and alloys are frequently used as a foundation of the coated elements to maintain structural rigidity and strength. Whereas, ceramic materials are considered as good outer layers for the tribological components, to ensure high hardness, as well as thermal and wear resistance. Miscellaneous particles such as zirconium dioxide (Zr_2O), aluminum oxide (Al_2O_3), silicon carbide (SiC), tungsten carbide (WC), and yttria-stabilized zirconia (YSZ) are employed for coatings of the friction elements in braking systems [3–8]. Ceramic intrinsically possess low fracture toughness, so TBCs are susceptible to mechanical or thermal stress-induced failures such as delamination or spallation of the top coating under harsh thermal conditions [9–11]. Especially, cracks of the brittle ceramic coating may be initiated in the friction elements of braking systems, when the concentration of subsurface tensile stresses appear due to the frictional heating [12]. Many properties of ceramic TBC are very different from that of metallic substrate, so their direct application may lead to cracking due to the thermal expansion mismatch at the interface of the core and the outer layer [13]. The interface of the coating-substrate system has been proved to be the most critical location for failure [1].

In order to overcome the mentioned problems of conventional ceramic coatings, the concept of functionally graded material (FGM) has been applied to the TBCs, obtaining the functionally graded coatings (FGCs). These coatings are characterized by smooth, graded transition of composition from the outer surface to the inside of the element. This feature tend to reduce stress concentrations resulting from materials mismatch, reinforce the bond cohesion, and significantly increase resilience and fracture strength [14–19]. Recently published [20] thermo-mechanical coupled sliding contact analysis of functionally graded coating-substrate structure showed that controlling the gradient parameter of FGC can decrease the residual tensile stress near the friction surface. Additionally, it was found that the functionally graded coating exposed to thermal shock may suffer far less damage than the corresponding homogeneous ceramic cover [21]. The functionality of FGC can be optimized by selecting appropriate gradation of material properties, while an inappropriate gradation can have adverse effects [22]. Therefore, thermal behavior of the friction elements coated with FGCs is an ongoing research topic including heat conduction problems and thermal stress analyses [5,14,20,23–25]. The investigation of thermally induced stress is crucial for predicting the failure mechanism such as fracture and crack propagation in the functionally graded materials [26]. Therefore, it is important to develop computational models used to evaluate the thermal response of functionally graded coatings during the design stage of friction couple.

In general, analytical solutions of the sliding contact problems are difficult, or even impossible, to be found for nonhomogeneous materials with finite dimensions [27,28]. As a consequence, the numerical solutions are primarily used for analysis of FGCs, mostly by means of the finite element approximation [5,10,15,18,23]. For instance, solution of the transient one-dimensional heat conduction equation with the arbitrary type of time-dependent boundary conditions has been obtained in [29] employing the central finite difference method. An analysis of the thermo-mechanical coupled problem in functionally graded metal/ceramic plates has been conducted by means of graded finite elements in [23].

The closed-form, exact solutions of the heat conduction and thermoelasticity problems for FGMs are possible to achieve only for a few cases with particular types of boundary conditions, geometry restrictions, and distributions of material properties [23]. The main issue encountered is concerned with modeling continuously varying gradient of material structure. To overcome this problem, most FGM models use the multi-layered approximation approach [2,19,20,22,23,30]. This method relies on a model of graded material heterogeneity by a package of homogeneous layers, which leads to a stepwise change in coating properties along the gradient direction. Two-dimensional frictional contact problem involving rigid stamps and half-plane with functionally graded coating has been considered in [22], introducing a multi-layer method, the gradation of Young's modulus in coating has been modeled as a piecewise constant function. Using this method, the two-dimensional heat conduction problem for functionally graded circular hollow cylinders subjected to transient thermal boundary conditions with radially dependent properties has been solved in [31]. Another solution of three-dimensional problem of elasticity for functionally graded coated half-space has been achieved in [2].

Our previous research on mathematical modeling of the FGMs friction heating process, presented in articles [24,27,32,33], was based on the semi–infinite body (half–space) model of one or both elements of the friction pair. The main difference between the results presented in this manuscript is the inclusion in the model of the finite thickness of one element—a FGC. It is also important, from the point of view of practical applications, obtaining asymptotic solutions for small and large values of the Fourier number, which do not require numerical integration. It was shown that the form of the obtained exact solution of the problem with uniform heating of the FGM surface of the coating allows for taking into account other than constant time profiles of the heat flux intensity using the Duhamel's theorem. This is important when modeling the temperature mode of braking systems, when the specific friction power, i.e., also the intensity of the heat flux, significantly changes with the braking time [12].

In this study, a thermal problem of friction was considered for a finite coating made of two-component FGM deposited on the semi-infinite homogeneous substrate, subjected to heat flux with constant and time-dependent intensities. An exponential variation of the thermal conductivity of the FGC along its thickness was adopted in the model. Obtained analytical solutions allow to determine the spatial-temporal distribution of temperature generated due to friction. The effect of FGM application was investigated, by comparing the results calculated for a FGC and for corresponding homogeneous ceramic coating.

2. Statement to the Problem

Consider a coated body, with outer layer $0 \le z \le d$ made of two-component FGM deposited on the surface $z = d$ of the homogeneous foundation $z \ge d$ (Figure 1). Thermal conductivity of the coating material K_1 increased exponentially along its thickness [34,35]:

$$K_1(z) = K_{1,1} e^{\gamma^* z/d},\ 0 \le z \le d, \tag{1}$$

where $\gamma^* \ge 0$ is the dimensionless parameter of gradient FGM [36], $K_{1,1} \equiv K_1(0)$ and $K_{1,2} \equiv K_1(d)$ are thermal conductivity coefficients of the FGM components.

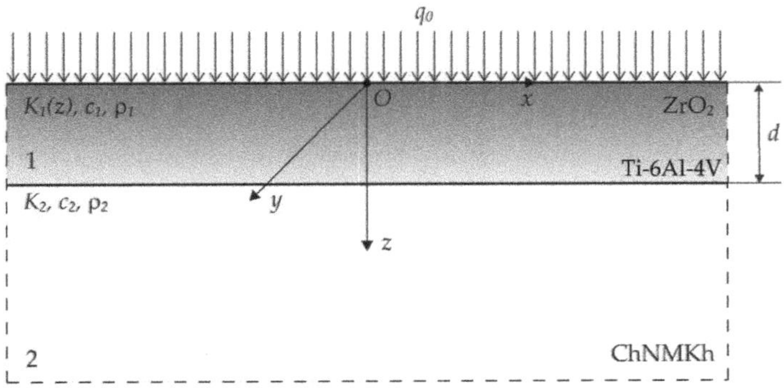

Figure 1. Scheme of heating the coated half-space.

At the initial time moment $t = 0$ the temperature T in the whole semi-infinite region $z \ge 0$ is constant $T = T_0$. Then, over the time $t > 0$ surface of the coating $z = 0$ is subjected to heating by the heat flux with constant intensity q_0. Assuming that the thermal contact between coating and foundation is perfect, the transient temperature field $T(z,t) = T_0 + \Theta(z,t), z \ge 0, t \ge 0$ caused by the heating, can be found from solution to the following boundary-value problem of heat conduction:

$$\frac{\partial}{\partial z}\left[K_1(z)\frac{\partial \Theta(z,t)}{\partial z}\right] = \rho_1 c_1 \frac{\partial \Theta(z,t)}{\partial t},\ 0 < z < d,\ t > 0, \tag{2}$$

$$K_2 \frac{\partial^2 \Theta(z,t)}{\partial z^2} = \rho_2 c_2 \frac{\partial \Theta(z,t)}{\partial t},\ z > d,\ t > 0, \tag{3}$$

$$K_1(z)\frac{\partial \Theta(z,t)}{\partial z}\bigg|_{z=0^+} = -q_0,\ t > 0, \tag{4}$$

$$\Theta(d^+,t) = \Theta(d^-,t),\ t > 0, \tag{5}$$

$$K_1(z)\frac{\partial \Theta(z,t)}{\partial z}\bigg|_{z=d^+} = K_2 \frac{\partial \Theta(z,t)}{\partial z}\bigg|_{z=d^-},\ t > 0, \tag{6}$$

$$\Theta(z,t) \to 0,\ z \to \infty,\ t > 0, \tag{7}$$

$$\Theta(z,0) = 0,\ z \geq 0, \tag{8}$$

where K_2 is the thermal conductivity coefficient of the substrate material; ρ_l, c_l are respectively the density and the specific heat capacities of the materials of coating $(l = 1)$ and substrate $(l = 2)$.

Introducing the dimensionless variables and parameters:

$$\zeta = \frac{z}{d},\ \tau = \frac{k_1 t}{d^2},\ K^* = \frac{K_2}{K_{1,1}},\ k^* = \frac{k_2}{k_1},\ \Theta^* = \frac{\Theta}{\Lambda}, \tag{9}$$

where

$$\Lambda = \frac{q_0 d}{K_{1,1}},\ k_1 = \frac{K_{1,1}}{c_1 \rho_1},\ k_2 = \frac{K_2}{c_2 \rho_2}, \tag{10}$$

the problems (2)–(8) can be written in the form:

$$\frac{\partial^2 \Theta^*(\zeta,\tau)}{\partial \zeta^2} + \gamma^* \frac{\partial \Theta^*(\zeta,\tau)}{\partial \zeta} - e^{-\gamma^* \zeta} \frac{\partial \Theta^*(\zeta,\tau)}{\partial \tau} = 0,\ 0 < \zeta < 1,\ \tau > 0, \tag{11}$$

$$\frac{\partial^2 \Theta^*(\zeta,\tau)}{\partial \zeta^2} - \frac{1}{k^*} \frac{\partial \Theta^*(\zeta,\tau)}{\partial \tau} = 0,\ \zeta > 1,\ \tau > 0, \tag{12}$$

$$\left.\frac{\partial \Theta^*(\zeta,\tau)}{\partial \zeta}\right|_{\zeta=0^+} = -1,\ \tau > 0, \tag{13}$$

$$\Theta^*(1^+,\tau) = \Theta^*(1^-,\tau),\ \tau > 0, \tag{14}$$

$$e^{\gamma^*} \left.\frac{\partial \Theta^*(\zeta,\tau)}{\partial \zeta}\right|_{\zeta=1^+} = K^* \left.\frac{\partial \Theta^*(\zeta,\tau)}{\partial \zeta}\right|_{\zeta=1^-},\ \tau > 0, \tag{15}$$

$$\Theta^*(\zeta,\tau) \to 0\,,\ \zeta \to \infty,\ \tau > 0, \tag{16}$$

$$\Theta^*(\zeta,0) = 0,\ \zeta \geq 0. \tag{17}$$

3. Solution to the Problem

Applying the integral Laplace transform [37]:

$$\overline{\Theta}^*(\zeta,p) \equiv L[\Theta^*(\zeta,\tau);p] = \int_0^\infty \Theta^*(\zeta,\tau) e^{-p\tau} d\tau,\ \mathrm{Re}p \geq 0, \tag{18}$$

into the problems (11)–(17), the following was obtained:

$$\frac{d^2 \overline{\Theta}^*(\zeta,p)}{d\zeta^2} + \gamma^* \frac{d\overline{\Theta}^*(\zeta,\tau)}{d\zeta} - p e^{-\gamma^* \zeta} \overline{\Theta}^*(\zeta,p) = 0,\ 0 < \zeta < 1, \tag{19}$$

$$\frac{d^2 \overline{\Theta}^*(\zeta,p)}{d\zeta^2} - \frac{p}{k^*} \overline{\Theta}^*(\zeta,p) = 0,\ \zeta > 1, \tag{20}$$

$$\left.\frac{d\overline{\Theta}^*(\zeta,p)}{d\zeta}\right|_{\zeta=0^+} = -\frac{1}{p}, \tag{21}$$

$$\overline{\Theta}^*(1^+,p) = \overline{\Theta}^*(1^-,p), \tag{22}$$

$$e^{\gamma^*} \left.\frac{d\overline{\Theta}^*(\zeta,p)}{d\zeta}\right|_{\zeta=1^+} = K^* \left.\frac{d\overline{\Theta}^*(\zeta,p)}{d\zeta}\right|_{\zeta=1^-}, \tag{23}$$

$$\overline{\Theta}^*(\zeta,p) \to 0\,,\ \zeta \to \infty, \tag{24}$$

The general solutions to the differential Equations (19) and (20) have the form:

$$\overline{\Theta}^*(\zeta,p) = \xi\sqrt{p}[A_1(p)\mathrm{I}_1(\xi\sqrt{p}) + B_1(p)\mathrm{K}_1(\xi\sqrt{p})],\ 0 \le \zeta \le 1, \tag{25}$$

$$\overline{\Theta}^*(\zeta,p) = A_2(p)e^{-\varsigma\sqrt{p}} + B_2(p)e^{\varsigma\sqrt{p}},\ \zeta \ge 1, \tag{26}$$

where

$$\xi = \frac{\alpha}{e^{\tilde{\alpha}\zeta}},\ \alpha = \frac{2}{\gamma^*},\ \tilde{\alpha} = \frac{1}{\alpha},\ \varsigma = \frac{\zeta - 1}{\sqrt{k^*}}, \tag{27}$$

$\mathrm{I}_n(x)$, $\mathrm{K}_n(x)$ here and further are modified Bessel functions of the nth order of the first and second kind, respectively [38]. Unknown functions $A_l(p)$ and $B_l(p)$, $l = 1,2$ in the solutions (25), (26) were found from the boundary conditions (21)–(24) in the form:

$$A_1(p) = \frac{\Delta_{A_1}(p)}{\alpha\, p^2\Delta(p)},\ B_1(p) = \frac{\Delta_{B_1}(p)}{\alpha\, p^2\Delta(p)},\ A_2(p) = \frac{e^{-\tilde{\alpha}}\Delta_{A_2}(p)}{p\sqrt{p}\Delta(p)},\ B_2(p) = 0, \tag{28}$$

where

$$\Delta(p) = \Delta_{A_1}(p)\mathrm{I}_0(\alpha\sqrt{p}) - \Delta_{B_1}(p)\mathrm{K}_0(\alpha\sqrt{p}), \tag{29}$$

$$\Delta_{A_1}(p) = \mathrm{K}_0(\beta\sqrt{p}) + \varepsilon\, e^{-\tilde{\alpha}}\mathrm{K}_1(\beta\sqrt{p}),\ \Delta_{B_1}(p) = \mathrm{I}_0(\beta\sqrt{p}) - \varepsilon\, e^{-\tilde{\alpha}}\mathrm{I}_1(\beta\sqrt{p}), \tag{30}$$

$$\Delta_{A_2}(p) = \mathrm{I}_0(\beta\sqrt{p})\mathrm{K}_1(\beta\sqrt{p}) + \mathrm{I}_1(\beta\sqrt{p})\mathrm{K}_0(\beta\sqrt{p}) \equiv (\beta\sqrt{p})^{-1}, \tag{31}$$

$$\beta = \frac{\alpha}{e^{\tilde{\alpha}}},\ \varepsilon = \frac{K^*}{\sqrt{k^*}}. \tag{32}$$

Considering Equations (28)–(32), solutions (25)–(27) took the form:

$$\overline{\Theta}^*(\zeta,p) = e^{-\tilde{\alpha}\zeta}\,\overline{\Theta}_0^*(p)\overline{\Theta}_1^*(\zeta,p), 0 \le \zeta \le 1,\ \overline{\Theta}^*(\zeta,p) = \tilde{\alpha}\,\overline{\Theta}_0^*(p)\overline{\Theta}_2^*(\zeta,p), \zeta \ge 1, \tag{33}$$

where

$$\overline{\Theta}_0^*(p) = \frac{1}{\sqrt{p}},\ \overline{\Theta}_1^*(\zeta,p) = \frac{\Delta_1(\zeta,p)}{p\Delta(p)},\ \overline{\Theta}_2^*(\zeta,p) = \frac{e^{-\varsigma\sqrt{p}}}{p\sqrt{p}\Delta(p)}, \tag{34}$$

$$\Delta_1(\zeta,p) = \Delta_{A_1}(p)\mathrm{I}_1(\xi\sqrt{p}) + \Delta_{B_1}(p)\mathrm{K}_1(\xi\sqrt{p}). \tag{35}$$

Based on the convolution of two functions theorem [39], Formulas (33)–(35) yield:

$$\Theta^*(\zeta,\tau) = e^{-\tilde{\alpha}\zeta}\int_0^{\tau}\Theta_0^*(\tau - s)\Theta_1^*(\zeta,s)ds,\ 0 \le \zeta \le 1,\ \tau \ge 0, \tag{36}$$

$$\Theta^*(\zeta,\tau) = \tilde{\alpha}\int_0^{\tau}\Theta_0^*(\tau - s)\Theta_2^*(\zeta,s)ds,\ \zeta \ge 1,\ \tau \ge 0, \tag{37}$$

where

$$\Theta_0^*(\tau) \equiv L^{-1}[\overline{\Theta}_0^*(p);\tau] = \frac{1}{\sqrt{\pi\tau}}, \tag{38}$$

$$\Theta_l^*(\zeta,\tau) \equiv L^{-1}[\overline{\Theta}_l^*(\zeta,p);\ \tau] = \frac{1}{2\pi i}\int_{\omega - i\infty}^{\omega + i\infty}\overline{\Theta}_l^*(\zeta,p)e^{p\tau}dp,\ l = 1,2,\ \omega \equiv \mathrm{Re}p > 0, i \equiv \sqrt{-1}. \tag{39}$$

Integration on the complex plane (Rep, Im p) in Equation (39) was performed along the closed curve Γ (Figure 2). It consists of the segment Γ_ω of a straight line Re $p = \omega$, the two arcs Γ_R and Γ_δ centered at a point $p = 0$ with radiuses R and δ, respectively, and a cut along the axis Re $p < 0$ with boundaries $\Gamma_\pm$. Within the contour Γ the integral functions $\overline{\Theta}_l^*(\zeta,p), l = 1,2$ (34) in the relation (39) are analytical and unambiguous, then by Cauchy's theorem [40] we have:

$$\frac{1}{2\pi i}\oint_\Gamma\overline{\Theta}_l^*(\zeta,p)e^{p\tau}dp = 0,\ l = 1,2. \tag{40}$$

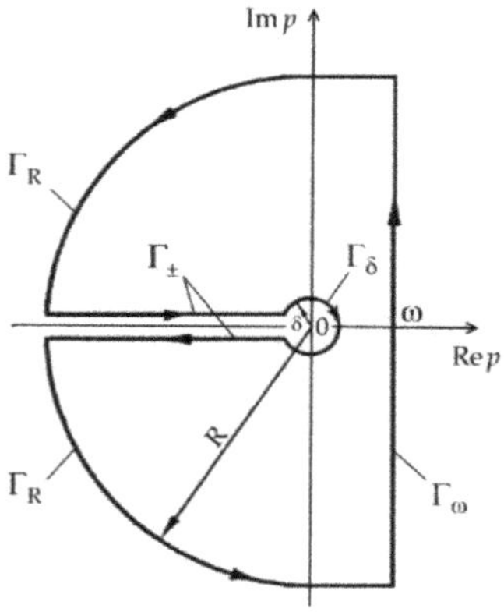

Figure 2. Integration contour.

Since the functions $\overline{\Theta}_l^*(\zeta, p), l = 1, 2$ (34) satisfy the Jordan's lemma conditions [37], integrals along the arc Γ_R tend to be zero for $R \to \infty$ and from Equations (39) and (40), follows that:

$$\Theta_l^*(\zeta, \tau) = -\Theta_{l,+}^*(\zeta, \tau) - \Theta_{l,-}^*(\zeta, \tau) - \Theta_{l,\delta}^*(\zeta, \tau), \ \tau \geq 0, \ l = 1, 2, \tag{41}$$

where

$$\Theta_{l,\pm}^*(\zeta, \tau) = \frac{1}{2\pi i} \int_{\Gamma_\pm} \overline{\Theta}_l^*(\zeta, p) e^{p\tau} dp, \ \Theta_{l,\delta}^*(\zeta, \tau) = \frac{1}{2\pi i} \int_{\Gamma_\delta} \overline{\Theta}_l^*(\zeta, p) e^{p\tau} dp, \ \tau \geq 0. \tag{42}$$

Proceeding to the polar system (r, ϕ) centered at a point $p = 0$, parameter of the Laplace transform $p = re^{i\phi}$, $r \geq 0$, $|\phi| \leq \pi$. Then, on the boundaries $\Gamma_\pm$ we received $p = re^{\pm i\pi} = -r$, $\sqrt{p} = \pm i\sqrt{r}$ and the first integrals from Equation (42) can be written as:

$$\Theta_{l,\pm}^*(\zeta, \tau) = \pm \frac{1}{2\pi i} \int_0^\infty \overline{\Theta}_{l,\pm}^*(\zeta, r) e^{-r\tau} dr, \ \tau \geq 0, \ l = 1, 2, \tag{43}$$

where $\overline{\Theta}_{l,\pm}^*(\zeta, r) \equiv \overline{\Theta}_l^*(\zeta, re^{\pm i\pi})$.

With consideration of the relations [38]:

$$\mathrm{I}_0(\pm ix) = \mathrm{J}_0(x), \ \mathrm{K}_0(\pm ix) = -0.5\pi[\mathrm{Y}_0(x) \pm i\mathrm{J}_0(x)], \tag{44}$$

$$\mathrm{I}_1(\pm ix) = \pm i\mathrm{J}_1(ix), \ \mathrm{K}_1(\pm ix) = -0.5\pi[\mathrm{J}_1(x) \mp i\mathrm{Y}_1(x)], \tag{45}$$

where $\mathrm{J}_n(x)$ and $\mathrm{Y}_n(x)$ are Bessel functions of the nth order of the first and second kind, respectively, from formulae (29)–(35) it was found:

$$\overline{\Theta}_{1,\pm}^*(\zeta, r) = \frac{\Delta_1^\pm(\zeta, r)}{r\,\Delta^\pm(r)}, \ 0 \leq \zeta \leq 1, \ \overline{\Theta}_{2,\pm}^*(\zeta, r) = \frac{\Delta_2^\pm(\zeta, r)}{r\sqrt{r}\,\Delta^\pm(r)}, \ \zeta \geq 1, \tag{46}$$

where

$$\Delta^\pm(r) = 0.5\pi[\Delta_\mathrm{R}(r) \mp i\varepsilon\, e^{-\tilde{\alpha}} \Delta_\mathrm{I}(r)], \tag{47}$$

$$\Delta_1^\pm(\zeta, r) = 0.5\pi[\varepsilon\, e^{-\tilde{\alpha}} \Delta_{1,\mathrm{R}}(\zeta, r) \pm i\Delta_{1,\mathrm{I}}(\zeta, r)], \tag{48}$$

$$\Delta_2^\pm(\zeta, r) = \sin(\varsigma\sqrt{r}) \pm i\cos(\varsigma\sqrt{r}), \tag{49}$$

$$\Delta_\mathrm{R}(r) = \mathrm{Y}_0(\alpha\sqrt{r})\mathrm{J}_0(\beta\sqrt{r}) - \mathrm{J}_0(\alpha\sqrt{r})\mathrm{Y}_0(\beta\sqrt{r}), \tag{50}$$

$$\Delta_{\mathrm{I}}(r) = \mathrm{Y}_0(\alpha\sqrt{r})\mathrm{J}_1(\beta\sqrt{r}) - \mathrm{J}_0(\alpha\sqrt{r})\mathrm{Y}_1(\beta\sqrt{r}), \tag{51}$$

$$\Delta_{1,\mathrm{R}}(\zeta, r) = \mathrm{Y}_1(\beta\sqrt{r})\mathrm{J}_1(\xi\sqrt{r}) - \mathrm{J}_1(\beta\sqrt{r})\mathrm{Y}_1(\xi\sqrt{r}), \tag{52}$$

$$\Delta_{1,\mathrm{I}}(\zeta, r) = \mathrm{Y}_0(\beta\sqrt{r})\mathrm{J}_1(\xi\sqrt{r}) - \mathrm{J}_0(\beta\sqrt{r})\mathrm{Y}_1(\xi\sqrt{r}), \tag{53}$$

parameters $\alpha, \widetilde{\alpha}, \xi, \varsigma$ are determined by Equations (27), and β, ε by (32).

On the circular arc Γ_δ there is $p = \delta e^{i\phi}$, $\sqrt{p} = \sqrt{\delta}e^{0.5i\phi}$, $|\phi| \leq \pi$. Approaching the limit $\delta \to 0$, the last integral (42) take the form:

$$\Theta^*_{l,\delta}(\zeta, \tau) = -\frac{1}{2\pi i}\lim_{\delta\to 0}\left(\int_{-\pi}^{\pi} \overline{\Theta}^*_{l,\delta}(\zeta, \delta e^{i\phi})e^{\delta e^{i\phi}\tau} i\delta e^{i\phi} d\phi\right), \ \tau \geq 0, \ l = 1,2, \tag{54}$$

where, with account of the solutions (34) we obtain:

$$\overline{\Theta}^*_{1,\delta}(\zeta, \delta e^{i\phi}) = \frac{\Delta_1(\zeta, \delta e^{i\phi})}{\delta e^{i\phi}\Delta(\delta e^{i\phi})}, 0 \leq \zeta \leq 1, \ \overline{\Theta}^*_{2,\delta}(\zeta, \delta e^{i\phi}) = \frac{e^{-\varsigma\sqrt{\delta}e^{0.5i\phi}}}{\delta\sqrt{\delta}e^{1.5i\phi}\Delta(\delta e^{i\phi})}, \zeta \geq 1, \tag{55}$$

and functions $\Delta(\delta e^{i\phi})$ and $\Delta_1(\zeta, \delta e^{i\phi})$ can be obtained from Equations (29)–(32) and (35).

Since for small values of the argument [38]:

$$\mathrm{I}_0(x) \cong 1, \ \mathrm{K}_0(x) \cong -\ln x, \ \mathrm{I}_1(x) \cong 0.5x, \ \mathrm{K}_1(x) \cong x^{-1}, \tag{56}$$

from formulae (29)–(32) and (35) the following was found:

$$\Delta(\delta e^{i\phi}) \cong \frac{\gamma^*}{2}\left(1 + \frac{\varepsilon}{\sqrt{\delta}e^{0.5i\phi}}\right), \Delta_1(\zeta, \delta e^{i\phi}) \cong \frac{\varepsilon}{2}\left(e^{-0.5\gamma^*\zeta} - e^{-\gamma^*(1-0.5\zeta)}\right) + \frac{\gamma^* e^{0.5\gamma^*\zeta}}{2\sqrt{\delta}e^{0.5i\phi}}, \tag{57}$$

Substituting the expressions (55)–(57) into the right side of the Equality (54) yields:

$$\Theta^*_{1,\delta}(\zeta, \tau) = -\frac{e^{0.5\gamma^*\zeta}}{\varepsilon}, \ 0 \leq \zeta \leq 1, \ \Theta^*_{2,\delta}(\zeta, \tau) = -\frac{2}{\varepsilon\gamma^*}, \ \zeta \geq 1, \ \tau \geq 0. \tag{58}$$

Applying functions $\Theta^*_{l,\pm}(\zeta, \tau)$ (43), (46)–(53), and $\Theta^*_{l,\delta}(\zeta, \tau)$ (58) to the Equation (41), with consideration of the designations $\sqrt{r} = x, r = x^2$, it was obtained:

$$\Theta^*_1(\zeta, \tau) = \frac{e^{0.5\gamma^*\zeta}}{\varepsilon} + \frac{2}{\pi}\int_0^\infty \frac{\Delta_1(\zeta, x)}{x\Delta(x)}e^{-x^2\tau}dx, \ 0 \leq \zeta \leq 1, \ \tau \geq 0, \tag{59}$$

$$\Theta^*_2(\zeta, \tau) = \frac{\alpha}{\varepsilon} - \frac{4}{\pi^2}\int_0^\infty \frac{\Delta_2(\zeta, x)}{x^2\Delta(x)}e^{-x^2\tau}dx, \ \zeta \geq 1, \ \tau \geq 0, \tag{60}$$

where

$$\Delta_1(\zeta, x) = \Delta_{\mathrm{R}}(x)\Delta_{1,\mathrm{I}}(\zeta, x) + \varepsilon^2 e^{-\gamma^*}\Delta_{\mathrm{I}}(x)\Delta_{1,\mathrm{R}}(\zeta, x), \tag{61}$$

$$\Delta_2(\zeta, x) = \Delta_{\mathrm{R}}(x)\cos(\varsigma\, x) + \varepsilon\, e^{-0.5\gamma^*}\Delta_{\mathrm{I}}(x)\sin(\varsigma\, x), \tag{62}$$

$$\Delta(x) = \Delta^2_{\mathrm{R}}(x) + \varepsilon^2 e^{-\gamma^*}\Delta^2_{\mathrm{I}}(x)], \tag{63}$$

$$\Delta_{\mathrm{R}}(x) = \mathrm{Y}_0(\alpha x)\mathrm{J}_0(\beta x) - \mathrm{J}_0(\alpha x)\mathrm{Y}_0(\beta x), \tag{64}$$

$$\Delta_{\mathrm{I}}(x) = \mathrm{Y}_0(\alpha x)\mathrm{J}_1(\beta x) - \mathrm{J}_0(\alpha x)\mathrm{Y}_1(\beta x), \tag{65}$$

$$\Delta_{1,\mathrm{R}}(\zeta, x) = \mathrm{J}_1(\beta x)\mathrm{Y}_1(\xi x) - \mathrm{Y}_1(\beta x)\mathrm{J}_1(\xi x), \tag{66}$$

$$\Delta_{1,\mathrm{I}}(\zeta, x) = \mathrm{J}_0(\beta x)\mathrm{Y}_1(\xi x) - \mathrm{Y}_0(\beta x)\mathrm{J}_1(\xi x). \tag{67}$$

Taking into account the functions $\Theta_0^*(\tau)$ (38) and $\Theta_l^*(\zeta,\tau)$, $l = 1,2$ (59), (60) in the relations (36) and (37), after integration, the sought dimensionless temperature rise was found in the form:

$$\Theta^*(\zeta,\tau) = 2\sqrt{\frac{\tau}{\pi}}\left[\frac{1}{\varepsilon} + \frac{1}{\sqrt{\pi}}e^{-0.5\gamma^*\zeta}\int_0^\infty \frac{\Delta_1(\zeta,x)}{x\Delta(x)}F(x\sqrt{\tau})dx\right],\ 0 < \zeta \leq 1,\ \tau \geq 0, \tag{68}$$

$$\Theta^*(\zeta,\tau) = 2\sqrt{\frac{\tau}{\pi}}\left[\frac{1}{\varepsilon} - \frac{\gamma^*}{\pi\sqrt{\pi}}\int_0^\infty \frac{\Delta_2(\zeta,x)}{x^2\Delta(x)}F(x\sqrt{\tau})dx\right],\ \zeta \geq 1,\ \tau \geq 0, \tag{69}$$

where

$$F(x) = \frac{2e^{-x^2}}{\sqrt{\pi}x}\int_0^x e^{s^2}ds. \tag{70}$$

To calculate function $F(x)$ (70), we used the following approximation formulae [41]:

$$F(x) = \frac{2}{\sqrt{\pi}}\sum_{n=0}^{\infty}(-1)^n\frac{(2x^2)^n}{(2n+1)!!},\ 0 < x < 3,\ F(x) = \frac{2}{\sqrt{\pi}}\sum_{n=0}^{N}\frac{(2n-1)!!}{(2x^2)^{n+1}},\ x \geq 3, \tag{71}$$

where $(-1)!! = 1$, $(2n+1)!! = 1\cdot 3\cdot 5\cdot \ldots \cdot (2n+1)$.

Besides the exact (in quadratures) solutions (68)–(71), the corresponding asymptotic solutions were also obtained for small and large values of the Fourier number (dimensionless time) τ.

Small values of τ (large values of the parameter p). Taking into account in Equations (29), (30), and (35) asymptotes of the modified Bessel functions for large values of the argument [38]:

$$\mathrm{I_n}(x) \cong \frac{e^x}{\sqrt{2\pi x}},\ \mathrm{K_n}(x) \cong \sqrt{\frac{\pi}{2x}}\,e^{-x},\ n = 0,\ 1,\ \ldots., \tag{72}$$

transforms (33) and (34) were obtained in the form:

$$\overline{\Theta}^*(\zeta,p) \cong e^{-0.25\gamma^*\zeta}\frac{e^{-(\alpha-\xi)\sqrt{p}}}{p\sqrt{p}},\ 0 \leq \zeta < 1, \tag{73}$$

$$\overline{\Theta}^*(\zeta,p) \cong \frac{2e^{-0.25\gamma^*}}{(1+\varepsilon\, e^{-0.5\gamma^*})}\frac{e^{-(\alpha-\beta+\varsigma)\sqrt{p}}}{p\sqrt{p}},\ \zeta \geq 1, \tag{74}$$

where, based on the definitions (27) and (32), we have:

$$\alpha - \xi = \frac{2}{\gamma^*}(1 - e^{-0.5\gamma^*\zeta}) > 0,\ \alpha - \beta + \varsigma = \frac{2}{\gamma^*}(1 - e^{-0.5\gamma^*}) + \frac{\zeta - 1}{\sqrt{k^*}} > 0. \tag{75}$$

Proceeding from the transforms (73) and (74) to the originals [39], asymptotes of dimensionless rise of temperature at the initial moments of the heating process were obtained in the form:

$$\Theta^*(\zeta,\tau) \cong 2\sqrt{\tau}\,e^{-0.25\gamma^*\zeta}\mathrm{ierfc}\left(\frac{\alpha-\xi}{2\sqrt{\tau}}\right),\ 0 \leq \zeta < 1,\ 0 \leq \tau << 1, \tag{76}$$

$$\Theta^*(\zeta,\tau) \cong \frac{4\sqrt{\tau}\,e^{-0.25\gamma^*}}{(1+\varepsilon\, e^{-0.5\gamma^*})}\,\mathrm{ierfc}\left(\frac{\alpha-\beta+\varsigma}{2\sqrt{\tau}}\right),\ \zeta \geq 1,\ 0 \leq \tau << 1, \tag{77}$$

where, $\mathrm{ierfc}(x) = \pi^{-0.5}e^{-x^2} - x\mathrm{erfc}(x)$ $\mathrm{erfc}(x) = 1 - \mathrm{erf}(x)$, $\mathrm{erf}(x)$—Gauss error function [38].

Large values of τ *(small values of the parameter p)*. For small values of the argument of modified Bessel functions with account of the expressions (56) and (57) Laplace transforms (33) and (34) were presented as:

$$\overline{\Theta}^*(\zeta,p) \cong \frac{\varepsilon\chi}{p(\sqrt{p}+\varepsilon)} + \frac{1}{p\sqrt{p}(\sqrt{p}+\varepsilon)},\ 0 \le \zeta < 1,\ \chi = \frac{1}{\gamma^*}(e^{-\gamma^*\zeta} - e^{-\gamma^*}), \tag{78}$$

$$\overline{\Theta}^*(\zeta,p) \cong \frac{e^{-\varsigma\sqrt{p}}}{p\sqrt{p}(\sqrt{p}+\varepsilon)},\ \zeta \ge 1, \tag{79}$$

where the coefficient ς is defined in Equation (27). Proceeding to the space of originals [39], the following asymptotes of dimensionless temperature rise for large values of the Fourier number τ:

$$\Theta^*(\zeta,\tau) \cong \frac{2}{\varepsilon}\sqrt{\frac{\tau}{\pi}} + \left(\chi - \frac{1}{\varepsilon^2}\right)\left[1 - e^{\varepsilon^2\tau}\mathrm{erfc}(\varepsilon\sqrt{\tau})\right],\ 0 \le \zeta < 1,\ \tau >> 1, \tag{80}$$

$$\Theta^*(\zeta,\tau) \cong \frac{2}{\varepsilon}\sqrt{\frac{\tau}{\pi}}\, e^{-\frac{\varsigma^2}{4\tau}} - \frac{1}{\varepsilon}\left(\varsigma + \frac{1}{\varepsilon}\right)\mathrm{erfc}\left(\frac{\varsigma}{2\sqrt{\tau}}\right) + \frac{1}{\varepsilon^2}e^{\varepsilon\varsigma+\varepsilon^2\tau}\mathrm{erfc}\left(\frac{\varsigma}{2\sqrt{\tau}} + \varepsilon\sqrt{\tau}\right),\ \zeta \ge 1,\ \tau >> 1. \tag{81}$$

On the interface $\zeta = 1$, coefficients are $\chi = \varsigma = 0$ and from the Equations (80) and (81) it was obtained:

$$\Theta^*(1^+,\tau) = \Theta^*(1^-,\tau) \cong \frac{2}{\varepsilon}\sqrt{\frac{\tau}{\pi}} - \frac{1}{\varepsilon^2}\left[1 - e^{\varepsilon^2\tau}\mathrm{erfc}(\varepsilon\sqrt{\tau})\right]. \tag{82}$$

4. Verification of the Solution

We will prove that the derived solution (68)–(70) satisfies the boundary conditions (13)–(16) and the initial condition (17). After differentiating the solution (68) with respect to the spatial variable ζ, it was obtained:

$$\frac{\partial\Theta^*(\zeta,\tau)}{\partial\zeta} = 2\sqrt{\frac{\tau}{\pi}}\left[\frac{1}{\sqrt{\pi}}e^{-0.5\gamma^*\zeta}\int_0^\infty \frac{[\Delta_1{}'(\zeta,x) - 0.5\gamma^*\Delta_1(\zeta,x)]}{x\Delta(x)}F(x\sqrt{\tau})dx\right],\ 0 \le \zeta \le 1,\ \tau \ge 0, \tag{83}$$

where derivative of the function $\Delta_1(\zeta,x)$ (61) has the form:

$$\Delta_1{}'(\zeta,x) = \Delta_{\mathrm{R}}(x)\Delta_{1,\mathrm{I}}{}'(\zeta,x) + \varepsilon^2 e^{-\gamma^*}\Delta_{\mathrm{I}}(x)\Delta_{1,\mathrm{R}}{}'(\zeta,x). \tag{84}$$

Considering derivatives [38]:

$$\mathrm{J}_1'(x) = \mathrm{J}_0(x) - x^{-1}\mathrm{J}_1(x),\ \mathrm{Y}_1'(x) = \mathrm{Y}_0(x) - x^{-1}\mathrm{Y}_1(x), \tag{85}$$

and Equations (64)–(67) for $\zeta = 0$ it was found:

$$\Delta_{1,\mathrm{R}}{}'(0,x) = 0.5\gamma^*\Delta_{1,\mathrm{R}}(0,x) - x\Delta_{\mathrm{I}}(x),\ \Delta_{1,\mathrm{I}}{}'(0,x) = 0.5\gamma^*\Delta_{1,\mathrm{I}}(0,x) - x\Delta_{\mathrm{R}}(x), \tag{86}$$

$$\Delta_{1,\mathrm{R}}(0,x) = \mathrm{Y}_1(\alpha\, x)\mathrm{J}_1(\beta\, x) - \mathrm{J}_1(\alpha\, x)\mathrm{Y}_1(\beta x),\ \Delta_{1,\mathrm{I}}(0,x) = \mathrm{Y}_1(\alpha\, x)\mathrm{J}_0(\beta x) - \mathrm{J}_1(\alpha\, x)\mathrm{Y}_0(\beta x), \tag{87}$$

Substituting the Equations (86) and (87) to the expression (84), it was obtained:

$$\Delta_1{}'(0,x) = 0.5\gamma^*\Delta_1(0,x) - x\Delta(x), \tag{88}$$

where functions $\Delta_1(0,x)$ and $\Delta(x)$ were determined from Equations (61) and (63), respectively. Next, from the Equation (83), with consideration of the derivative (88), the following was found:

$$\left.\frac{\partial\Theta^*(\zeta,\tau)}{\partial\zeta}\right|_{\zeta=0^+} = -\frac{2}{\sqrt{\pi}}\sqrt{\frac{\tau}{\pi}}\int_0^\infty F(x\sqrt{\tau})dx,\ \tau \ge 0, \tag{89}$$

where function $F(x)$ has form (69), (70). Taking into account the value of the integral [41]:

$$\int_0^\infty F(x)dx = \frac{\pi}{2}, \tag{90}$$

from the Equation (88) it follows that boundary condition (13) is met.

Comparing the forms of solutions (68) and (69) it can be concluded that the boundary condition (14) is satisfied, when:

$$e^{-0.5\gamma^*}\Delta_1(1,x) = -\frac{\gamma^*\Delta_2(1,x)}{\pi x}. \tag{91}$$

Taking into consideration that for $\zeta = 1$ yields $\xi = \beta$ and $\varsigma = 0$, from Equations (61)–(67) it was obtained:

$$\Delta_{1,\mathrm{R}}(1,x) = 0,\ \Delta_1(1,x) = \Delta_\mathrm{R}(x)\Delta_{1,\mathrm{I}}(1,x),\ \Delta_2(1,x) = \Delta_\mathrm{R}(x). \tag{92}$$

Then, using relations [38]:

$$\Delta_{1,\mathrm{I}}(1,x) = \mathrm{J}_0(\beta x)\mathrm{Y}_1(\beta\, x) - \mathrm{Y}_0(\beta x)\mathrm{J}_1(\beta\, x) = -\frac{2}{\pi\beta x}, \tag{93}$$

definition of coefficient β (27), (32) and Equation (92), the Equality (91) takes the form:

$$\frac{\gamma^*}{\pi x}\Delta_\mathrm{R}(x) \equiv \frac{\gamma^*}{\pi x}\Delta_\mathrm{R}(x), \tag{94}$$

which confirm the fulfillment of the boundary condition (14).

For $\zeta = 1$ from Equations (83) and (84), it was found:

$$\left.\frac{\partial\Theta^*(\zeta,\tau)}{\partial\zeta}\right|_{\zeta=1^+} = 2\sqrt{\frac{\tau}{\pi}}\left[\frac{1}{\sqrt{\pi}}e^{-0.5\gamma^*}\int_0^\infty \frac{[\Delta_1{}'(1,x) - 0.5\gamma^*\Delta_1(1,x)]}{x\Delta(x)}F(x\sqrt{\tau})dx\right],\ \tau \geq 0, \tag{95}$$

$$\Delta_1{}'(1,x) = \Delta_\mathrm{R}(x)\Delta_{1,\mathrm{I}}{}'(1,x) + \varepsilon^2 e^{-\gamma^*}\Delta_\mathrm{I}(x)\Delta_{1,\mathrm{R}}{}'(1,x), \tag{96}$$

where the corresponding values of the derivatives of the function $\Delta_{1,\mathrm{R}}(\zeta,x)$ (66) and $\Delta_{1,\mathrm{I}}(\zeta,x)$ (67) with respect to the variable ζ have the form:

$$\Delta_{1,\mathrm{R}}{}'(1,x) = -\frac{\gamma^*}{\pi},\ \Delta_{1,\mathrm{I}}{}'(1,x) = -\frac{\gamma^*}{\pi\beta x}, \tag{97}$$

Substituting the above Equation (97) to the right side of Equality (96), and next, considering in this manner the obtained derivative $\Delta_1{}'(1,x)$ and functions $\Delta_1(1,x)$ (92), (93) in the integrand (95), it was received:

$$e^{\gamma^*}\left.\frac{\partial\Theta^*(\zeta,\tau)}{\partial\zeta}\right|_{\zeta=1^+} = -\frac{2}{\pi^2}\varepsilon^2\gamma^* e^{-0.5\gamma^*}\sqrt{\tau}\int_0^\infty \frac{\Delta_\mathrm{I}(x)}{x\Delta(x)}F(x\sqrt{\tau})dx,\ \tau \geq 0. \tag{98}$$

On the other hand, differentiating the solution (69) with respect to the variable ζ yields:

$$K^*\left.\frac{\partial\Theta^*(\zeta,\tau)}{\partial\zeta}\right|_{\zeta=1^-} = -\frac{2}{\pi^2}\gamma^* K^*\sqrt{\tau}\int_0^\infty \frac{\Delta_2{}'(1,x)}{x^2\Delta(x)}F(x\sqrt{\tau})dx,\ \tau \geq 0, \tag{99}$$

where value of the derivative function $\Delta_2(\zeta,x)$ (61) for $\zeta = 1$ is equal to:

$$\Delta_2{}'(1,x) = \frac{\varepsilon\, e^{-0.5\gamma^*}}{\sqrt{k^*}}x\Delta_\mathrm{I}(x). \tag{100}$$

Taking into consideration of derivative (100) and definition (32) of thermal activity ε in the right side of the equality (99), it was established that it is the same as the right side of Equation (98). In this way, it was proved that the obtained solutions (68)–(70) meet the boundary condition (15). Fulfillment of the condition (16) of temperature rise disappearance (69) for $\zeta \to \infty$ is guaranteed by omitting the function $B_2(p)$ (28) in Equation (26). Additionally, it was checked during the numerical calculations. Obviously, the solutions (68)–(70) satisfy also the initial condition (17).

5. Heating the Coating Surface by a Heat Flux with Linearly Decreasing Intensity in Time

Presented above exact solutions (68)–(70) were developed for constant intensity of heat flux q_0 over time. In this chapter, the heating process of the FGC surface deposited on the homogeneous substrate, with time-dependent heat flux intensity:

$$q(t) = q_0 q^*(t),\ q^*(t) = 1 - t\, t_s^{-1},\ 0 \le t \le t_s, \tag{101}$$

where t_s—the final moment of heating process. It should be noted that the evolution of the heat flux intensity in the form (101) is characteristic for thermal problems of friction formulated for braking process with constant deceleration [42]. Dimensionless temperature rise $\hat{\Theta}^*(\zeta,\tau)$, corresponding to the heat flux intensity (101) can be found based on Duhamel's theorem [43]:

$$\hat{\Theta}^*(\zeta,\tau) = \frac{\partial}{\partial\tau}\int_0^{\tau} q^*(\tau - s)\Theta^*(\zeta,s)ds,\ \zeta \ge 0,\ 0 \le \tau \le \tau_s, \tag{102}$$

where $\Theta^*(\zeta,\tau)$ is dimensionless temperature rise (68)–(70), and function $q^*(\tau)$ has the form:

$$q^*(\tau) = 1 - \tau\, \tau_s^{-1},\ 0 \le \tau \le \tau_s,\ \tau_s = k_1 t_s d^{-2}. \tag{103}$$

Substituting the solution (68) and function $q^*(\tau)$ (103) to the integrand in the expression (102), it was found:

$$\hat{\Theta}^*(\zeta,\tau) = \frac{\partial}{\partial\tau}\left[\frac{2}{\sqrt{\pi\varepsilon}}Q_1(\tau) + \frac{2}{\pi}e^{-0.5\gamma^*\zeta}\int_0^{\infty}\frac{\Delta_1(\zeta,x)}{x\Delta(x)}Q_2(\tau,x)dx\right],0 \le \zeta \le 1,\ 0 \le \tau \le \tau_s, \tag{104}$$

where

$$Q_1(\tau) = \int_0^{\tau}\left(1 - \frac{\tau - s}{\tau_s}\right)\sqrt{s}ds = \frac{2}{3}\tau\sqrt{\tau}\left(1 - \frac{2\tau}{5\tau_s}\right), \tag{105}$$

$$Q_2(\tau,x) = \int_0^{\tau}\left(1 - \frac{\tau - s}{\tau_s}\right)\sqrt{s}F(x\sqrt{s})ds = \left(1 - \frac{\tau}{\tau_s}\right)Q_{21}(\tau,x) + \frac{1}{\tau_s}Q_{22}(\tau,x), \tag{106}$$

$$Q_{21}(\tau,x) = \int_0^{\tau}\sqrt{s}F(x\sqrt{s})ds,\ Q_{22}(\tau,x) = \int_0^{\tau}s\sqrt{s}F(x\sqrt{s})ds. \tag{107}$$

Taking into consideration the function $F(x)$ (70), integrals (107) were written in the form:

$$Q_{21}(\tau,x) = \frac{2}{x^2}\sqrt{\frac{\tau}{\pi}}\left[1 - \frac{\sqrt{\pi}}{2}F(x\sqrt{\tau})\right], \tag{108}$$

$$Q_{22}(\tau,x) = \frac{2}{x^4}\sqrt{\frac{\tau}{\pi}}\left[1 + \frac{1}{3}x^2\tau - \frac{\sqrt{\pi}}{2}(1 + x^2\tau)F(x\sqrt{\tau})\right]. \tag{109}$$

Applying functions $Q_{21}(\tau,x)$ (108) and $Q_{22}(\tau,x)$ (109) into Equation (106), it was received:

$$Q_2(\tau,x) = \frac{2}{x^2}\sqrt{\frac{\tau}{\pi}}\left[1 - \frac{2\tau}{3\tau_s} + \frac{1}{x^2\tau_s} - \frac{\sqrt{\pi}}{2}\left(1 + \frac{1}{x^2\tau_s}\right)F(x\sqrt{\tau})\right]. \quad (110)$$

Next, differentiating the functions $Q_1(\tau)$ (105) and $Q_2(\tau,x)$ (110), it was found:

$$P_1(\tau) \equiv \frac{dQ_1(\tau)}{d\tau} = \sqrt{\tau}\,P_1^*(\tau),\ P_1^*(\tau) = 1 - \frac{2\tau}{3\tau_s}, \quad (111)$$

$$P_2(\tau,x) \equiv \frac{\partial Q_2(\tau,x)}{\partial\tau} = \sqrt{\tau}\,P_2^*(\tau,x),\ P_2^*(\tau,x) = \left(1 + \frac{1}{x^2\tau_s}\right)F(x\sqrt{\tau}) - \frac{2}{\sqrt{\pi}x^2\tau_s}. \quad (112)$$

From the Duhamel's formula (104) with account of derivatives (111) and (112), it was obtained:

$$\hat{\Theta}^*(\zeta,\tau) = 2\sqrt{\frac{\tau}{\pi}}\left[\frac{1}{\varepsilon}P_1^*(\tau) + \frac{1}{\sqrt{\pi}}e^{-0.5\gamma^*\zeta}\int_0^\infty \frac{\Delta_1(\zeta,x)}{x\Delta(x)}P_2^*(\tau,x)dx\right],\ 0 \le \zeta \le 1, 0 \le \tau \le \tau_s. \quad (113)$$

Noting that solution (113) is structurally close (with accuracy to the functions $P_1^*(\tau)$ (111) i $P_2^*(\tau,x)$ (112)) to the solution (68), based on the expression (69) the following can be written:

$$\hat{\Theta}^*(\zeta,\tau) = 2\sqrt{\frac{\tau}{\pi}}\left[\frac{1}{\varepsilon}P_1^*(\tau) - \frac{\gamma^*}{\pi\sqrt{\pi}}\int_0^\infty \frac{\Delta_2(\zeta,x)}{x^2\Delta(x)}P_2^*(\tau,x)dx\right],\ \zeta \ge 1,\ 0 \le \tau \le \tau_s. \quad (114)$$

It should be noted that for $\tau_s \to \infty$ from the Equations (111) and (112) follows that $P_1^*(\tau) = 1, P_2^*(\tau,x) = F(x\sqrt{\tau})$ and solutions (113), (114) become the same, as well as the previously obtained solutions (68), (69) at constant heat flux intensity.

6. Numerical Analysis

Calculations were performed for a coating made of two-component FGM, applied on the homogeneous substrate. On the outer surface of FGC is pure zirconium dioxide ZrO_2 that smoothly transforms into a titanium alloy Ti-6Al-4V in the structure of the coating material. The element substrate is made of cast iron ChNMKh. Essential for calculations, properties of these materials at the ambient temperature $T_0 = 20\ ^\circ C$ are included in Table 1.

Table 1. Materials properties [34,44].

Material	Thermal Conductivity $Wm^{-1}K^{-1}$	Specific Heat Capacity $J\ kg^{-1}K^{-1}$	Density $kg\ m^{-3}$
ZrO_2	$K_{1,1} = 1.94$	$c_{1,1} = 452.83$	$\rho_{1,1} = 6102.16$
Ti-6Al-4V	$K_{1,2} = 6.87$	$c_{1,2} = 538.08$	$\rho_{1,2} = 4431.79$
ChNMKh	$K_2 = 52.17$	$c_2 = 444.6$	$\rho_2 = 7100$

Specific heat capacity and density of functionally graded coating material was determined according to the mixture law:

$$c_1 = c_{1,1}\mathrm{v} + (1-\mathrm{v})c_{1,2},\ \rho_1 = \rho_{1,1}\mathrm{v} + (1-\mathrm{v})\rho_{1,2}, \quad (115)$$

where $0 \le \mathrm{v} \le 1$ is the volume fracture of the FGM components. For the same proportion of both components ($\mathrm{v} = 0.5$) it was established that $c_1 = 495.55\ J\ kg^{-1}K^{-1}$, $\rho_1 = 5266.98 kg\ m^{-3}$. Moreover from the Equation (9) the dimensionless values of thermal conductivity $K^* = 26.89$ and diffusivity $k^* = 22.23$ were found, and the thermal activity $\varepsilon = 5.7$ from the Equation (32). Gradient parameter of considered FGM was found based on the following relation [45]:

$$\gamma^* = \ln(K_{1,2}K_{1,1}^{-1}) = 1.26. \quad (116)$$

Main objects of the numerical analysis are dimensionless temperature rises $\Theta^*(\zeta,\tau)$ (68), (69) and $\hat{\Theta}^*(\zeta,\tau)$ (113), (114) initiated by heating the coating surface by heat fluxes with constant and linearly decreasing intensities over time. Numerical integration in the Equations (68), (69) and (113), (114) were performed utilizing the QAGI procedure from the package QUADPACK [46].

Variations of dimensionless temperature rise $\Theta^*(\zeta,\tau)$ (68), (69) over time of heating (Fourier number τ) in the coating and substrate is presented in Figures 3 and 4. Deposition of a FGC on a substrate causes a drop of temperature compared to a homogeneous coating made entirely of zirconium dioxide (Figure 3a). It is valuable that this effect becomes more noticeable with the passage of heating time, and thus with increasing the temperature of the protective layer. So, the Ti-6Al-4V titanium alloy, having a thermal conductivity three times higher than zirconium dioxide ZrO_2, fulfills well the assumed *a priori* role of dissipating heat from the heated surface of the FGC and thus lowering its temperature. It should be noted that the described effect of lowering the temperature as a result of using FGM is maintained inside the layer ($0 \leq \zeta < 1$). However, starting from the interface and further into the substrate ($\zeta \geq 1$), we observe the opposite behavior of the temperature evolution—the temperature of the substrate with the FGC is higher than when using a homogeneous coating (Figure 3b). This is due to the significantly higher thermal conductivity of cast iron compared to the titanium alloy (Table 1). The temperature of the substrate is much (more than an order of magnitude) lower than the temperature of the coating.

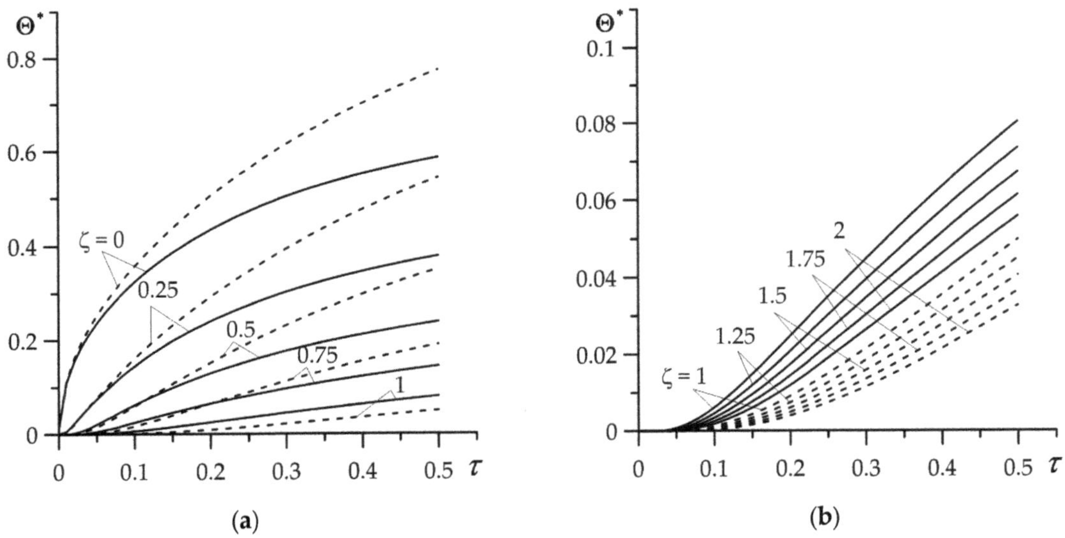

Figure 3. Evolutions of dimensionless temperature rise $\Theta^*(\zeta,\tau)$ for selected values of dimensionless spatial variable ζ in: (**a**) coating, (**b**) substrate. Solid lines—FGC, dashed lines—homogeneous ZrO_2 coating.

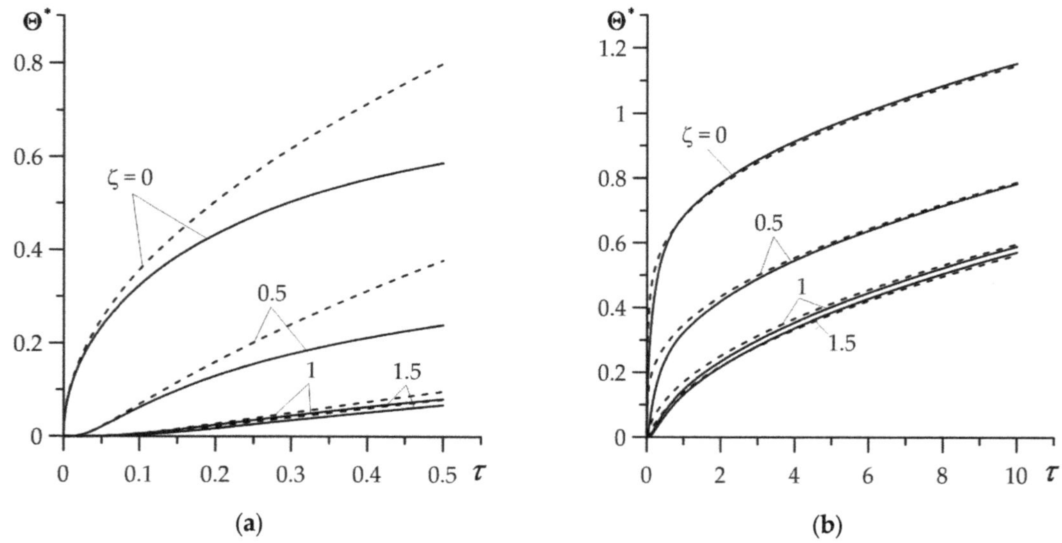

Figure 4. Comparison of time profiles of dimensionless temperature rise $\Theta^*(\zeta,\tau)$ in the FGC, obtained using the exact (68), (69) (solid lines) and asymptotic (dashed lines) solutions: (**a**) for small (76), (77), (**b**) for large (80), (81) Fourier number τ, for selected values of dimensionless spatial variable ζ.

Results demonstrated in Figure 3 were obtained based on exact solutions for coating (68) and substrate (69), whereas Figure 4 illustrates the outcomes established by means of the asymptotic solutions for small (76), (77) and large (80), (81) values of dimensionless time (Fourier number) τ. Calculations were performed only for the FGC.

Satisfactory agreement of exact and asymptotic (for small τ) solutions takes place in the range $0 \leq \tau \leq 0.1$ (Figure 4a). On the other hand, the good agreement of the results obtained with the use of exact and asymptotic (for large τ) solutions occurs at $\tau \geq 1$ (Figure 4b). The advantage of asymptotic solutions is their analytical form, which eliminates the need for laborious procedures of numerical integration. Thus, the developed asymptotic solutions can be used for expressing the estimation of the temperature mode of the considered coating–substrate system.

As shown in Figures 3 and 4, the outcomes were obtained for FGM gradient parameter $\gamma^* = 1.26$ determined based on the Equation (116). On the other hand, γ^* can be considered as independent input parameter, responsible for the speed of "transition" of the thermal conductivity coefficient K_1 of coating material, from the value $K_{1,1}$ of the first component to the value $K_{1,2}$ of the second component, according to the formula (1). Influence of the gradient γ^* on the temperature of heated coating surface is presented in Figure 5, where results obtained for $\gamma^* = 0$ correspond to the coating made entirely from the zirconium dioxide. Increase in FGM gradient causes a drop in temperature, which becomes more noticeable as the heating process progresses (Figure 5a). At the final moment $\tau = 0.5$, the decrease in the temperature of the heated FGC surface is almost linear (Figure 5b).

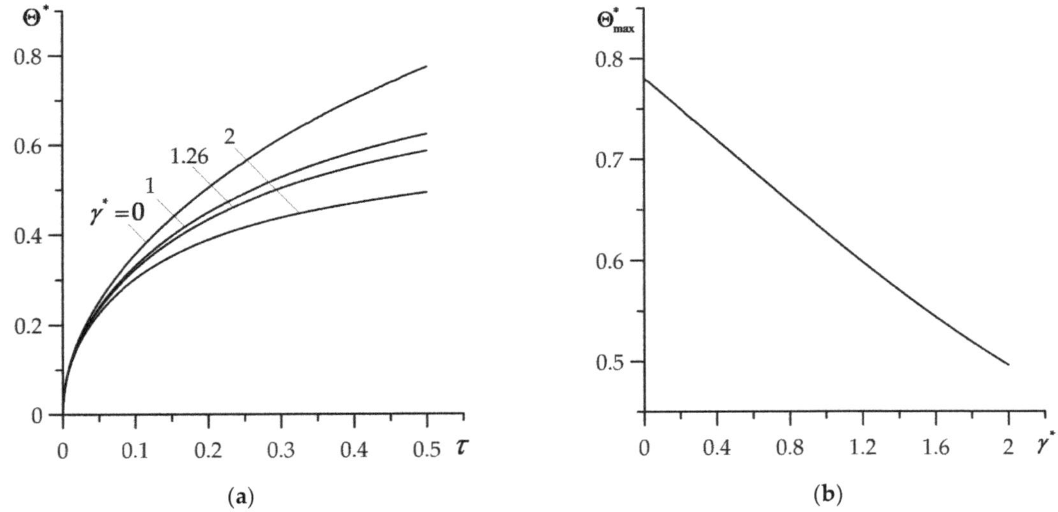

Figure 5. Effect of FGC gradient γ^* on dimensionless temperature rise: (**a**) evolution $\Theta^*(0,\tau)$ (68), (69) for selected values of γ^*, (**b**) dependency $\Theta^*_{\max} \equiv \Theta^*(0,\ 0.5)$ on γ^*.

Investigation of the impact of linearly decreasing time profile of the heat flux intensity (103) was carried out based on solutions (113) and (114) for $\tau_s = 0.5$, $\gamma^* = 1.26$ (Figure 6). For a constant intensity of the heat flux, the temperature of the coating and substrate increases monotonically with the heating time (Figure 3). However, in the considered case of a linear heat flux intensity profile, the temperature of the coating reaches its maximum value $\hat{\Theta}^*_{\max}$ at the moment of time $0 < \tau_{\max} < \tau_s$, and then decreases (Figure 6a). The maximum temperature value $\hat{\Theta}^*_{\max}$ is achieved the earliest on the FGC outer surface $\zeta = 0$. The time to reach $\hat{\Theta}^*_{\max}$ increases with the distance from heated surface.

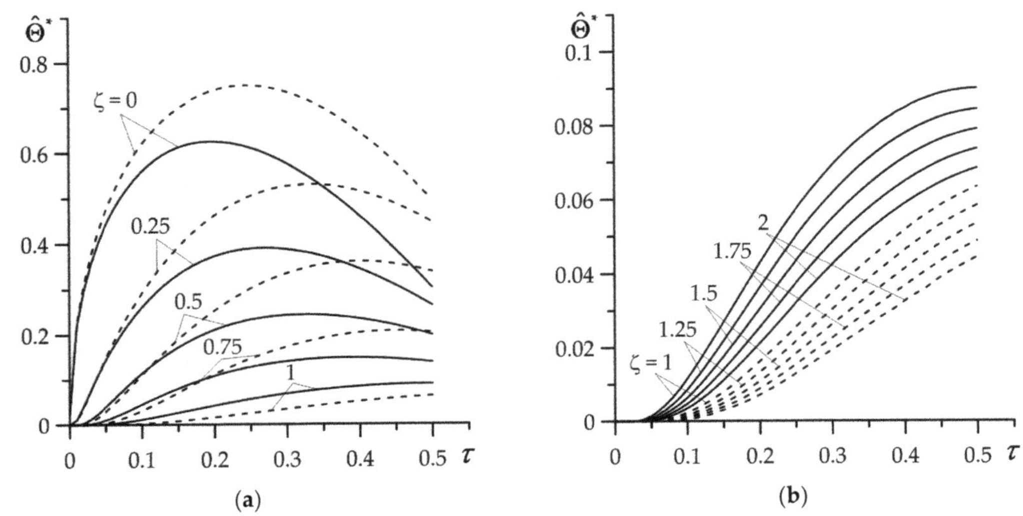

Figure 6. Evolution of dimensionless temperature rise $\hat{\Theta}^*(\zeta,\tau)$ (113), (114) for selected dimensionless values of spatial variable ζ in: (**a**) coating, (**b**) substrate. Solid lines—FGC, dashed lines—homogeneous coating made of ZrO_2.

Starting from the interface $\zeta = 1$ and further into the substrate $\zeta > 1$, the temperature increases throughout the heating process, reaching the highest value at the moment $\tau_{\max} = \tau_s$ (Figure 6b). Both at a constant and at the time-dependent intensity of the heat flux, consideration of the FGC gradient causes a drop in the temperature value in relation to the temperature of the coating made of a homogeneous material. In the substrate, the situation is opposite—at fixed value of ζ, the temperature is lower in case of a homogeneous coating material. Regardless of the coating material (FGM or homogeneous), the substrate temperature is an order lower than in the coating.

Spatial-temporal distributions of temperature are presented in Figure 7, for constant and time-dependent heat flux intensities. Visible differences between the shape of the corresponding isotherms occur only in the more heated element of the system—the coating that absorbs the main part of the thermal load. In the case of a constant intensity of the heat flux, the isotherms with a fixed temperature level go deeper inside the FGC during the whole heating process. On the other hand, during heating by a heat flux with linearly decreasing intensity over time, the isotherms in the coating reach the maximum distance from the heated surface before the end of the process.

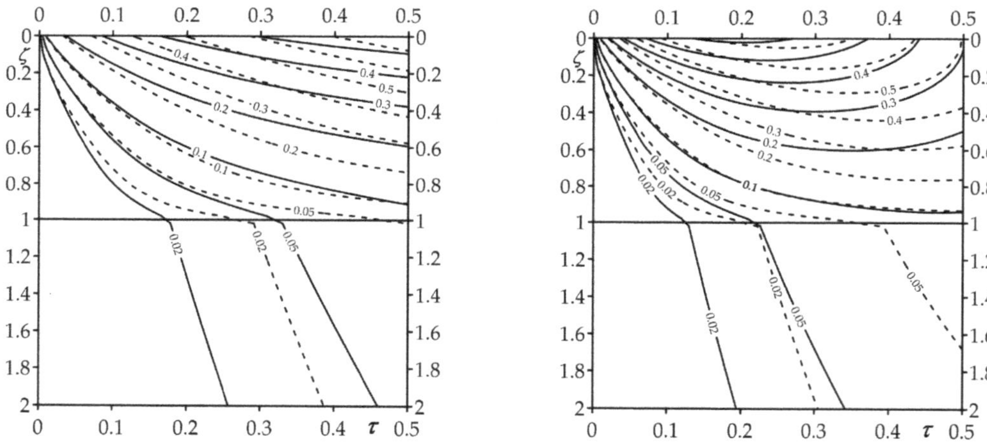

(a) (b)

Figure 7. Dimensionless isolines of the temperature rises: (**a**) $\Theta^*(\zeta,\tau)$ (68), (69), (**b**) $\hat{\Theta}^*(\zeta,\tau)$(113), (114). Solid lines—FGC, dashed lines—homogeneous coating made of ZrO_2.

7. Conclusions

As a result of the performed analysis, it was found that:

1. Deposition of functionally graded coating on the homogeneous substrate allow to effectively lower the temperature on the heated surface;
2. FGC is the main adsorbent of frictional heat generated. As a result, values of temperature achieved in the substrate are much lower than that obtained in the coating temperature level;
3. The temporal profile of the heat flux intensity has a noticeable impact on the spatial-temporal distribution of isotherms only in the coating;
4. Gradient parameter of the FGC has a crucial influence on the maximum temperature for the selected coating–substrate system;
5. Obtained asymptotic solutions are useful for the express estimation of the temperature of the FGC-substrate system at small and large values of the Fourier number;
6. The proposed mathematical model can be utilized as an effective tool for simulating the temperature mode of homogeneous bodies with functionally graded coating.

In summary, it should be noted that the developed methodology for obtaining accurate and asymptotic solutions can be successfully applied also to the FGM class, not only with increasing but also decreasing thermal conductivity along the thickness, as well as in the problems of thermal conduction, considering the generation of heat due to friction on the contact surface of the protective layer with the counterbody. For example, such a pair may be formed by a brake disc coated with FGC in combination with a pad.

Author Contributions: Conceptualization and methodology, A.Y.; software, P.Z.; validation, A.Y. and K.T.; formal analysis, A.Y. and P.Z.; investigation, A.Y., K.T. and P.Z.; writing—original draft preparation, A.Y. and K.T.; writing—review and editing, K.T.; visualization and figures preparation, P.Z.; supervision, A.Y.; project administration, K.T. All authors have read and agreed to the published version of the manuscript.

Funding: This research was funded by the National Science Centre in Poland, grant number 2017/27/B/ST8/01249 and the APC was funded by the program of the Minister of Education and Science of Poland named "Regional Initiative of Excellence" in 2019–2022, project number 011/RID/2018/19.

Institutional Review Board Statement: Not applicable.

Informed Consent Statement: Not applicable.

Data Availability Statement: No new data were created or analyzed in this study. Data sharing is not applicable to this article.

Conflicts of Interest: The authors declare no conflict of interest.

Nomenclature

c	Specific heat capacity (J kg^{-1}K^{-1})
d	Thickness of FGC (m)
$\mathrm{I_n}(\cdot)$	Modified Bessel functions of the nth order of the first kind
$\mathrm{J_n}(\cdot)$	Bessel functions of the nth order of the first kind
$\mathrm{K_n}(\cdot)$	Modified Bessel functions of the nth order of the second kind
k	Thermal diffusivity (m^{2}s^{-1})
K	Thermal conductivity (W m^{-1}K^{-1})
q	Intensity of heat flux (W m^{-2})
q_0	Nominal value of the heat flux intensity (W m^{-2})
t	Time (s)
t_s	Final moment of the heating process (s)
T	Temperature (°C)
T_0	Initial temperature (°C)
v	Volume fraction of the material phases (dimensionless)
$\mathrm{Y_n}(\cdot)$	Bessel functions of the nth order of the second kind
z	Spatial coordinate in axial direction (m)
γ	Parameter of FGM gradient (m^{-1})
γ^*	Dimensionless parameter of FGM gradient
Λ	Temperature rise scaling factor (°C)
ε	Dimensionless coefficient of thermal activity of friction couple
Θ	Temperature rise (°C)
Θ^*	Dimensionless temperature rise
ρ	Density (kg m^{-3})
τ	Dimensionless time
τ_s	Dimensionless final time of heating
ζ	Dimensionless spatial coordinate in axial direction

References

1. Alinia, Y.; Beheshti, A.; Guler, M.A.; El-Borgi, S.; Polycarpou, A.A. Sliding contact analysis of functionally graded coating/substrate system. *Mech. Mater.* **2016**, *94*, 142–155. [CrossRef]
2. Kulchytsky-Zhyhailo, R.; Bajkowski, A. Analytical and numerical methods of solution of three-dimensional problem of elasticity for functionally graded coated half-space. *Int. J. Mech. Sci.* **2012**, *54*, 105–112. [CrossRef]

3. Bishnoi, D. Pressure exertion and heat dissipation analysis on uncoated and ceramic (Al_2O_3, TiO_2 and ZrO_2) coated braking pads. *Mater. Today Proc.* **2023**, *74*, 774–787. [CrossRef]
4. Yevtushenko, A.; Kuciej, M.; Och, E. Influence of thermal sensitivity of the materials on temperature and thermal stresses of the brake disc with thermal barrier coating. *Int. Commun. Heat Mass Transf.* **2017**, *87*, 288–294. [CrossRef]
5. Varecha, D.; Brončeк, J.; Kohar, R.; Nový, F.; Vicen, M.; Radek, N. Research of friction materials applicable to the multi-disc brake concept. *J. Mater. Res. Technol.* **2021**, *14*, 647–661. [CrossRef]
6. Strojny-Nędza, A.; Pietrzak, K.; Gili, F.; Chmielewski, M. FGM based on copper–alumina composites for brake disc applications. *Arch. Civ. Mech. Eng.* **2020**, *20*, 83. [CrossRef]
7. Govindaraju, M.; Megalingam, A.; Murugasan, J.; Vignesh, R.V.; Kota, P.K.; Ram, A.S.; Lakshana, P.; Kumar, V.N. Investigations on the tribological behavior of functionally gradient iron-based brake pad material. *Proc. Inst. Mech. Eng. Part C J. Mech. Eng. Sci.* **2020**, *234*, 2474–2486. [CrossRef]
8. Babu, K.V.; Marichamy, S.; Ganesan, P.; Madan, D.; Uthayakumar, M.; Rajan, T.P. Processing of functionally graded aluminum composite brake disc and machining parameters optimization. *Mater. Today Proc.* **2020**, *21*, 563–567. [CrossRef]
9. Pakseresht, A.H.; Rahimipour, M.R.; Alizadeh, M.; Hadavi, S.M.M.; Shahbazkhan, A. Concept of advanced thermal barrier functional coatings in high temperature engineering components. In *Research Perspectives on Functional Micro- and Nanoscale Coatings*; Zuzuarregui, A., Morant-Minana, M.C., Eds.; IGI Global: Hershey, PA, USA, 2016; pp. 396–419. [CrossRef]
10. Buyukkaya, E. Thermal analysis of functionally graded coating AlSi alloy and steel pistons. *Surf. Coat. Technol.* **2008**, *202*, 3856–3865. [CrossRef]
11. Mondal, K.; Nuñez III, L.; Downey, C.M.; Van Rooyen, I.J. Recent advances in the thermal barrier coatings for extreme environments. *Mater. Sci. Energy Technol.* **2021**, *4*, 208–210. [CrossRef]
12. Topczewska, K. Analytical model for investigation of the effect of friction power on the thermal stresses in friction elements of brakes. *J. Theor. Appl. Mech.* **2018**, *56*, 1017–1027. [CrossRef]
13. Zhao, J.; Li, Y.; Ai, X. Analysis of transient thermal stress in sandwich plate with functionally graded coatings. *Thin Solid Film.* **2008**, *516*, 7581–7587. [CrossRef]
14. Sathish, M.; Radhika, N.; Saleh, B.A. critical review on functionally graded coatings: Methods, properties, and challenges. *Compos. Part B Eng.* **2021**, *225*, 109278. [CrossRef]
15. Pasupuleti, K.T.; Dsouza, S.; Thejaraju, R.; Venkataraman, S.; Ramaswamy, P.; Murty, S.N. Performance and steady state heat transfer analysis of functionally graded thermal barrier coatings systems. *Mater. Today Proc.* **2018**, *5*, 27936–27945. [CrossRef]
16. Suresh, S. Graded materials for resistance to contact deformation and damage. *Science* **2001**, *292*, 2447–2451. [CrossRef]
17. Khor, K.A.; Gu, Y.W. Thermal properties of plasma-sprayed functionally graded thermal barrier coatings. *Thin Solid Film.* **2000**, *372*, 104–113. [CrossRef]
18. Balci, M.N.; Dag, S.; Yildirim, B. Subsurface stresses in graded coatings subjected to frictional contact with heat generation. *J. Therm. Stress.* **2017**, *40*, 517–534. [CrossRef]
19. Balci, M.N.; Dag, S. Solution of the dynamic frictional contact problem between a functionally graded coating and a moving cylindrical punch. *Int. J. Solids Struct.* **2019**, *161*, 267–281. [CrossRef]
20. Fu, P.; Zhao, J.; Zhang, X.; Kang, G.; Wang, P.; Kan, Q. Thermo-mechanically coupled sliding contact shakedown analysis of functionally graded coating-substrate structures. *Int. J. Mech. Sci.* **2022**, *222*, 107241. [CrossRef]
21. Jin, Z.H. An asymptotic solution of temperature field in a strip a functionally graded material. *Int. Commun. Heat Mass Transf.* **2002**, *29*, 887–895. [CrossRef]
22. Jobin, K.J.; Abhilash, M.N.; Murthy, H. A simplified analysis of 2D sliding frictional contact between rigid indenters and FGM coated substrates. *Tribol. Int.* **2017**, *108*, 174–185. [CrossRef]
23. Burlayenko, V.N.; Altenbach, H.; Sadowski, T.; Dimitrova, S.D.; Bhaskar, A. Modelling functionally graded materials in heat transfer and thermal stress analysis by means of graded finite elements. *Appl. Math. Model.* **2017**, *45*, 422–438. [CrossRef]
24. Yevtushenko, A.; Topczewska, K.; Zamojski, P. Temperature during Repetitive Short-Term Operation of a Brake with Functionally Graded Friction Element. *Materials* **2023**, *16*, 881. [CrossRef]
25. Jojith, R.; Sam, M.; Radhika, N. Recent advances in tribological behavior of functionally graded composites: A review. *Eng. Sci. Technol. Int. J.* **2022**, *25*, 100999. [CrossRef]
26. Bhandari, M.; Purohit, K. Dynamic fracture analysis of functionally graded material structures—A critical review. *Compos. Part C Open Access* **2022**, *7*, 100227. [CrossRef]
27. Yevtushenko, A.; Topczewska, K.; Zamojski, P. The Effect of Functionally Graded Materials on Temperature During Frictional Heating: Under Uniform Sliding. *Materials* **2021**, *14*, 4285. [CrossRef]
28. Topczewska, K.; Gerlici, J.; Yevtushenko, A.; Kuciej, M.; Kravchenko, K. Analytical Model of the Frictional Heating in a Railway Brake Disc at Single Braking with Experimental Verification. *Materials* **2022**, *15*, 6821. [CrossRef] [PubMed]
29. Rahmati Nezhad, Y.; Asemi, K.; Akhlaghi, M. Transient solution of temperature field in functionally graded hollow cylinder with finite length using multi layered approach. *Int. J. Mech. Mater. Des.* **2011**, *7*, 71–82. [CrossRef]
30. Kulchytsky-Zhyhailo, R.; Bajkowski, A.S. Axisymmetrical problem of thermoelasticity for half-space with gradient coating. *Int. J. Mech. Sci.* **2016**, *106*, 62–71. [CrossRef]
31. Kiani, Y.; Eslami, M.R. Geometrically non-linear rapid heating of temperature-dependent circular FGM plates. *J. Therm. Stress.* **2014**, *37*, 1495–1518. [CrossRef]

32. Yevtushenko, A.; Topczewska, K.; Zamojski, P. The Heat Partition Ratio during Braking in a Functionally Graded Friction Couple. *Materials* **2022**, *15*, 4623. [CrossRef]
33. Yevtushenko, A.; Kuciej, M.; Topczewska, K.; Zamojski, P. Temperature in the Friction Couple Consisting of Functionally Graded and Homogeneous Materials. *Materials* **2022**, *15*, 3600. [CrossRef] [PubMed]
34. Lee, S.W.; Jang, Y.H. Frictionally excited thermoelastic instability in a thin layer of functionally graded material sliding between two half-planes. *Wear* **2009**, *267*, 1715–1722. [CrossRef]
35. Liu, J.; Ke, L.L.; Wang, Y.S. Two-dimensional thermoelastic contact problem of functionally graded materials involving frictional heating. *Int. J. Solids Struct.* **2011**, *48*, 2536–2548. [CrossRef]
36. Suresh, S.; Mortensen, A. *Fundamentals of Functionally Graded Materials*; The Institute of Materials, IOM Communications Ltd.: London, UK, 1998.
37. Sneddon, I.N. *The Use of Integral Transforms*; McGraw-Hill: New York, NY, USA, 1972.
38. Abramowitz, M.; Stegun, I. *Handbook of Mathematical Functions with Formulas, Graphs, and Mathematical Tables*; United States Department of Commerce, National Bureau of Standards (NBS): Washington, DC, USA, 1964.
39. Bateman, H.; Erdelyi, A. *Tables of Integrals Transforms*; McGraw-Hill: New York, NY, USA, 1954; Volume 1.
40. Luikov, A.V. *Analitycal Heat Diffusion Theory*; Academic Press: New York, NY, USA, 1968.
41. Barber, J.R.; Martin-Moran, C.J. Green's functions for transient thermoelastic contact problems for the half-plane. *Wear* **1982**, *79*, 11–19. [CrossRef]
42. Yevtushenko, A.A.; Kuciej, M.; Topczewska, K. Analytical model for investigation of the effect of friction power on temperature in the disc brake. *Adv. Mech. Eng.* **2017**, *9*, 1–12. [CrossRef]
43. Özişik, N.M. *Heat Conduction*; John Wiley: New York, NY, USA, 1993.
44. Chichinadze, A.V. *Polymers in Friction Assembles of Machines and Devices: A Handbook*; Allerton Press Inc.: New York, NY, USA, 1984.
45. Mao, J.J.; Ke, L.L.; Yang, J.; Kitipornchai, S.; Wang, Y.S. The coupled thermoelastic instability of FGM coatings with arbitrarily varying properties: In-plane sliding. *Acta Mech.* **2018**, *229*, 2979–2995. [CrossRef]
46. Piessens, R.; De Doneker-Kapenga, E.; Überhuber, C.W.; Kahaner, D.K. *Quadpack: A Subroutine for Automatic Integration*; Springer Science & Business Media: Berlin/Heidelberg, Germany, 2012; Volume 1.

materials

MDPI

Article

Influence of Functionally Graded Protective Coating on the Temperature in a Braking System

Aleksander Yevtushenko, Katarzyna Topczewska * and Przemysław Zamojski

Faculty of Mechanical Engineering, Bialystok University of Technology (BUT), 45C Wiejska Street, 15-351 Bialystok, Poland; a.yevtushenko@pb.edu.pl (A.Y.); p.zamojski@pb.edu.pl (P.Z.)
* Correspondence: k.topczewska@pb.edu.pl

Abstract: A mathematical model of heat generation due to friction in a disc–pad braking system was developed with consideration of a thermal barrier coating (TBC) on the friction surface of the disc. The coating was made of functionally graded material (FGM). The three-element geometrical scheme of the system consisted of two homogeneous half-spaces (pad and disc) and a functionally graded coating (FGC) deposited on the friction surface of the disc. It was assumed that the frictional heat generated on the coating-pad contact surface was absorbed to the insides of friction elements along the normal to this surface. Thermal contact of friction between the coating and the pad as well as the heat contact between the coating and the substrate were perfect. On the basis of such assumptions, the thermal friction problem was formulated, and its exact solution was obtained for constant and linearly descending specific friction power over time. For the first case, the asymptotic solutions for small and large values of time were also found. A numerical analysis was performed on an example of the system containing a metal ceramic (FMC-11) pad, sliding on the surface of a FGC (ZrO_2–Ti-6Al-4V) applied on a cast iron (ChNMKh) disc. It was established that the application of a TBC made of FGM on the surface of a disc could effectively reduce the level of temperature achieved during braking.

Keywords: braking; frictional heating; functionally graded material; temperature; thermal barrier coating

1. Introduction

Frictional heating occurs when two contacting elements slide against each other, resulting in the conversion of mechanical energy into heat [1,2]. Modern braking systems operate based on this phenomenon [3]. Therefore, the main requirement for friction elements is resistance to elevated temperature, which can be improved by thermal barrier coatings applied on friction surfaces [4,5]. Due to their high thermal and wear resistance, ceramic materials are mostly used in the production of protective coatings for friction components [6,7]. However, in conventional coated elements, cracking may appear on the interface between the layer and a substrate, because of material properties mismatch [8]. To overcome this problem, functionally graded coatings have been introduced that possess smooth gradation of properties to reduce stress concentration on the interface and to reinforce the bond cohesion [9,10].

Accurate models of the frictional heating process are critical in the design of brakes, because they can provide insights into the temperature and thermal stress distributions initiated in the friction elements during braking actions [11]. Such models are developed on the basis of the thermal problems of friction which, in most cases, are obtainable only for bodies bounded by parallel planes (e.g., semi-spaces and strips) [12]. The simplest model used to formulate thermal problems of friction constitutes a single body, obtained after a virtual separation of the friction pair elements. Then, the friction interaction on the contact surface of the elements is replaced by a heat flux with an intensity proportional to the specific friction power. For this purpose, the heat partition ratio is introduced a priori to the model in order to determine the amount of heat absorbed by each element of the friction pair [13]. The most commonly used in the literature is the two-element geometric scheme, which represents both elements of the friction pair and considers heat generation on the contact area due to friction. Another scheme is a three-element model of a tribosystem, consisting of a semi-space sliding on the outer surface of a strip deposited on a semi-infinite substrate. Such a scheme can be adopted to simulate a frictional disc/pad/caliper system, consisting of a coated disc paired with a brake pad. A comparative analysis of the solutions to the thermal problems of friction for a disc brake system, obtained by means of two- and three-element models, was demonstrated in [14]. The three-element scheme considered the influence of the material properties of the caliper, on which the brake pad was placed, on the distribution of temperature and thermal stresses. Other transient thermal problems of friction for braking systems on the basis of a three-element scheme have been studied in [15–22]. The analytical solution to a boundary value problem of heat conduction for such a system was obtained under uniform sliding in a study by [8]. A three-element scheme to formulate thermal problems of friction for single braking with constant deceleration was used in [17,18]. The achieved solutions determined the temperature distribution and corresponding quasi-static thermal stresses in the disc, pad, and caliper of a tribosystem, both in the sliding phase during braking as well as in the cooling phase after braking action. The same scheme was used to consider a three-element braking system, i.e., a top semi-space (the disc) and a strip (the pad) deposited on a substrate (the caliper), in a study by [19]. The heat conduction problem was formulated and solved with time-dependent pressure and the assumption of imperfect thermal contact conditions on the disc–pad interface, in order to study the influence of heat resistance on the thermal behavior of the system. A generalization of this solution for a case considering fluctuations in pressure on the contact surface was presented in [20]. Such a temporal profile of contact pressure during braking with consideration of its oscillations was also considered in [23]. Asymptotic solutions (at large and small values of time) of the heat conduction problem for a three-element tribosystem with generalized boundary conditions on the sliding surface were obtained in [21]. All the mentioned studies concern the thermal problems of friction formulated for homogeneous materials. The thermal problems of friction for a three-element system with a brake pad made of periodic composite material was considered analytically in [22]. The assumption was made that the composite contained four sub-cells with rectangular cross sections, with different thermo-physical properties. The influence of geometrical dimensions of composite sub-cells on the maximum temperature in the system was investigated. However, modern friction elements can have far more complicated internal structure, such as functionally graded materials (FGMs), which have continuously changing properties throughout the volume of material [24]. Some analytical solutions to the heat conduction problems formulated for braking systems with functionally graded friction materials have already been obtained in [25–30]. However, these studies concerned simpler geometrical schemes, including a two FGM semi-spaces system [25,26]; functionally graded semi-infinite body coupled with homogeneous element [27,28]; and a one-element system consisting of a heating semi-space with deposited functionally graded coating, heated by the frictional heat flux on the friction surface [29]. A comparative analysis of frictional heating models formulated for FGMs by means of two-element and one-element schemes was performed in [30]. Based on the obtained results, a new heat partition ratio formula was proposed for a functionally graded friction couple.

In our previous paper [29], the problem we considered was a coated brake disc simulated using a homogeneous substrate with a deposited coating made of functionally graded material. The outer surface of the coating was assumed to be heated by the frictional heat flux. The current study is a continuation of that study, by introducing a second element in a friction pair, i.e., a brake pad, as the counterbody in a tribosystem. In this study, the thermal problem of friction is considered for a three-element system consisting of two homogeneous semi-spaces (pad and disc) and a functionally graded coating (FGC) is deposited on the friction surface of the disc, considering the heat generation on the disc–pad interface due to the friction during braking. The mathematical model obtained was based on the following algorithm:

(1) Formulation of the proper boundary value problem of heat conduction.
(2) Transition of the problem to the Laplace integral space.
(3) Finding the solutions in the form of Laplace transforms.
(4) Proceeding from the transforms of the solutions to the originals.
(5) Verification of the obtained solutions.
(6) Designating the asymptotic solutions for small and high values of the Fourier number.
(7) Performance of a numerical analysis for a selected case.

2. Problem Formulation

In this study, we consider the sliding contact of two semi-infinite bodies (half-spaces) taking into consideration the generation of heat due to friction (Figure 1). The upper half-space consists of a protective coating applied on the surface of the substrate. The materials of the substrate and lower semi-space (counterbody) are homogeneous, so their properties, i.e., thermal conductivity (K_l), specific heat (c_l), and density (ρ_l) are uniform throughout volumes of the elements, whereas the coating is made of a two-element functionally gradient material (FGM) with a thermal conductivity coefficient K_1 increasing exponentially along its thickness [31]:

$$K_1(z) = K_{1,1}e^{\gamma^* z/d},\ \gamma^* = \ln(K_{1,2}\,K_{1,1}^{-1}),\ 0 \le z \le d, \tag{1}$$

where d is the thickness of the coating, $\gamma^* \geq 0$ is the gradient parameter FGM [32], $K_{1,1} \equiv K_1(0)$ and $K_{1,2} \equiv K_1(d)$ are the thermal conductivity coefficients of the FGM constituents, and z is the spatial coordinate in the axial direction.

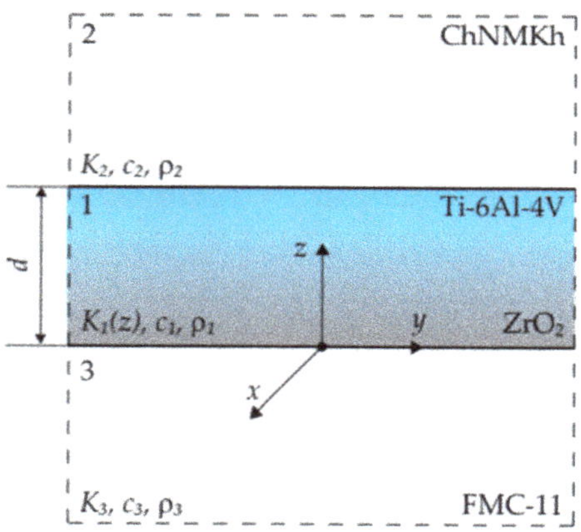

Figure 1. Scheme of the considered three-element tribosystem.

At the initial moment $t = 0$, the temperature T of all bodies in the system are constant and equal to T_0. Next, both semi-spaces, under the effect of pressure p_0 and acting parallel to the axis Oz, come into contact and simultaneously begin to slide in the positive direction of the axis Ox with constant velocity V_0. Due to friction on the contact surface $z = 0$, heat is generated, and the bodies heat up. Assuming that the friction thermal contact is perfect, i.e., at a set moment of time $t > 0$ the friction surfaces of the coating and counterbody are heated to the same temperature, and the sum of the intensities of heat fluxes, directed to the insides of the coating and counterbody along the normal to the contact surface, is equal to the specific friction power $q_0 = f p_0 V_0$, where f is the friction coefficient. The thermal connection of the coating with the substrate is perfect, i.e., the temperature and intensity of heat fluxes of these elements at the interface $z = d$ are the same. Thermal sensitivity of materials and wear are neglected. All assumptions in more detail are listed in our previous papers [25,26]. The aim of this study is to develop a mathematical model for the analytical determination of the temperature T of the considered three-element system at a fixed location $|z| < \infty$ at the selected time moment $t > 0$:

$$\frac{\partial}{\partial z}\left[K_1(z)\frac{\partial\Theta(z,t)}{\partial z}\right] = \rho_1 c_1 \frac{\partial\Theta(z,t)}{\partial t},\ 0 < z < d,\ t > 0, \tag{2}$$

$$K_2\frac{\partial^2\Theta(z,t)}{\partial z^2} = \rho_2 c_2 \frac{\partial\Theta(z,t)}{\partial t},\ z > d,\ t > 0, \tag{3}$$

$$K_3\frac{\partial^2\Theta(z,t)}{\partial z^2} = \rho_3 c_3 \frac{\partial\Theta(z,t)}{\partial t},\ z < d,\ t > 0, \tag{4}$$

$$\Theta(0^+,t) = \Theta(0^-,t),\ t > 0, \tag{5}$$

$$K_1(z)\frac{\partial\Theta(z,t)}{\partial z}\bigg|_{z=0^+} - K_3\frac{\partial\Theta(z,t)}{\partial z}\bigg|_{z=0^-} = -q_0,\ t > 0, \tag{6}$$

$$\Theta(d^+,t) = \Theta(d^-,t),\ t > 0, \tag{7}$$

$$K_1(z)\frac{\partial\Theta(z,t)}{\partial z}\bigg|_{z=d^+} = K_2\frac{\partial\Theta(z,t)}{\partial z}\bigg|_{z=d^-},\ t > 0, \tag{8}$$

$$\Theta(z,t) \to 0,\ |z| \to \infty,\ t > 0, \tag{9}$$

$$\Theta(z,0) = 0,\ |z| < \infty, \tag{10}$$

where K_2 and K_3 are the thermal conductivities of the substrate and counterbody materials, respectively; ρ_l and c_l, respectively, are the density and specific heat of materials of the coating $(l = 1)$, the substrate $(l = 2)$, and the counterbody $(l = 3)$.

Incorporating the following dimensionless variables and parameters:

$$\zeta = \frac{z}{d},\ \tau = \frac{k_1 t}{d^2},\ K_3^* = \frac{K_3}{K_{1,1}},\ k_2^* = \frac{k_2}{k_1},\ k_3^* = \frac{k_2}{k_1},\ \Theta^* = \frac{\Theta}{\Lambda}, \tag{11}$$

where

$$k_1 = \frac{K_{1,1}}{c_1\rho_1},\ k_2 = \frac{K_2}{c_2\rho_2},\ k_3 = \frac{K_3}{c_3\rho_3},\ \Lambda = \frac{q_0 d}{K_{1,1}}, \tag{12}$$

Problem (2)–(10) was written in the form:

$$\frac{\partial^2\Theta^*(\zeta,\tau)}{\partial\zeta^2} + \gamma^*\frac{\partial\Theta^*(\zeta,\tau)}{\partial\zeta} - e^{-\gamma^*\zeta}\frac{\partial\Theta^*(\zeta,\tau)}{\partial\tau} = 0,\ 0 < \zeta < 1,\ \tau > 0, \tag{13}$$

$$\frac{\partial^2\Theta^*(\zeta,\tau)}{\partial\zeta^2} - \frac{1}{k_2^*}\frac{\partial\Theta^*(\zeta,\tau)}{\partial\tau} = 0,\ \zeta > 1,\ \tau > 0, \tag{14}$$

$$\frac{\partial^2\Theta^*(\zeta,\tau)}{\partial\zeta^2} - \frac{1}{k_3^*}\frac{\partial\Theta^*(\zeta,\tau)}{\partial\tau} = 0,\ \zeta < 0,\ \tau > 0, \tag{15}$$

$$\Theta^*(0^+,\tau) = \Theta^*(0^-,\tau),\ \tau > 0, \tag{16}$$

$$\left.\frac{\partial\Theta^*(\zeta,\tau)}{\partial\zeta}\right|_{\zeta=0^+} - K_3^*\left.\frac{\partial\Theta^*(\zeta,\tau)}{\partial\zeta}\right|_{\zeta=0^-} = -1,\ \tau > 0, \tag{17}$$

$$\Theta(1^+,\tau) = \Theta(1^-,\tau),\ \tau > 0, \tag{18}$$

$$e^{\gamma^*}\left.\frac{\partial\Theta^*(\zeta,\tau)}{\partial\zeta}\right|_{\zeta=1^+} = K_2^*\left.\frac{\partial\Theta^*(\zeta,\tau)}{\partial\zeta}\right|_{\zeta=1^-},\ \tau > 0, \tag{19}$$

$$\Theta^*(\zeta,\tau) \to 0\,,\ |\zeta| \to \infty,\ \tau > 0, \tag{20}$$

$$\Theta^*(\zeta,0) = 0,\ |\zeta| < \infty. \tag{21}$$

3. Exact Solution

By means of the Laplace integral transform [33]:

$$\overline{\Theta}^*(\zeta,p) \equiv L[\Theta^*(\zeta,\tau);p] = \int_0^\infty \Theta^*(\zeta,\tau)e^{-p\tau}d\tau,\ \mathrm{Re}p \geq 0, \tag{22}$$

the boundary value problem (13)–(21) was reduced to the following boundary problem for a system of three ordinary differential equations of the second order:

$$\frac{d^2\overline{\Theta}^*(\zeta,p)}{d\zeta^2} + \gamma^*\frac{d\overline{\Theta}^*(\zeta,\tau)}{d\zeta} - pe^{-\gamma^*\zeta}\overline{\Theta}^*(\zeta,p) = 0,\ 0 < \zeta < 1, \tag{23}$$

$$\frac{d^2\overline{\Theta}^*(\zeta,p)}{d\zeta^2} - \frac{p}{k_2^*}\overline{\Theta}^*(\zeta,p) = 0,\ \zeta > 1, \tag{24}$$

$$\frac{d^2\overline{\Theta}^*(\zeta,p)}{d\zeta^2} - \frac{p}{k_3^*}\overline{\Theta}^*(\zeta,p) = 0,\ \zeta < 0, \tag{25}$$

$$\overline{\Theta}^*(0^+,p) = \overline{\Theta}^*(0^-,p), \tag{26}$$

$$\left.\frac{d\overline{\Theta}^*(\zeta,p)}{d\zeta}\right|_{\zeta=0^+} - K_3^*\left.\frac{d\overline{\Theta}^*(\zeta,p)}{d\zeta}\right|_{\zeta=0^-} = -\frac{1}{p}, \tag{27}$$

$$\overline{\Theta}^*(1^+,p) = \overline{\Theta}^*(1^-,p), \tag{28}$$

$$e^{\gamma^*}\left.\frac{d\overline{\Theta}^*(\zeta,p)}{d\zeta}\right|_{\zeta=1^+} = K_2^*\left.\frac{d\overline{\Theta}^*(\zeta,p)}{d\zeta}\right|_{\zeta=1^-}, \tag{29}$$

$$\overline{\Theta}^*(\zeta,p) \to 0\,,\ |\zeta| \to \infty. \tag{30}$$

The solution to problem (23)–(30) has the following form:

$$\overline{\Theta}^*(\zeta,p) = e^{-\tilde{\alpha}\zeta}\,\overline{\Theta}_0^*(p)\overline{\Theta}_1^*(\zeta,p),\ 0 \leq \zeta \leq 1, \tag{31}$$

$$\overline{\Theta}^*(\zeta,p) = \tilde{\alpha}\,\overline{\Theta}_0^*(p)\overline{\Theta}_2^*(\zeta,p),\ \zeta \geq 1, \tag{32}$$

$$\overline{\Theta}^*(\zeta,p) = \overline{\Theta}_0^*(p)\overline{\Theta}_3^*(\zeta,p),\ \zeta < 0, \tag{33}$$

where

$$\overline{\Theta_0^*}(p) = \frac{1}{\sqrt{p}},\ \overline{\Theta_1^*}(\zeta, p) = \frac{\Delta_1(\zeta, p)}{p\Delta(p)},\ \overline{\Theta_2^*}(\zeta, p) = \frac{e^{-\zeta_2\sqrt{p}}}{p\sqrt{p}\Delta(p)},\ \overline{\Theta_3^*}(\zeta, p) = \frac{\Delta_3(p)e^{-\zeta_3\sqrt{p}}}{p\Delta(p)}, \tag{34}$$

$$\Delta_1(\zeta, p) = A(p)\mathrm{I}_1(\zeta_1\sqrt{p}) + B(p)\mathrm{K}_1(\zeta_1\sqrt{p}),\ \Delta_3(p) = A(p)\mathrm{I}_1(\alpha\sqrt{p}) + B(p)\mathrm{K}_1(\alpha\sqrt{p}), \tag{35}$$

$$\Delta(p) = A(p)[\mathrm{I}_0(\alpha\sqrt{p}) + \varepsilon_3\mathrm{I}_1(\alpha\sqrt{p})] - B(p)[\mathrm{K}_0(\alpha\sqrt{p}) - \varepsilon_3\mathrm{K}_1(\alpha\sqrt{p})], \tag{36}$$

$$A(p) = \mathrm{K}_0(\beta\sqrt{p}) + \varepsilon_2\, e^{-\tilde{\alpha}}\mathrm{K}_1(\beta\sqrt{p}),\ B(p) = \mathrm{I}_0(\beta\sqrt{p}) - \varepsilon_2\, e^{-\tilde{\alpha}}\mathrm{I}_1(\beta\sqrt{p}), \tag{37}$$

$$\alpha = \frac{2}{\gamma^*}, \tilde{\alpha} = \frac{1}{\alpha}, \beta = \frac{\alpha}{e^{\tilde{\alpha}}}, \varepsilon_2 = \frac{K_2^*}{\sqrt{k_2^*}},\ \varepsilon_3 = \frac{K_3^*}{\sqrt{k_3^*}},\ \zeta_1 = \frac{\alpha}{e^{\tilde{\alpha}\zeta}},\ \zeta_2 = \frac{\zeta - 1}{\sqrt{k_2^*}}, \zeta_3 = \frac{|\zeta|}{\sqrt{k_3^*}}, \tag{38}$$

where $\mathrm{I}_n(x)$ and $\mathrm{K}_n(x)$ are the modified Bessel functions of the nth order of the first and second kind, respectively [34].

Considering the forms of the transformed solutions (31)–(33) and based on the convolution theorem of the two functions, the dimensionless temperature rises were found [33]:

$$\Theta^*(\zeta, \tau) = e^{-\tilde{\alpha}\zeta}\int_0^\tau \Theta_0^*(\tau - s)\Theta_1^*(\zeta, s)ds,\ 0 \le \zeta \le 1,\ \tau \ge 0, \tag{39}$$

$$\Theta^*(\zeta, \tau) = \tilde{\alpha}\int_0^\tau \Theta_0^*(\tau - s)\Theta_2^*(\zeta, s)ds,\ \zeta \ge 1,\ \tau \ge 0, \tag{40}$$

$$\Theta^*(\zeta, \tau) = \int_0^\tau \Theta_0^*(\tau - s)\Theta_3^*(\zeta, s)ds,\ \zeta \le 0,\ \tau \ge 0, \tag{41}$$

where

$$\Theta_0^*(\tau) \equiv L^{-1}[\overline{\Theta_0^*}(p); \tau] = \frac{1}{\sqrt{\pi\tau}}, \tag{42}$$

$$\Theta_l^*(\zeta, \tau) \equiv L^{-1}[\overline{\Theta_l^*}(\zeta, p);\ \tau] = \frac{1}{2\pi i}\int_{\omega - i\,\infty}^{\omega + i\,\infty} \overline{\Theta_l^*}(\zeta, p)e^{p\tau}dp,\ l = 1,2,3,\ \omega \equiv \mathrm{Re}p > 0,\ i \equiv \sqrt{-1}. \tag{43}$$

Integration on the complex plane ($\mathrm{Re}p$, $\mathrm{Im}p$) in Equation (43) was carried out according to the methodology described in detail in [29] and using the following relations [34]:

$$\mathrm{I}_0(\pm ix) = \mathrm{J}_0(x),\ \mathrm{K}_0(\pm ix) = -0.5\pi[\mathrm{Y}_0(x) \pm i\mathrm{J}_0(x)], \tag{44}$$

$$\mathrm{I}_1(\pm ix) = \pm i\mathrm{J}_1(ix),\ \mathrm{K}_1(\pm ix) = -0.5\pi[\mathrm{J}_1(x) \mp i\mathrm{Y}_1(x)], \tag{45}$$

where $\mathrm{J}_n(x)$ and $\mathrm{Y}_n(x)$ are Bessel functions of the nth order of the first and second kind, respectively. As a result, it was obtained:

$$\Theta_1^*(\zeta, \tau) = \frac{e^{\tilde{\alpha}\zeta}}{\varepsilon_2 + \varepsilon_3} + \frac{2}{\pi}\int_0^\infty \frac{\Phi_1(\zeta, x)}{x\Psi(x)}e^{-x^2\tau}dx,\ 0 \le \zeta \le 1,\ \tau \ge 0, \tag{46}$$

$$\Theta_2^*(\zeta, \tau) = \frac{\alpha}{\varepsilon_2 + \varepsilon_3} - \frac{4}{\pi^2}\int_0^\infty \frac{\Phi_2(\zeta, x)}{x^2\Psi(x)}e^{-x^2\tau}dx,\ \zeta \ge 1,\ \tau \ge 0, \tag{47}$$

$$\Theta_3^*(\zeta,\tau)=\frac{1}{\varepsilon_2+\varepsilon_3}+\frac{2}{\pi}\int_0^{\infty}\frac{\Phi_3(\zeta,x)}{x\Psi(x)}e^{-x^2\tau}dx,\ \zeta\leq 0,\ \tau\geq 0, \tag{48}$$

where

$$\Phi_1(\zeta,x)=\Delta_{\mathrm{R}}(x)\Delta_{1,\mathrm{I}}(\zeta_1,x)-\Delta_{\mathrm{I}}(x)\Delta_{1,\mathrm{R}}(\zeta_1,x), \tag{49}$$

$$\Phi_2(\zeta,x)=\Delta_{\mathrm{R}}(x)\cos(\zeta_2 x)-\Delta_{\mathrm{I}}(x)\sin(\zeta_2\, x), \tag{50}$$

$$\begin{aligned}\Phi_3(\zeta,x)=&[\Delta_{\mathrm{R}}(x)\Delta_{3,\mathrm{I}}(x)-\Delta_{\mathrm{I}}(x)\Delta_{3,\mathrm{R}}(x)]\cos(\zeta_3 x)-\\&[\Delta_{\mathrm{R}}(x)\Delta_{3,\mathrm{R}}(x)+\Delta_{\mathrm{I}}(x)\Delta_{3,\mathrm{I}}(x)]\sin(\zeta_3 x)\,,\end{aligned} \tag{51}$$

$$\Psi(x)=\Delta_{\mathrm{R}}^2(x)+\Delta_{\mathrm{I}}^2(x), \tag{52}$$

$$\Delta_{\mathrm{R}}(x)=\mathrm{Y}_0(\alpha x)\mathrm{J}_0(\beta x)-\mathrm{J}_0(\alpha x)\mathrm{Y}_0(\beta x)+\varepsilon_3\Delta_{3,\mathrm{R}}(x), \tag{53}$$

$$\Delta_{\mathrm{I}}(x)=\varepsilon_3\Delta_{3,\mathrm{I}}(x)-\varepsilon_2 e^{-\tilde{\alpha}}[\mathrm{Y}_0(\alpha x)\mathrm{J}_1(\beta x)-\mathrm{J}_0(\alpha x)\mathrm{Y}_1(\beta x)], \tag{54}$$

$$\Delta_{1,\mathrm{R}}(\zeta,x)=\varepsilon_2 e^{-\tilde{\alpha}}[\mathrm{J}_1(\beta x)\mathrm{Y}_1(\zeta_1 x)-\mathrm{Y}_1(\beta x)\mathrm{J}_1(\zeta_1 x)], \tag{55}$$

$$\Delta_{1,\mathrm{I}}(\zeta,x)=\mathrm{J}_0(\beta x)\mathrm{Y}_1(\zeta_1 x)-\mathrm{Y}_0(\beta x)\mathrm{J}_1(\zeta_1 x), \tag{56}$$

$$\Delta_{3,\mathrm{R}}(x)=\varepsilon_2 e^{-\tilde{\alpha}}[\mathrm{Y}_1(\alpha x)\mathrm{J}_1(\beta x)-\mathrm{J}_1(\alpha x)\mathrm{Y}_1(\beta x)], \tag{57}$$

$$\Delta_{3,\mathrm{I}}(x)=\mathrm{Y}_1(\alpha x)\mathrm{J}_0(\beta x)-\mathrm{J}_1(\alpha x)\mathrm{Y}_0(\beta x). \tag{58}$$

Taking into consideration the functions $\Theta_0^*(\tau)$ (42) and $\Theta_l^*(\zeta,\tau), l=1,\,2,\,3$ (46)–(48) in Equations (39)–(41), after performing the integration, the sought dimensionless temperature rise was found in the form:

$$\Theta^*(\zeta,\tau)=2\sqrt{\frac{\tau}{\pi}}\left[\frac{1}{\varepsilon_2+\varepsilon_3}+\frac{2}{\pi}e^{-0.5\gamma^*\zeta}\int_0^{\infty}\frac{\Phi_1(\zeta,x)}{x\Psi(x)}\mathrm{F}(x\sqrt{\tau})dx\right],\ 0\leq\zeta\leq 1,\ \tau\geq 0, \tag{59}$$

$$\Theta^*(\zeta,\tau)=2\sqrt{\frac{\tau}{\pi}}\left[\frac{1}{\varepsilon_2+\varepsilon_3}-\frac{2\gamma^*}{\pi^2}\int_0^{\infty}\frac{\Phi_2(\zeta,x)}{x^2\Psi(x)}\mathrm{F}(x\sqrt{\tau})dx\right],\ \zeta\geq 1,\ \tau\geq 0, \tag{60}$$

$$\Theta^*(\zeta,\tau)=2\sqrt{\frac{\tau}{\pi}}\left[\frac{1}{\varepsilon_2+\varepsilon_3}+\frac{2}{\pi}\int_0^{\infty}\frac{\Phi_3(\zeta,x)}{x\Psi(x)}\mathrm{F}(x\sqrt{\tau})dx\right],\ \zeta\leq 0,\ \tau\geq 0, \tag{61}$$

where

$$\mathrm{F}(x)=\frac{e^{-x^2}}{x}\int_0^{x}e^{s^2}ds. \tag{62}$$

To calculate the function $\mathrm{F}(x)$ (62) the following approximation formulas were used [35]:

$$\mathrm{F}(x)=\sum_{n=0}^{\infty}(-1)^n\frac{(2x^2)^n}{(2n+1)!!},\ 0<x<3,\ \mathrm{F}(x)=\sum_{n=0}^{N}\frac{(2n-1)!!}{(2x^2)^{n+1}},\ x\geq 3. \tag{63}$$

where $(-1)!!=1$, $(2n+1)!!=1\times 3\times 5\times\ldots\times(2n+1)$.

4. Verification of the Solution

Correctness of solutions (59)–(61) will be shown by proving that they satisfy the boundary conditions (16)–(20) and the initial condition (21). By comparing the forms of solutions (59) and (61), it follows that the Equation condition (16) of the temperature of the coating and counterbody on the contact surface $\zeta = 0$ will be met if:

$$\Phi_1(0^+, x) = \Phi_3(0^-, x). \quad (64)$$

Substituting in Formulas (55)–(58) $\zeta = 0$ ($\zeta_- = \alpha$) it was found that

$$\Delta_{1,R}(0, x) = \Delta_{3,R}(x), \; \Delta_{1,I}(0, x) = \Delta_{3,I}(x), \quad (65)$$

from where, on the basis of Equations (49) and (51), we obtain the Equation (64).

Comparing solutions (59) and (60), it can be seen that condition (18) of the temperature equality at the interface $\zeta = 1$ will be met, when

$$e^{-0.5\gamma^*} \pi x \, \Phi_1(1^+, x) + \gamma^* \Phi_2(1^-, x) = 0. \quad (66)$$

For $\zeta = 1$ from Equation (38), it follows that $\zeta_1 = \beta$ and $\zeta_2 = 0$. Then, Formulas (55) and (56) yield [34]:

$$\Delta_{1,R}(1, x) = 0, \; \Delta_{1,I}(1, x) = J_0(\beta x) Y_1(\beta x) - Y_0(\beta x) J_1(\beta x) \equiv -2(\pi \beta x)^{-1}. \quad (67)$$

Including definition (38) of parameter β, from Equations (49), (50) and (67) it was found:

$$\Phi_1(1^+, x) = -\frac{\gamma^* e^{0.5\gamma^*}}{\pi x} \Delta_R(x), \; \Phi_2(1^-, x) = \Delta_R(x), \quad (68)$$

which proves that Equation (66) is fulfilled, which means that the boundary condition (18) is met.

After differentiating solution (59) with respect to a spatial variable, ζ (the corresponding derivatives hereinafter are denoted by the symbol « ′ ») was obtained:

$$\Theta'^*(\zeta, \tau) = \frac{2}{\pi} e^{-0.5\gamma^* \zeta} \sqrt{\tau} \int_0^\infty \frac{[\Phi_1'(\zeta, x) - 0.5\gamma^* \Phi_1(\zeta, x)]}{x \Psi(x)} F(x\sqrt{\tau}) dx, \; 0 \le \zeta \le 1, \; \tau \ge 0, \quad (69)$$

where, from Equation (49), it yields:

$$\Phi_1'(\zeta, x) = \Delta_R(x) \Delta_{1,I}'(\zeta, x) - \Delta_I(x) \Delta_{1,R}'(\zeta, x). \quad (70)$$

Considering derivatives [34]:

$$J_1'(x) = J_0(x) - x^{-1} J_1(x), \; Y_1'(x) = Y_0(x) - x^{-1} Y_1(x), \quad (71)$$

from Formulas (53)–(56) it was found:

$$\Delta_{1,R}'(\zeta, x) = 0.5\gamma^* \Delta_{1,R}(\zeta, x) - x e^{-0.5\gamma^* \zeta} \hat{\Delta}_I(\zeta, x), \quad (72)$$

$$\Delta_{1,I}'(\zeta, x) = 0.5\gamma^* \Delta_{1,I}(\zeta, x) - x e^{-0.5\gamma^* \zeta} \hat{\Delta}_R(\zeta, x), \quad (73)$$

where

$$\hat{\Delta}_R(\zeta, x) = J_0(\beta x) Y_0(\zeta_1 x) - Y_0(\beta x) J_0(\zeta_1 x), \quad (74)$$

$$\hat{\Delta}_I(\zeta, x) = \varepsilon_2 e^{-0.5\gamma^*} [J_1(\beta x) Y_0(\zeta_1 x) - Y_1(\beta x) J_0(\zeta_1 x)]. \quad (75)$$

Substituting derivatives (72)–(75) to the right side of Equation (70), it was obtained:

$$\Phi_1'(\zeta,x) = 0.5\gamma^*\Phi_1(\zeta,x) - xe^{-0.5\gamma^*\zeta}\hat{\Phi}_1(\zeta,x), \tag{76}$$

where

$$\hat{\Phi}_1(\zeta,x) = \Delta_R(x)\hat{\Delta}_R(\zeta,x) - \Delta_I(x)\hat{\Delta}_I(\zeta,x). \tag{77}$$

Taking into consideration the Formulas (76) and (77), the derivative of function $\Phi_1(\zeta,x)$ in Equation (69) gives:

$$\Theta'^*(\zeta,\tau) = -\frac{2}{\pi}e^{-\gamma^*\zeta}\sqrt{\tau}\int_0^\infty \frac{\hat{\Phi}_1(\zeta,x)}{\Psi(x)}\,\mathrm{F}(x\sqrt{\tau})dx,\ 0 \le \zeta \le 1,\ \tau \ge 0. \tag{78}$$

Next, differentiating solutions (60) and (61), it was found:

$$\Theta'^*(\zeta,\tau) = -\frac{2\gamma^*}{\pi^2}\sqrt{\tau}\int_0^\infty \frac{\Phi_2'(\zeta,x)}{x^2\Psi(x)}\,\mathrm{F}(x\sqrt{\tau})dx,\ \zeta \ge 1,\ \tau \ge 0, \tag{79}$$

$$\Theta'^*(\zeta,\tau) = \frac{2}{\pi}\sqrt{\tau}\int_0^\infty \frac{\Phi_3'(\zeta,x)}{x\Psi(x)}\,\mathrm{F}(x\sqrt{\tau})dx,\ \zeta \le 0,\ \tau \ge 0, \tag{80}$$

where, based on relations (50)–(54) and (57) and (58), the following was determined:

$$\Phi_2'(\zeta,x) = -\frac{x}{\sqrt{k_2^*}}[\Delta_R(x)\sin(\zeta_2 x) + \Delta_I(x)\cos(\zeta_2\, x)], \tag{81}$$

$$\begin{aligned}\Phi_3'(\zeta,x) = -\tfrac{x}{\sqrt{k_3^*}}\{&[\Delta_R(x)\Delta_{3,I}(x) - \Delta_I(x)\Delta_{3,R}(x)]\sin(\zeta_3 x) - \\ &[\Delta_R(x)\Delta_{3,R}(x) + \Delta_I(x)\Delta_{3,I}(x)]\cos(\zeta_3 x)\}.\end{aligned} \tag{82}$$

Substituting the derivatives (78), (80), and (82) for $\zeta = 0$ ($\zeta_1 = \alpha$,$\zeta_3 = 0$) to the left side of boundary condition (17), it was achieved:

$$\Theta'^*(0^+,\tau) - K_3^*\Theta'^*(0^-,\tau) = -\frac{2}{\pi}\sqrt{\tau}\int_0^\infty \frac{\{\hat{\Phi}_1(0,x) + \varepsilon_3[\Delta_R(x)\Delta_{3,R}(x) + \Delta_I(x)\Delta_{3,I}(x)]\}}{\Psi(x)}\,\mathrm{F}(x\sqrt{\tau})dx,\ \tau \ge 0. \tag{83}$$

Considering relation (77), the following was written:

$$\begin{aligned}&\hat{\Phi}_1(0,x) + \varepsilon_3[\Delta_R(x)\Delta_{3,R}(x) + \Delta_I(x)\Delta_{3,I}(x)] = \\ &\quad \Delta_R(x)[\hat{\Delta}_R(0,x) + \varepsilon_3\Delta_{3,R}(x)] - \Delta_I(x)[\hat{\Delta}_I(0,x) - \varepsilon_3\Delta_{3,I}(x)].\end{aligned} \tag{84}$$

Then, taking into consideration the forms of functions $\Delta_{3,R}(x)$ (57) and $\Delta_{3,I}(x)$ (58) and the following functions from Equations (74) and (75):

$$\hat{\Delta}_R(0,x) = \mathrm{Y}_0(\alpha x)\mathrm{J}_0(\beta x) - \mathrm{J}_0(\alpha x)\mathrm{Y}_0(\beta x), \tag{85}$$

$$\hat{\Delta}_I(0,x) = \varepsilon_2 e^{-0.5\gamma^*}[\mathrm{Y}_0(\alpha x)\mathrm{J}_1(\beta x) - \mathrm{J}_0(\alpha x)\mathrm{Y}_1(\beta x)], \tag{86}$$

it was established that

$$\hat{\Delta}_R(0,x) + \varepsilon_3\Delta_{3,R}(x) = \Delta_R(x),\ \hat{\Delta}_I(0,x) - \varepsilon_3\Delta_{3,I}(x) = -\Delta_I(x), \tag{87}$$

where the functions $\Delta_R(x)$ and $\Delta_I(x)$ have the Formulas (54) and (55), respectively. Considering the results (87) in the right side of Equation (84), as well as the function $\Psi(x)$ (52), the Equation (83) can be written in the form:

$$\Theta'^*(0^+,\tau) - K_3^*\Theta'^*(0^-,\tau) = -\frac{2}{\pi}\sqrt{\tau}\int_0^\infty F(x\sqrt{\tau})dx,\ \tau \geq 0. \quad (88)$$

Bearing in mind that [35]

$$\int_0^\infty F(x)dx = \frac{\pi}{2}, \quad (89)$$

from the Equation (88), it follows that the boundary condition (17) has been fulfilled.

Substituting $\zeta = 1$ ($\zeta_1 = \beta,\ \zeta_2 = 0$) in Equations (78) and (79) gives:

$$e^{\gamma^*}\Theta'^*(1^+,\tau) = -\frac{2}{\pi}\sqrt{\tau}\int_0^\infty \frac{\hat{\Phi}_1(1,x)}{\Psi(x)}F(x\sqrt{\tau})dx,\ \tau \geq 0, \quad (90)$$

$$K_2^*\Theta'^*(1^-,\tau) = -\frac{2\gamma^* K_2^*}{\pi^2}\sqrt{\tau}\int_0^\infty \frac{\Phi_2'(1,x)}{x^2\Psi(x)}F(x\sqrt{\tau})dx,\ \tau \geq 0, \quad (91)$$

where, based on relations (77) and (81), it was found:

$$\hat{\Phi}_1(1^+,x) = \Delta_R(x)\hat{\Delta}_R(1,x) - \Delta_I(x)\hat{\Delta}_I(1,x),\ K_2^*\Phi_2'(1^-,x) = -\varepsilon_2 x\Delta_I(x). \quad (92)$$

From (74) and (75), it follows that [34]

$$\hat{\Delta}_R(1,x) = 0,\ \hat{\Delta}_I(1,x) = \varepsilon_2 e^{-0.5\gamma^*}[J_1(\beta x)Y_0(\beta x) - Y_1(\beta x)J_0(\beta x)] \equiv \varepsilon_2\gamma^*(\pi x)^{-1}, \quad (93)$$

and the first Equation (92) takes the form

$$\hat{\Phi}_1(1^+,x) = -\varepsilon_2\gamma^*(\pi x)^{-1}\Delta_I(x). \quad (94)$$

Considering Equations (93) and (94) in the right sides of Equations (90) and (91), it was determined:

$$e^{\gamma^*}\Theta'^*(1^+,\tau) = \frac{2}{\pi^2}\varepsilon_2\gamma^*\sqrt{\tau}\int_0^\infty \frac{\Delta_I(x)}{x\Psi(x)}F(x\sqrt{\tau})dx = K_2^*\Theta'^*(1^-,\tau), \quad (95)$$

which confirms that the obtained solutions meet the boundary condition (19).

Fulfillment of condition (20) of the disappearance of dimensionless temperature rises (60) and (61) for $\zeta \to \infty$ was considered by rejecting the terms $e^{\zeta_2\sqrt{p}}$ and $e^{\zeta_3\sqrt{p}}$ while solving the boundary problem (23)–(33) and was verified by numerical calculations. Finally, it should be noted that solutions (59)–(61) also satisfy the initial condition (21).

5. Some Special Cases of Solution

In addition to the exact (in quadrature) solutions (59)–(62), appropriate asymptotic solutions were also obtained for small and large values of the Fourier number (dimensionless time) τ (11).

Small values of τ *(large values of parameter* p*).* Including Formulas (34)–(37) asymptotes of the modified Bessel functions for large argument values [34]:

$$I_n(x) \cong \frac{e^x}{\sqrt{2\pi x}},\ K_n(x) \cong \sqrt{\frac{\pi}{2x}}e^{-x},\ n = 0,\ 1,\ \ldots, \quad (96)$$

transformed solutions (31)–(33) were written in the forms:

$$\overline{\Theta}^{*}(\zeta,p) \cong \frac{e^{-0.25\gamma^{*}\zeta}}{(1+\varepsilon_3)} \frac{e^{-(\alpha-\zeta_1)\sqrt{p}}}{p\sqrt{p}},\ 0 \le \zeta < 1, \quad (97)$$

$$\overline{\Theta}^{*}(\zeta,p) \cong \frac{2e^{-0.25\gamma^{*}}}{(1+\varepsilon_2 e^{-0.5\gamma^{*}})(1+\varepsilon_3)} \frac{e^{-(\alpha-\beta+\zeta_2)\sqrt{p}}}{p\sqrt{p}},\ \zeta \ge 1, \quad (98)$$

$$\overline{\Theta}^{*}(\zeta,p) \cong \frac{e^{-\zeta_3\sqrt{p}}}{(1+\varepsilon_3)p\sqrt{p}},\ \zeta \le 0, \quad (99)$$

where, based on definition (38), it was obtained:

$$\alpha-\zeta_1 = \frac{2}{\gamma^{*}}(1-e^{-0.5\gamma^{*}\zeta}) \ge 0,\ \alpha-\beta+\zeta_2 = \frac{2}{\gamma^{*}}(1-e^{-0.5\gamma^{*}}) + \frac{\zeta-1}{\sqrt{k_2^{*}}} > 0,\ \zeta_3 \ge 0. \quad (100)$$

Proceeding from transforms (97)–(100) to the originals by means of the relation [36]:

$$L^{-1}\left[\frac{e^{-\lambda\sqrt{p}}}{p\sqrt{p}};\tau\right] = 2\sqrt{\tau}\,\mathrm{ierfc}\left(\frac{\lambda}{2\sqrt{\tau}}\right),\ \lambda \ge 0, \quad (101)$$

asymptotes of the dimensionless temperature rise in the initial moments of the heating process were found in the forms:

$$\Theta^{*}(\zeta,\tau) \cong \frac{2e^{-0.25\gamma^{*}\zeta}\sqrt{\tau}}{(1+\varepsilon_3)}\mathrm{ierfc}\left(\frac{\alpha-\zeta_1}{2\sqrt{\tau}}\right),\ 0 \le \zeta < 1,\ 0 \le \tau << 1, \quad (102)$$

$$\Theta^{*}(\zeta,\tau) \cong \frac{4e^{-0.25\gamma^{*}}\sqrt{\tau}}{(1+\varepsilon_2 e^{-0.5\gamma^{*}})(1+\varepsilon_3)}\ \mathrm{ierfc}\left(\frac{\alpha-\beta+\zeta_2}{2\sqrt{\tau}}\right),\ \zeta \ge 1,\ 0 \le \tau << 1, \quad (103)$$

$$\Theta^{*}(\zeta,\tau) \cong \frac{2\sqrt{\tau}}{(1+\varepsilon_3)}\mathrm{ierfc}\left(\frac{\zeta_3}{2\sqrt{\tau}}\right),\ \zeta \le 0,\ 0 \le \tau << 1, \quad (104)$$

where $\mathrm{ierfc}(x) = \pi^{-0.5}e^{-x^2} - x\,\mathrm{erfc}(x)$, $\mathrm{erfc}(x) = 1 - \mathrm{erf}(x)$, and $\mathrm{erf}(x)$ are Gauss error functions [34].

Large values of τ (small values of parameter p). For small argument values, the modified Bessel functions behave as follow [34]:

$$\mathrm{I}_0(x) \cong 1,\ \mathrm{I}_1(x) \cong 0.5x,\ \mathrm{K}_0(x) \cong -\ln(x),\ \mathrm{K}_1(x) \cong x^{-1}. \quad (105)$$

Considering asymptotes (105) in Equations (34)–(37), Laplace transforms (31)–(33) were presented in the forms:

$$\overline{\Theta}^{*}(\zeta,p) \cong \frac{1}{b}\left[\frac{1}{p\sqrt{p}(\sqrt{p}+c)} + \frac{\varsigma}{p(\sqrt{p}+c)}\right],\ 0 \le \zeta \le 1, \quad (106)$$

$$\overline{\Theta}^{*}(\zeta,p) \cong \frac{e^{-\zeta_2\sqrt{p}}}{b\,p\sqrt{p}(\sqrt{p}+c)},\ \zeta \ge 1, \quad (107)$$

$$\overline{\Theta}^{*}(\zeta,p) \cong \frac{e^{-\zeta_3\sqrt{p}}}{b}\left[\frac{1}{p\sqrt{p}(\sqrt{p}+c)} + \frac{a}{p(\sqrt{p}+c)}\right],\ \zeta \ge 0, \quad (108)$$

where

$$a = \varepsilon_2\frac{(1-e^{-\gamma^{*}})}{\gamma^{*}},\ b = 1+\varepsilon_3 a,\ c = \frac{\varepsilon_2+\varepsilon_3}{b},\ \varsigma = \varepsilon_2\frac{(e^{-\gamma^{*}\zeta}-e^{-\gamma^{*}})}{\gamma^{*}}. \quad (109)$$

Bearing in mind that [36]

$$L^{-1}\left[\frac{c\,e^{-\lambda\sqrt{p}}}{p(\sqrt{p}+c)};\tau\right]=\mathrm{erfc}\left(\frac{\lambda}{2\sqrt{\tau}}\right)-e^{\lambda c+c^2\tau}\mathrm{erfc}\left(\frac{\lambda}{2\sqrt{\tau}}+c\sqrt{\tau}\right),\ c>0,\ \lambda\geq 0, \quad (110)$$

$$L^{-1}\left[\frac{c^2e^{-\lambda\sqrt{p}}}{p\sqrt{p}(\sqrt{p}+c)};\tau\right]=2c\sqrt{\frac{\tau}{\pi}}\,e^{-\left(\frac{\lambda}{2\sqrt{\tau}}\right)^2}-(1+\lambda c)\mathrm{erfc}\left(\frac{\lambda}{2\sqrt{\tau}}\right)+e^{\lambda c+c^2\tau}\mathrm{erfc}\left(\frac{\lambda}{2\sqrt{\tau}}+c\sqrt{\tau}\right), \quad (111)$$

the following asymptotes of the dimensionless temperature rise at high values of the Fourier number τ were obtained:

$$\Theta^*(\zeta,\tau)\cong\frac{1}{(\varepsilon_2+\varepsilon_3)}\left\{2\sqrt{\frac{\tau}{\pi}}-\left(\frac{1}{c}-\varsigma\right)[1-e^{c^2\tau}\mathrm{erfc}(c\sqrt{\tau})]\right\},\ 0\leq\zeta\leq 1,\tau>>1, \quad (112)$$

$$\Theta^*(\zeta,\tau)\cong\frac{1}{(\varepsilon_2+\varepsilon_3)}\left\{2\sqrt{\frac{\tau}{\pi}}\,e^{-\left(\frac{\zeta_2}{2\sqrt{\tau}}\right)^2}-\left(\frac{1}{c}+\zeta_2\right)\mathrm{erfc}\left(\frac{\zeta_2}{2\sqrt{\tau}}\right)+\frac{1}{c}e^{c\zeta_2+c^2\tau}\mathrm{erfc}\left(\frac{\zeta_2}{2\sqrt{\tau}}+c\sqrt{\tau}\right)\right\},\ \zeta\geq 1,\ \tau>>1, \quad (113)$$

$$\begin{aligned}\Theta^*(\zeta,\tau)\cong\tfrac{1}{(\varepsilon_2+\varepsilon_3)}\Big\{&2\sqrt{\tfrac{\tau}{\pi}}\,e^{-\left(\frac{\zeta_3}{2\sqrt{\tau}}\right)^2}-\left(\tfrac{1}{c}-a+\zeta_3\right)\mathrm{erfc}\left(\tfrac{\zeta_3}{2\sqrt{\tau}}\right)+\\&\left(\tfrac{1}{c}-a\right)e^{c\zeta_3+c^2\tau}\mathrm{erfc}\left(\tfrac{\zeta_3}{2\sqrt{\tau}}+c\sqrt{\tau}\right)\Big\},\ \zeta\leq 0,\ \tau>>1.\end{aligned} \quad (114)$$

On the friction surface $\zeta=0$ and from Equations (38) and (109), it follows that $\zeta_3=0$ and $\varsigma=a$, and from solutions (112) and (114) it was determined:

$$\Theta^*(0^+,\tau)=\Theta^*(0^-,\tau)\cong\frac{1}{(\varepsilon_2+\varepsilon_3)}\left\{2\sqrt{\frac{\tau}{\pi}}-\left(\frac{1}{c}-a\right)[1-e^{c^2\tau}\mathrm{erfc}(c\sqrt{\tau})]\right\},\ \tau>>1, \quad (115)$$

In a similar way, considering that on the interface $\zeta=1$ we have $\zeta_2=0$ and $\varsigma=0$, from solutions (112) and (113) it was found:

$$\Theta^*(1^+,\tau)=\Theta^*(1^-,\tau)\cong\frac{1}{(\varepsilon_2+\varepsilon_3)}\left\{2\sqrt{\frac{\tau}{\pi}}-\frac{1}{c}[1-e^{c^2\tau}\mathrm{erfc}(c\sqrt{\tau})]\right\},\ \tau>>1. \quad (116)$$

Linearly descending temporal profile of specific friction power. The exact solutions (59)–(61) presented above were obtained with the specific friction power q_0 remaining constant over time. Whereas, for modeling the frictional heating process in disc brake systems, the time profile of the specific friction power is most often used in the form [37]:

$$q(t)=q_0q^*(t),\ q^*(t)=1-t\,t_s^{-1},\ 0\leq t\leq t_s,\ \tau>>1, \quad (117)$$

where t_s is the moment of stopping the vehicle, and thus the final moment of the heating process. The dimensionless temperature rise $\hat{\Theta}^*(\zeta,\tau)$, corresponding to the specific friction power (117) was found based on the Duhamel's theorem [38,39]:

$$\hat{\Theta}^*(\zeta,\tau)=\frac{\partial}{\partial\tau}\int_0^{\tau}q^*(\tau-s)\Theta^*(\zeta,s)ds,\ \zeta\geq 0,\ 0\leq\tau\leq\tau_s, \quad (118)$$

where $\Theta^*(\zeta,\tau)$ is dimensionless temperature increase (59)–(61), and function $q^*(\tau)$ has the following form:

$$q^*(\tau)=1-\tau\,\tau_s^{-1},\ 0\leq\tau\leq\tau_s,\ \tau_s=k_1t_sd^{-2}. \quad (119)$$

Substituting solutions (59)–(61) and function $q^*(\tau)$ (119) as the integrand in the right side of Equation (118), after performing the integration according to the methodology from [29], it was found:

$$\Theta^*(\zeta,\tau) = 2\sqrt{\frac{\tau}{\pi}}\left[\frac{P(\tau)}{\varepsilon_2+\varepsilon_3} + \frac{2}{\pi}e^{-0.5\gamma^*\zeta}\int_0^\infty \frac{\Phi_1(\zeta,x)}{x\Psi(x)}Q(\tau,x)dx\right],\ 0 \le \zeta \le 1,\ 0 \le \tau \le \tau_s, \quad (120)$$

$$\Theta^*(\zeta,\tau) = 2\sqrt{\frac{\tau}{\pi}}\left[\frac{P(\tau)}{\varepsilon_2+\varepsilon_3} - \frac{2\gamma^*}{\pi^2}\int_0^\infty \frac{\Phi_2(\zeta,x)}{x^2\Psi(x)}Q(\tau,x)dx\right],\ \zeta \ge 1,\ 0 \le \tau \le \tau_s, \quad (121)$$

$$\Theta^*(\zeta,\tau) = 2\sqrt{\frac{\tau}{\pi}}\left[\frac{P(\tau)}{\varepsilon_2+\varepsilon_3} + \frac{2}{\pi}\int_0^\infty \frac{\Phi_3(\zeta,x)}{x\Psi(x)}Q(\tau,x)dx\right],\ \zeta \le 0,\ 0 \le \tau \le \tau_s, \quad (122)$$

where

$$P(\tau) = 1 - \frac{2\tau}{3\tau_s},\ Q(\tau,x) = \left(1 + \frac{1}{x^2\tau_s}\right)\mathrm{F}(x\sqrt{\tau}) - \frac{2}{\sqrt{\pi}x^2\tau_s}, \quad (123)$$

and functions $\Phi_k(\zeta,x), k = 1,2,3$ are determined from (49)–(51) and (53)–(58).

6. Numerical Analysis

Calculations were carried out for a friction pair consisting of two half-spaces; one has a two-component FGM coating applied on the substrate, and the other (counterbody) slides on the working surface of the FGC with constant or linearly decreasing velocity. The base and core of the FGM are, respectively, zirconium dioxide ZrO_2 and titanium alloy Ti-6Al-4V. The substrate is ChNMKh gray cast iron, and the counterbody is cermet FMC-11. The properties of these materials, at initial temperature $T_0 = 20°C$, are given in Table 1.

Table 1. Materials properties [23,29].

Material	Thermal Conductivity $Wm^{-1}K^{-1}$	Specific Heat $J\,kg^{-1}K^{-1}$	Density $kg\,m^{-3}$
ZrO_2	$K_{1,1} = 1.94$	$c_{1,1} = 452.83$	$\rho_{1,1} = 6102.16$
Ti-6Al-4V	$K_{1,2} = 6.87$	$c_{1,2} = 538.08$	$\rho_{1,2} = 4431.79$
ChNMKh	$K_2 = 52.17$	$c_2 = 444.6$	$\rho_2 = 7100$
FMC-11	$K_3 = 35.0$	$c_3 = 478.9$	$\rho_3 = 4700$

The specific heat and density of the coating material were determined according to the mixture law, based on the data for ZrO_2 and Ti-6Al-4V (Table 1). For equal participation of the volume fractions of the base and core components, it was established that $c_1 = 495.55\ J\,kg^{-1}K^{-1}$ and $\rho_1 = 5266.98 kg\,m^{-3}$. The dimensionless gradient parameter of the FGM and the time of braking were equal to $\gamma^* = \ln(K_{1,2}\,K_{1,1}^{-1}) = 1.26$ and $\tau_s = 1$, respectively.

The results of the calculations for dimensionless temperature rises $\Theta^*(\zeta,\tau)$ (59)–(63) for constant, and $\hat{\Theta}^*(\zeta,\tau)$ (120)–(123) for linearly decreasing specific power of friction are presented in Figure 2 (evolutions) and Figure 3 (isotherms). For numerical integration, the QAGI procedure from the QUADPACK library was implemented [40]. A numerical analysis was carried out to compare the results obtained for the applied FGC on the substrate (solid lines) with the corresponding data found for a homogeneous coating made entirely of zirconium dioxide (dashed lines).

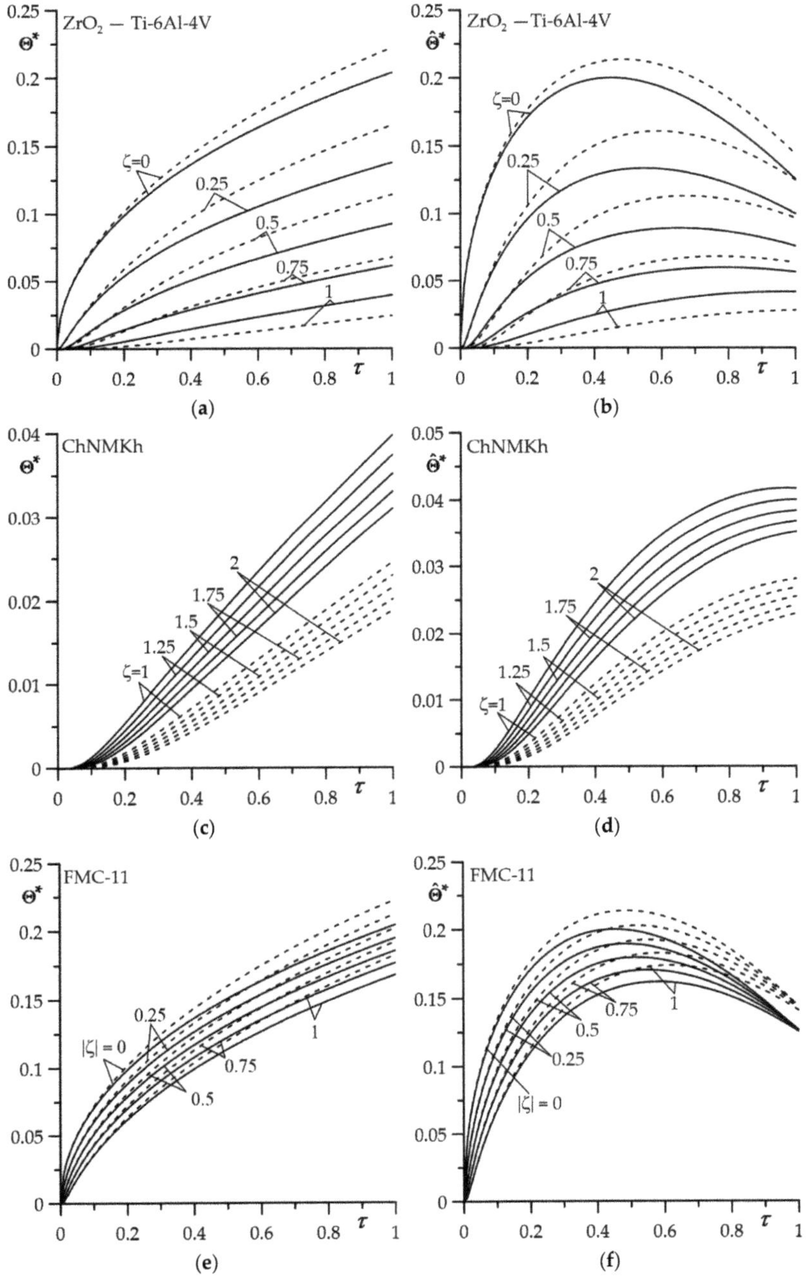

Figure 2. Evolutions of dimensionless temperature rises $\Theta^*(\zeta,\tau)$ (59)–(63) (**a**,**c**,**e**) and $\hat{\Theta}^*(\zeta,\tau)$ (120)–(123) (**b**,**d**,**f**) for selected values of dimensionless distance from the contact surface ζ in: (**a**,**b**) The coating; (**c**,**d**) the substrate; (**e**,**f**) the counterbody. Solid lines—FGC, dashed lines—coating made of ZrO_2.

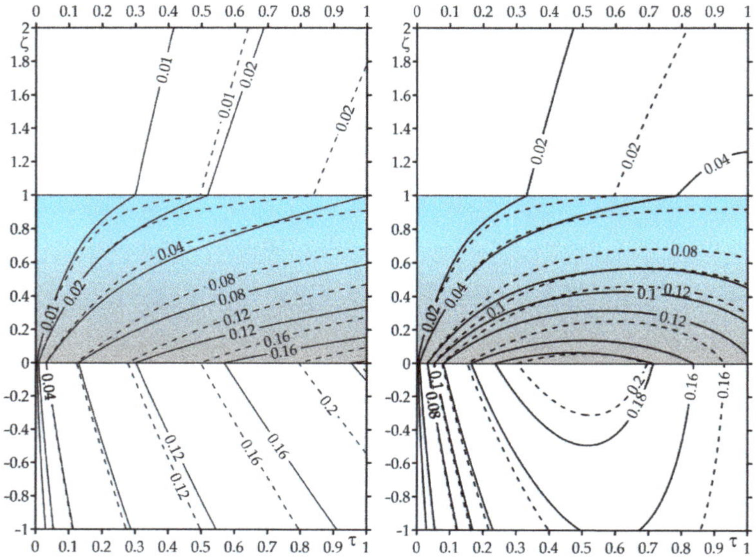

(**a**) (**b**)

Figure 3. Isotherms: (**a**) $\Theta^*(\zeta, \tau)$ (59)–(63); (**b**) $\hat{\Theta}^*(\zeta, \tau)$ (120)–(123). Solid lines—FGC, dashed lines—coating made of ZrO_2.

The Ti-6Al-4V titanium alloy, with its thermal conductivity 3.5 times greater than zirconium dioxide ZrO_2, effectively dissipates heat from the contact surface. As a result, the temperature of the FGC is lower compared to that determined using a homogeneous material (Figure 2a,d). Such a temperature mode changes to the opposite, starting from the interface ($\zeta = 1$) and further into the substrate ($\zeta > 1$) (Figure 2b,e). The ChNMKh cast iron used for the disc substrate has significantly (about 7.8 times) higher thermal conductivity than the Ti-6Al-4V titanium alloy. As a result, the temperature of the substrate at a fixed distance from the interface during the entire heat generation process is lower in the case of a homogeneous coating made of zirconium dioxide. In both cases, the level of substrate temperature is much (by an order of magnitude) lower than the coating temperature level. The change in the counterbody temperature over time (FMC-11, Figure 2c,f) is quantitatively and qualitatively similar to the evolution of the coating temperature, shown in Figure 2a,d. However, there are some features of the temporal profiles of the counterbody temperature that differ from the corresponding time courses of the coating temperature. Firstly, the effect of the FGM on the temperature of the counterbody is much lower than in the coating itself. Finally, the drop in the temperature in the counterbody (pad) with the distance from the contact surface is much slower than in the coating.

The spatial-temporal distributions of the dimensionless temperature rises for the constant and time-dependent intensity of specific friction power are demonstrated in Figure 3. They confirm the result from Figure 2 that the effective absorber of heat generated by friction on the contact surface is the coating made of the considered two-component FGM. It is clearly visible here that it plays the role of a thermal barrier, effectively protecting the substrate against overheating.

The asymptotic solutions for small (102)–(104) and large (112)–(114) values of the Fourier number τ (11) were an effective tool for estimating the temperature of the considered system in the case of constant specific friction power. The calculation results presented in Figure 4 show that the satisfactory convergence of the exact and asymptotic (at small τ) solutions takes place in the range $0 \leq \tau \leq 0.5$ (Figure 4a). However, it is surprising that the results obtained using the exact and asymptotic (for large τ) solutions show terrific agreement in almost the entire range of Fourier number changes (Figure 4b). It is all the more valuable due to the fact that asymptotic solutions, unlike exact solutions, do not require numerical integration.

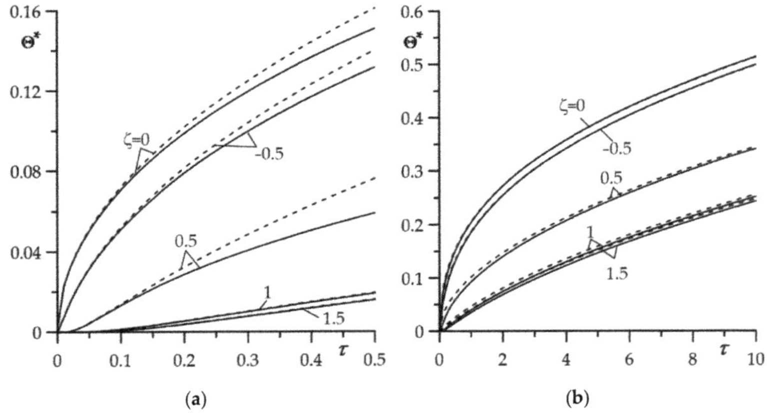

Figure 4. Evolution of the dimensionless temperature rises $\Theta^*(\zeta,\tau)$ obtained using the exact (59)–(63) (solid lines) and asymptotic (dashed lines) solutions: (**a**) For small (102)–(104); (**b**) for large (112)–(114) Fourier numbers τ for selected values of dimensionless spatial variable ζ.

Based on the Fourier law, dimensionless intensities of heat fluxes for constant specific power of friction, directed perpendicularly to the contact surface $\zeta = 0$ to the insides of the FGC and homogeneous counterbody were written in the forms, respectively:

$$q_1^*(\tau) = \Theta'^*(0^+,\tau),\ q_3^*(\tau) = -K_3^*\Theta'^*(0^-,\tau),\ \tau \geq 0, \tag{124}$$

where derivatives were determined from Formulas (78) and (80). Calculations carried out on the basis of Equation (124) demonstrated that the greater part of the heat generated on the contact area was absorbed by the cermet pad, which had much better heat conduction capabilities compared to the zirconium dioxide (Figure 5). At the initial moments of the heating process, about 80% of the heat is absorbed by the pad, and only 20% by the FGC. With the elapse of heating time, the amount of heat directed to the pad (coating) decreases (increases) and, for $\tau = 1$, it is equal to 70% (30%).

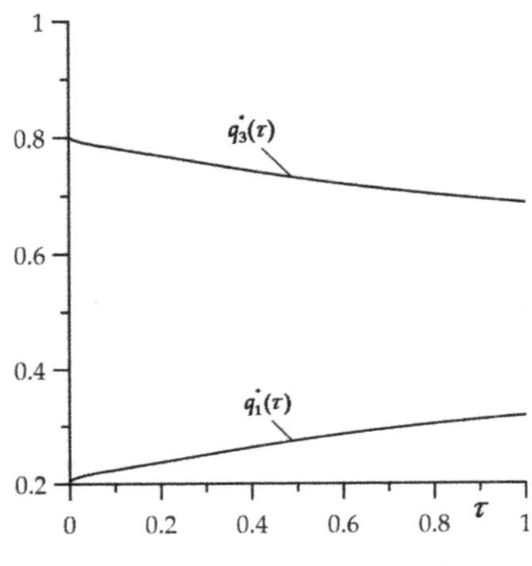

Figure 5. Evolutions of the dimensionless intensities of heat fluxes $q_l^*(\tau)$, l = 1, 3, defined by Equation (124).

7. Conclusions

An analytical model to determining the temperature field of a three-element friction system is developed, consisting of a substrate with protective coating deposited on the surface and a counterbody. The materials of the substrate and counterbody are homogeneous, while the coating is made of a functionally gradient material with an exponentially increasing thermal conductivity coefficient along the thickness. The counterbody slides on the surface of the coating, as a result of which heat is generated, and the elements of the system heat up. This type of system is used to simulate the frictional heating process in the pair, i.e., the pad (counterbody) and the coated disc (substrate with a coating). The crucial element of the model is the boundary value problem of heat conduction, considering the generation of heat due to friction at a constant and linearly decreasing specific friction power. Exact and asymptotic solutions of such a problem were obtained. Based on the achieved solutions, a numerical analysis was carried out for a cermet (FMK-11) brake pad sliding on the surface of the FGC (ZrO_2–Ti-6Al-4V) perfectly connected to a cast iron (ChNMKh) disc. It is concluded that:

1. Application of the selected functionally graded coating on the friction surfaces of the disc is an effective tool to lower the maximum temperature of the system.
2. The temporal profile of the specific friction power has a significant influence on the spatial-temporal distribution of the isotherms in the coating and the pad. The temperature of the disc is by an order of magnitude lower than the temperatures of the coating and the pad.
3. Asymptotic solutions for small and large values of the Fourier number can be used for quick estimation of temperature with high accuracy for all elements of the system.
4. The higher part of the heat ($\approx 3/4$) generated on the contact surface due to friction is absorbed by the pad, and the smaller part ($\approx 1/4$) is absorbed by the FGC.

It should be noted that protective coatings made of functionally graded materials are applied in braking systems for the following reasons: On the one hand, to reduce wear on the friction surface of the disc it is desirable to use a component with high wear resistance in the FGC (in the analyzed case—ZrO2). On the other hand, these materials usually have a low capacity to dissipate the heat generated by friction on the contact surface. The problem of lowering the possibly high temperature on the friction surface is solved by simultaneously using in the FGC a component with a much higher thermal conductivity coefficient than the previous one (in our case, Ti-6Al-4V). The numerical analysis carried out on the basis of the proposed analytical model showed that the last task of the FGC of the selected type is completely solved.

In the future, authors intend to expand research on the heating process in brakes with a functionally graded friction element and to develop models to investigate the influence of imperfect thermal contact of friction, convective cooling with the environment, as well as the finite thickness of friction pair elements.

Author Contributions: Conceptualization and methodology, A.Y.; software, P.Z.; validation, A.Y. and K.T., formal analysis, A.Y., K.T. and P.Z.; investigation, A.Y., K.T. and P.Z.; writing—original draft preparation A.Y. and K.T.; writing–review and editing, K.T.; visualization and figures preparation, P.Z.; supervision, A.Y.; project administration, K.T. All authors have read and agreed to the published version of the manuscript.

Funding: This research was funded by the National Science Centre in Poland, grant No. 2017/27/B/ST8/01249 and research project No. WZ/WM-IIM/4/2023 at the Bialystok University of Technology in Poland.

Institutional Review Board Statement: Not applicable.

Informed Consent Statement: Not applicable.

Data Availability Statement: No new data were created or analyzed in this study. Data sharing is not applicable to this article.

Conflicts of Interest: The authors declare no conflict of interest.

Nomenclature

c	Specific heat (J $kg^{-1}K^{-1}$)
d	Thickness of coating (m)
f	Coefficient of friction (dimensionless)
Imp	Imaginary part of a complex Laplace parameter
$I_n(\cdot)$	Modified Bessel functions of the nth order of the first kind
$J_n(\cdot)$	Bessel functions of the nth order of the first kind
$K_n(\cdot)$	Modified Bessel functions of the nth order of the second kind
k	Thermal diffusivity (m^2s^{-1})
K	Thermal conductivity (W $m^{-1}K^{-1}$)
p	Parameter of the Laplace integral transform (dimensionless)
p_0	Nominal pressure on the contact surface (Pa)
Rep	Real part of a complex Laplace parameter
q	Specific power of friction (W m^{-2})
q_0	Nominal value of the specific friction power (W m^{-2})
t	Time (s)
t_s	Stop moment of the process (s)
T	Temperature (°C)
T_0	Initial temperature (°C)
V_0	Sliding velocity (m s^{-1})
$Y_n(\cdot)$	Bessel functions of the nth order of the second kind
x	Spatial coordinate in tangential direction (m)
y	Spatial coordinate in radial direction (m)
z	Spatial coordinate in axial direction (m)
γ^*	Gradient parameter of FGM
Λ	Temperature rise scaling factor (°C)
ε	Dimensionless coefficient of thermal activity
Θ	Temperature rise (°C)
Θ^*	Dimensionless temperature rise
ρ	Density (kg m^{-3})
τ	Dimensionless time
τ_s	Dimensionless stop time
ζ	Dimensionless spatial coordinate in axial direction
subscript l	Number of the element of the system

References

1. Li, D.; Wang, S.; Fang, X.; Guo, Y.; Hu, N.; Zhang, D. Research progress of temperature field calculation of disc brake braking interface based on numerical analysis. *Proc. Inst. Mech. Eng. Part D J. Automob. Eng.* **2022**, 09544070221128288. [CrossRef]
2. Gerlici, J.; Gorbunov, M.; Kravchenko, K.; Prosvirova, O.; Lack, T. Noise and temperature reduction in the contact of tribological elements during braking. *MATEC Web Conf. Mach. Model. Simul.* **2018**, *157*, 02010. [CrossRef]
3. Borawski, A.; Szpica, D.; Mieczkowski, G.; Borawska, E.; Awad, M.M.; Shalaby, R.M.; Sallah, M. Theoretical Analysis of the Motorcycle Front Brake Heating Process during High Initial Speed Emergency Braking. *J. Appl. Comput. Mech.* **2020**, *6*, 1431–1437. [CrossRef]
4. Öz, A.; Gürbüz, H.; Yakut, A.K.; Sağiroğlu, S. Braking performance and noise in excessive worn brake discs coated with HVOF thermal spray process. *J. Mech. Sci. Technol.* **2017**, *31*, 535–543. [CrossRef]
5. Holubets, V.M.; Pashechko, M.I.; Borc, J.; Tisov, O.V.; Shpuliar, Y.S. Wear resistance of electrospark-deposited coatings in dry sliding friction conditions. *Powder Metall. Metal Ceram.* **2021**, *60*, 90–96. [CrossRef]
6. Łatka, L.; Pawłowski, L.; Winnicki, M.; Sokołowski, P.; Małachowska, A.; Kozerski, S. Review of Functionally Graded Thermal Sprayed Coatings. *Appl. Sci.* **2020**, *10*, 5153. [CrossRef]
7. Bishnoi, D. Pressure exertion and heat dissipation analysis on uncoated and ceramic (Al_2O_3, TiO_2 and ZrO_2) coated braking pads. *Mater. Today Proc.* **2023**, *74*, 774–787. [CrossRef]
8. Zhao, J.; Li, Y.; Ai, X. Analysis of transient thermal stress in sandwich plate with functionally graded coatings. *Thin Solid Film* **2008**, *516*, 7581–7587. [CrossRef]

9. Alinia, Y.; Beheshti, A.; Guler, M.A.; El-Borgi, S.; Polycarpou, A.A. Sliding contact analysis of functionally graded coating/substrate system. *Mech. Mater.* **2016**, *94*, 142–155. [CrossRef]
10. Sathish, M.; Radhika, N.; Saleh, B.A. critical review on functionally graded coatings: Methods, properties, and challenges. *Compos. Part B Eng.* **2021**, *225*, 109278. [CrossRef]
11. Ungureanu, M.; Medan, N.; Ungureanu, N.S.; Pop, N.; Nadolny, K. Tribological Aspects Concerning the Study of Overhead Crane Brakes. *Materials* **2022**, *15*, 6549. [CrossRef]
12. Topczewska, K.; Gerlici, J.; Yevtushenko, A.; Kuciej, M.; Kravchenko, K. Analytical Model of the Frictional Heating in a Railway Brake Disc at Single Braking with Experimental Verification. *Materials* **2022**, *15*, 6821. [CrossRef] [PubMed]
13. Loizou, A.; Sheng Qi, H.; Day, A.J. A fundamental study on the heat partition ratio of vehicle disk brakes. *J. Heat Transf.* **2013**, *135*. [CrossRef]
14. Yevtushenko, A.A.; Kuciej, M. Two heat conduction problems with frictional heating during braking. *J. Theor. Appl. Mech.* **2010**, *48*, 367–380.
15. Mao, J.J.; Ke, L.L.; Yang, J.; Kitipornchai, S.; Wang, Y.S. Thermoelastic instability of functionally graded coating with arbitrarily varying properties considering contact resistance and frictional heat. *Appl. Math. Model.* **2017**, *43*, 521–537. [CrossRef]
16. Yevtushenko, A.A.; Kuciej, M. Influence of the protective strip properties on distribution of the temperature at transient frictional heating. *Int. J. Heat Mass Transf.* **2009**, *52*, 376–384. [CrossRef]
17. Yevtushenko, A.A.; Kuciej, M. Frictional heating during braking in a threeelement tribosystem. *Int. J. Heat Mass Transf.* **2009**, *52*, 2942–2948. [CrossRef]
18. Yevtushenko, A.A.; Kuciej, M. Influence of the pad's material properties on the thermal stresses during braking. *Numer. Heat Transf. A-Appl.* **2009**, *56*, 931–945. [CrossRef]
19. Yevtushenko, A.A.; Kuciej, M.; Yevtushenko, O.O. Three-element model of frictional heating during braking with contact thermal resistance and timedependent pressure. *Int. J. Therm. Sci.* **2011**, *50*, 1116–1124. [CrossRef]
20. Kuciej, M. Accounting changes of pressure in time in one-dimensional modeling the process of friction heating of disc brake. *Int. J. Heat Mass Transf.* **2011**, *54*, 468–474. [CrossRef]
21. Yevtushenko, A.A.; Kuciej, M.; Yevtushenko, O. The asymptotic solutions of heat problem of friction for a three-element tribosystem with generalized boundary conditions on the surface of sliding. *Int. J. Heat Mass Transf.* **2014**, *70*, 128–136. [CrossRef]
22. Kuciej, M. The thermal problem of friction during braking for a three-element tribosystem with a composite pad. *Int. J. Heat Mass Transf.* **2011**, *38*, 1322–1329. [CrossRef]
23. Topczewska, K.; Zamojski, P. Effect of pressure fluctuations on the temperature during braking. *Acta Mech. Autom.* **2020**, *14*, 103–107. [CrossRef]
24. Burlayenko, V.N.; Altenbach, H.; Sadowski, T.; Dimitrova, S.D.; Bhaskar, A. Modelling functionally graded materials in heat transfer and thermal stress analysis by means of graded finite elements. *Appl. Math. Model.* **2017**, *45*, 422–438. [CrossRef]
25. Yevtushenko, A.; Topczewska, K.; Zamojski, P. The Effect of Functionally Graded Materials on Temperature during Frictional Heating at Single Braking. *Materials* **2021**, *14*, 6241. [CrossRef] [PubMed]
26. Yevtushenko, A.; Topczewska, K.; Zamojski, P. Influence of Thermal Sensitivity of Functionally Graded Materials on Temperature during Braking. *Materials* **2022**, *15*, 963. [CrossRef]
27. Yevtushenko, A.; Kuciej, M.; Topczewska, K.; Zamojski, P. Temperature in the Friction Couple Consisting of Functionally Graded and Homogeneous Materials. *Materials* **2022**, *15*, 3600. [CrossRef]
28. Yevtushenko, A.; Topczewska, K.; Zamojski, P. Temperature during Repetitive Short-Term Operation of a Brake with Functionally Graded Friction Element. *Materials* **2023**, *16*, 881. [CrossRef]
29. Yevtushenko, A.; Topczewska, K.; Zamojski, P. Use of Functionally Gradem Material to Decrease Maximum Temperature of a Coating–Substrate System. *Materials* **2023**, *16*, 265. [CrossRef]
30. Yevtushenko, A.; Topczewska, K.; Zamojski, P. The Heat Partition Ratio during Braking in a Functionally Graded Friction Couple. *Materials* **2022**, *15*, 4623. [CrossRef]
31. Mao, J.J.; Ke, L.L.; Yang, J.; Kitipornchai, S.; Wang, Y.S. The coupled thermoelastic instability of FGM coatings with arbitrarily varying properties: In-plane sliding. *Acta Mech.* **2018**, *229*, 2979–2995. [CrossRef]
32. Suresh, S.; Mortensen, A. *Fundamentals of Functionally Graded Materials*; The Institute of Materials, IOM Communications Ltd.: London, UK, 1998.
33. Sneddon, I.N. *The Use of Integral Transforms*; McGraw-Hill: New York, NY, USA, 1972.
34. Abramowitz, M.; Stegun, I. *Handbook of Mathematical Functions with Formulas, Graphs, and Mathematical Tables*; United States Department of Commerce, National Bureau of Standards (NBS): Washington, DC, USA, 1964.
35. Barber, J.R.; Martin-Moran, C.J. Green's functions for transient thermoelastic contact problems for the half-plane. *Wear* **1982**, *79*, 11–19. [CrossRef]
36. Bateman, H.; Erdelyi, A. *Tables of Integrals Transforms*; McGraw-Hill: New York, NY, USA, 1954; Volume 1.
37. Yevtushenko, A.A.; Kuciej, M.; Topczewska, K. Analytical model for investigation of the effect of friction power on temperature in the disc brake. *Adv. Mech. Eng.* **2017**, *9*, 1–12. [CrossRef]
38. Luikov, A.V. *Analitycal Heat Diffusion Theory*; Academic Press: New York, NY, USA, 1968.

39. Özişik, N.M. *Heat Conduction;* John Wiley: New York, NY, USA, 1993.
40. Piessens, R.; De Doneker-Kapenga, E.; Überhuber, C.W.; Kahaner, D.K. *Quadpack: A Subroutine for Automatic Integration;* Springer Science & Business Media: Berlin/Heidelberg, Germany, 2012; Volume 1.

materials

Article

Effect of Convective Cooling on the Temperature in a Friction System with Functionally Graded Strip

Aleksander Yevtushenko, Michał Kuciej, Katarzyna Topczewska * and Przemysław Zamojski

Faculty of Mechanical Engineering, Bialystok University of Technology (BUT), 45C Wiejska Street, 15-351 Bialystok, Poland; a.yevtushenko@pb.edu.pl (A.Y.); m.kuciej@pb.edu.pl (M.K.); p.zamojski@pb.edu.pl (P.Z.)
* Correspondence: k.topczewska@pb.edu.pl

Abstract: An exact solution of the boundary-value problem of heat conduction was obtained with consideration of heat generation due to friction and convective cooling for the strip/semi-space system. Analytical solutions to this problem are known for the case with both friction elements made of homogeneous materials or a composite layer with a micro-periodic structure. However, in this study, the strip is made of a two-component functionally gradient material (FGM). In addition, the exact, asymptotic solutions were also determined at small and large values of the Fourier number. By means of Duhamel's theorem, it was shown that the developed solution for a constant friction power allows to obtain appropriate solutions with a changing time profile of this value during heating. Numerical analysis in dimensionless form was carried out for the FGM (ZrO_2—Ti-6Al-4V) strip in combination with the cast iron semi-space. The influence of the convective cooling intensity (Biot number) on the temperature field in the considered friction system was investigated. The developed mathematical model allows for a quick estimation of the maximum temperature of systems, in which one of the elements (FGM strip) is heated on the friction surface and cooled by convection on the free surface.

Keywords: frictional heating; convective cooling; temperature; functionally graded material

1. Introduction

Analytical models of the frictional heating process in the pad/disc system are based on the experimentally confirmed assumption that the main part of the heat generated by friction is absorbed inside both sliding elements in the direction perpendicular to the contact surface [1–3]. This makes it possible to ignore changes in the temperature gradient in the other two directions, parallel to this surface [4]. As a consequence, the temperature field of the friction element is determined based on an analytical or analytical–numerical solution of the one-dimensional boundary-value problem of heat conduction for a two-element semi-space/semi-space, strip/semi-space and strip/strip systems, made of homogeneous materials [5,6]. The choice of one of these geometric schemes depends on the size of the effective depth of heat penetration d_{eff} to the insides of each element of the friction pair [7]. Various formulas are used to estimate this parameter at the design stage, and one of the most commonly used is $d_{eff} = \sqrt{3kt_s}$, where k is the coefficient of the thermal diffusivity of the considered element and t_s is the time of the frictional heating process [8]. If it turns out that $d_{eff} < d$, where d is the thickness of the element (pad, disc, etc.), during the formulation of the thermal problem of friction, then this element was considered as a semi-bounded body (half-space); otherwise, it was modeled with a strip.

In this article, the strip/semi-space friction system is considered, so we focused on the most important results directly related to it. First, the process of heat generation during braking with a constant deceleration under perfect thermal conditions of friction between the layer and the half-space was considered. During the entire heating process, the free surface of the strip was maintained at a constant (initial) temperature, or it was adiabatic (thermally insulated) [9]. The exact solutions of the corresponding boundary-value thermal problems of friction formulated in this way were obtained by the superposition method using the mathematical apparatus of the Laplace integral transform. Calculations were made for a metal–ceramic (FMC-11) layer and a cast iron (ChNMKh) half-space.

The solutions obtained in article [9] were generalized by taking into consideration the heat exchange with the surrounding environment on the free surface of the strip, according to Newton's law [10]. The transition from the space of transforms to the space of originals was made using the integration technique in the plane of the complex parameter of the Laplace transform. Numerical analysis was carried out for the steel strip and semi-space made of aluminum. Using the Biot number, the influence of the intensity of convective cooling of the free surface of the steel strip during its sliding with a constant velocity on the surface of the aluminum semi-space was investigated.

The next studies concerned the modeling of imperfect thermal contact between the strip and the semi-space [11,12]. Such contact takes place in most cases during sliding and accompanying frictional heating of rough rubbing surfaces [13,14]. They assume that the resulting temperature jump on the contact surface is proportional to the differences in the intensity of heat fluxes directed along the normal from this surface to the insides of the sliding bodies. The coefficient of proportionality is the coefficient of the contact heat transfer—a parameter inversely proportional to the thermal resistance of the contact surface. An exact solution of the thermal problem of friction was obtained with uniform sliding of the strip over the surface of the semi-space, taking into account the imperfect thermal contact and the preservation of the initial temperature on the free surface of the strip [11]. The subsequent numerical analysis for the layer made of FMC-11 and the cast iron (ChNMKh) half-space made it possible, in particular, to determine the applicability ranges of the conditions of the perfect and imperfect thermal contact of friction for the considered pair. Another analytical model was developed to simulate the temperature field of the pad/disc system during braking with a constant deceleration under constant contact pressure [12] based on the strip/semi-space scheme. This model takes into consideration the thermal resistance of the contact surface with simultaneous convective cooling on the free surface of the strip.

It should be noted that all the mentioned works concerned the case of homogeneous materials of both elements of the friction couple. However, nowadays with the frictional elements of modern tribosystems, the nonhomogeneous materials are more likely to be used such as functionally graded materials (FGMs) [15–17]. Their usage is attractive to engineers and material scientists given the continuous grading and tailoring capabilities, when compared to traditional monolithic counterparts [18,19]. Usually, functionally graded friction materials are two-component composites made of ceramic and metal. So, the friction elements have high heat and wear resistance of ceramic on the outer zone, as well as the mechanical strength of the metal inside elements at the same time [20,21]. A smooth transition of composition from the outer surface to the inside ensure a reduction in stress concentration and an increase in fracture strength [21,22]. Research shows that controlling the gradient parameter of FGM can reduce residual tensile stress and improve thermal shock resistance [23]. Many computational models have been recently developed to evaluate the thermal response of FGMs, which is crucial for predicting failure mechanisms and designing friction couples [24–33]. Comprehensive reviews of the literature on thermal and thermoelastic problems of friction for functionally graded materials are given in our previous articles [34–36]. In these studies, the authors obtained the solutions of the boundary-value problems of heat conduction for a coating (layer) made of FGM with a heated surface that was ideally thermally connected to the surface of a homogeneous substrate (half-space) [35]. Then, the process of heat generation in a friction system consisting of a homogeneous semi-space sliding on the surface of the FGM strip applied to a homogeneous semi-space was investigated [36]. In both of these articles, the FGM layer performed the role of the thermal-barrier coating (TBC), allowing for more efficient heat dissipation from the heated surface. However, in this article, the FGM strip is one of the elements of the friction pair, absorbing the heat generated as a result of friction on the contact surface and cooled by convection on the free surface. The latter factor has not been taken into consideration in analytical models of friction heating involving FGM so far.

2. Statement to the Problem

Let the strip $0 \le z \le d$ at the initial moment $t = 0$ begin to slide with constant velocity V_0 in the positive direction of the axis Ox on the outer surface of the semi-space $z \le 0$ (Figure 1). The strip is made of a two-component functionally graded material, which has exponentially increasing thermal conductivity coefficient along the thickness, whereas the material of semi-space is homogeneous. Due to the friction on the contact surface $z = 0$, the heat is generated and absorbed by each body of the friction couple. It was assumed that the thermal contact of friction between the strip and the semi-space is full. This means that the temperature of the friction surfaces of the strip and the half-space is the same, and the sum of the intensities of heat fluxes directed along the normal from the contact surface to the insides of each body is equal to the specific friction power $q_0 = f p_0 V_0$, where f is the coefficient of friction, and p_0 is the contact pressure. The free surface of the strip $z = d$ is convectively cooled with a constant coefficient of heat transfer h. The initial temperature T_0 of the considered system is constant. A more detailed description of the assumptions of the mathematical model is given in our previous article [34].

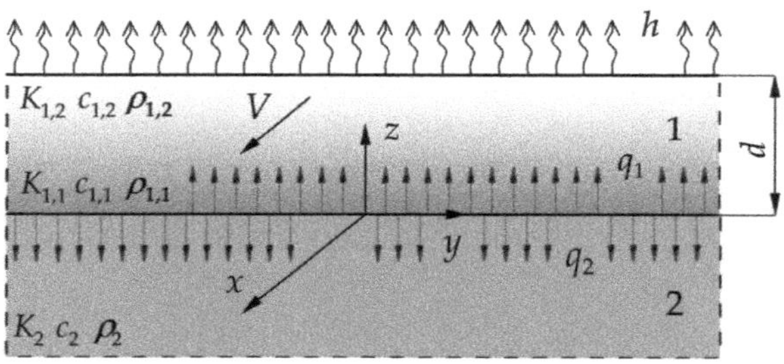

Figure 1. Scheme of the friction system strip/semi-space.

The aim of this study is to explain the effect of FGM on the transient temperature fields $T(z,t)$, $-\infty < z \le d$, $t \ge 0$ of the strip and the semi-space. For this purpose, based on the above assumptions, the following thermal problem of friction for a single braking process in relation to the temperature rise $\Theta(z,t) = T(z,t) - T_0$ was formulated:

$$\frac{\partial}{\partial z}\left[K_1(z)\frac{\partial\Theta(z,t)}{\partial z}\right] - \rho_1 c_1 \frac{\partial\Theta(z,t)}{\partial t} = 0,\ 0 < z < d,\ t > 0, \quad (1)$$

$$K_2\frac{\partial^2\Theta(z,t)}{\partial z^2} - \rho_2 c_2 \frac{\partial\Theta(z,t)}{\partial t} = 0,\ z < 0,\ t > 0, \quad (2)$$

$$\Theta(0^+,t) = \Theta(0^-,t),\ t > 0, \quad (3)$$

$$K_2\frac{\partial\Theta(z,t)}{\partial z}\bigg|_{z=0^-} - K_1(z)\frac{\partial\Theta(z,t)}{\partial z}\bigg|_{z=0^+} = q_0,\ t > 0, \quad (4)$$

$$K_1(z)\frac{\partial\Theta(z,t)}{\partial z}\bigg|_{z=d} = h[T_0 - T(d,t)],\ t > 0, \quad (5)$$

$$\Theta(z,t) \to 0\ ,\ z \to -\infty,\ t > 0, \quad (6)$$

$$\Theta(z,0) = 0, -\infty < z \le d, \quad (7)$$

where [37,38]

$$K_1(z) = K_{1,1}e^{\gamma^* z/d},\ \gamma^* = \ln(K_{1,2}K_{1,1}^{-1}),\ 0 \le z \le d, \tag{8}$$

and $K_{1,1}$, $K_{1,2}$ K_2 are the coefficients of thermal conductivity of FGM components and semi-space, respectively, ρ_l, c_l are the density and specific heat of materials of the strip $(l = 1)$ and the semi-space $(l = 2)$, and γ^* is the dimensionless gradient parameter.

Incorporating the following dimensionless parameters and variables:

$$\zeta = \frac{z}{d},\ \tau = \frac{k_1 t}{d^2},\ K^* = \frac{K_2}{K_{1,1}},\ k^* = \frac{k_2}{k_1},\ \Theta^* = \frac{\Theta}{\Theta_0},\ Bi = \frac{hd}{K_{1,1}}, \tag{9}$$

where

$$k_1 = \frac{K_{1,1}}{c_1\rho_1},\ k_2 = \frac{K_2}{c_2\rho_2},\ \Theta_0 = \frac{q_0 d}{K_{1,1}}, \tag{10}$$

Equations (1)–(8) were written in the form:

$$\frac{\partial^2\Theta^*(\zeta,\tau)}{\partial\zeta^2} + \gamma^*\frac{\partial\Theta^*(\zeta,\tau)}{\partial\zeta} - e^{-\gamma^*\zeta}\frac{\partial\Theta^*(\zeta,\tau)}{\partial\tau} = 0,\ 0 < \zeta < 1,\ \tau > 0, \tag{11}$$

$$\frac{\partial^2\Theta^*(\zeta,\tau)}{\partial\zeta^2} - \frac{1}{k_2^*}\frac{\partial\Theta^*(\zeta,\tau)}{\partial\tau} = 0,\ \zeta < 0,\ \tau > 0, \tag{12}$$

$$\Theta^*(0^+,\tau) = \Theta^*(0^-,\tau),\ \tau > 0, \tag{13}$$

$$K^*\left.\frac{\partial\Theta^*(\zeta,\tau)}{\partial\zeta}\right|_{\zeta=0^-} - \left.\frac{\partial\Theta^*(\zeta,\tau)}{\partial\zeta}\right|_{\zeta=0^+} = 1,\ \tau > 0, \tag{14}$$

$$e^{\gamma^*}\left.\frac{\partial\Theta^*(\zeta,\tau)}{\partial\zeta}\right|_{\zeta=1} + Bi\,\Theta^*(1,\tau) = 0,\ \tau > 0, \tag{15}$$

$$\Theta^*(\zeta,\tau) \to 0\,,\ \zeta \to -\infty,\ \tau > 0, \tag{16}$$

$$\Theta^*(\zeta,0) = 0,\ |\zeta| < \infty. \tag{17}$$

3. Exact Solution

Using the Laplace integral transform [39]:

$$\overline{\Theta}^*(\zeta,p) \equiv L[\Theta^*(\zeta,\tau);p] = \int_0^\infty \Theta^*(\zeta,\tau)e^{-p\tau}d\tau,\ \mathrm{Re}p \ge 0, \tag{18}$$

to the boundary-value problem (11)–(17), the following boundary problem was obtained with two ordinary differential equations:

$$\frac{d^2\overline{\Theta}^*(\zeta,p)}{d\zeta^2} + \gamma^*\frac{d\overline{\Theta}^*(\zeta,p)}{d\zeta} - pe^{-\gamma^*\zeta}\overline{\Theta}^*(\zeta,p) = 0,\ 0 < \zeta < 1, \tag{19}$$

$$\frac{d^2\overline{\Theta}^*(\zeta,p)}{d\zeta^2} - \frac{p}{k_2^*}\overline{\Theta}^*(\zeta,p) = 0,\ \zeta < 0, \tag{20}$$

$$\overline{\Theta}^*(0^+,p) = \overline{\Theta}^*(0^-,p), \tag{21}$$

$$K^*\left.\frac{d\overline{\Theta}^*(\zeta,p)}{d\zeta}\right|_{\zeta=0^-} - \left.\frac{d\overline{\Theta}^*(\zeta,p)}{d\zeta}\right|_{\zeta=0^+} = \frac{1}{p}, \tag{22}$$

$$e^{\gamma^*}\frac{d\overline{\Theta}^*(\zeta,p)}{d\zeta}\bigg|_{\zeta=1}+Bi\,\overline{\Theta}^*(1,p)=0, \tag{23}$$

$$\overline{\Theta}^*(\zeta,p)\to 0\,,\ \zeta\to-\infty. \tag{24}$$

The solution to the boundary problem (19)–(24) has the form:

$$\overline{\Theta}^*(\zeta,p)=\frac{e^{-\frac{1}{2}\gamma^*\zeta}\Delta_1(\zeta,p)}{p\sqrt{p}\,\Delta(p)},\ 0\le\zeta\le1,\ \overline{\Theta}^*(\zeta,p)=\frac{\Delta_2(\zeta,p)}{p\sqrt{p}\,\Delta(p)},\ \zeta\le0, \tag{25}$$

where

$$\Delta_1(\zeta,p)=A_1(p)\mathrm{I}_1\left(\frac{2}{\gamma^*}e^{-\frac{1}{2}\gamma^*\zeta}\sqrt{p}\right)+B_1(p)\mathrm{K}_1\left(\frac{2}{\gamma^*}e^{-\frac{1}{2}\gamma^*\zeta}\sqrt{p}\right),\ \Delta_2(\zeta,p)=A_2(p)e^{\sqrt{\frac{p}{k^*}}\zeta}, \tag{26}$$

$$\Delta(p)=A_1(p)\left[\mathrm{I}_0\left(\frac{2}{\gamma^*}\sqrt{p}\right)+\varepsilon\,\mathrm{I}_1\left(\frac{2}{\gamma^*}\sqrt{p}\right)\right]-B_1(p)\left[\mathrm{K}_0\left(\frac{2}{\gamma^*}\sqrt{p}\right)-\varepsilon\,\mathrm{K}_1\left(\frac{2}{\gamma^*}\sqrt{p}\right)\right], \tag{27}$$

$$A_1(p)=\mathrm{K}_0\left(\frac{2}{\gamma^*}e^{-\frac{1}{2}\gamma^*}\sqrt{p}\right)+Bi\,\frac{e^{-\frac{1}{2}\gamma^*}}{\sqrt{p}}\mathrm{K}_1\left(\frac{2}{\gamma^*}e^{-\frac{1}{2}\gamma^*}\sqrt{p}\right), \tag{28}$$

$$B_1(p)=\mathrm{I}_0\left(\frac{2}{\gamma^*}e^{-\frac{1}{2}\gamma^*}\sqrt{p}\right)-Bi\,\frac{e^{-\frac{1}{2}\gamma^*}}{\sqrt{p}}\mathrm{I}_1\left(\frac{2}{\gamma^*}e^{-\frac{1}{2}\gamma^*}\sqrt{p}\right), \tag{29}$$

$$A_2(p)=A_1(p)\mathrm{I}_1\left(\frac{2}{\gamma^*}\sqrt{p}\right)+B_1(p)\mathrm{K}_1\left(\frac{2}{\gamma^*}\sqrt{p}\right), \tag{30}$$

$$\varepsilon=\frac{K^*}{\sqrt{k^*}}, \tag{31}$$

where $\mathrm{I_n}(x)$, $\mathrm{K_n}(x)$ are modified Bessel's functions of the nth ($n=0,1$) order of the first and second kind, respectively [40].

Applying the inverse Laplace transform [39]:

$$\Theta^*(\zeta,\tau)\equiv L^{-1}[\overline{\Theta}^*(\zeta,p);\,\tau]=\frac{1}{2\pi i}\int\limits_{\omega-i\infty}^{\omega+i\infty}\overline{\Theta}^*(\zeta,p)e^{p\tau}dp,\ \omega\equiv\mathrm{Re}p>0,\ i\equiv\sqrt{-1}, \tag{32}$$

to the transformed solution in Equations (25)–(31), and by carrying out the integration on the plane of the complex variable p, according to the methodology described in the articles [10,35], with account of the following relations [40]:

$$\mathrm{I}_0(\pm ix)=\mathrm{J}_0(x),\ \mathrm{K}_0(\pm ix)=-0.5\pi[\mathrm{Y}_0(x)\pm i\mathrm{J}_0(x)], \tag{33}$$

$$\mathrm{I}_1(\pm ix)=\pm i\mathrm{J}_1(ix),\ \mathrm{K}_1(\pm ix)=0.5\pi[\mathrm{J}_1(x)\pm i\mathrm{Y}_1(x)], \tag{34}$$

where $\mathrm{J_n}(x)$ and $\mathrm{Y_n}(x)$ are the Bessel functions of the nth ($n=0,1$) order of the first and second kind, respectively, the dimensionless temperature rises in the strip and semi-space were obtained in the form:

$$\Theta^*(\zeta,\tau)=\vartheta(\zeta)-\frac{2}{\pi}\varepsilon\,e^{-\frac{1}{2}\gamma^*\zeta}\int\limits_0^\infty F(x)G_1(\zeta,x)e^{-x^2\tau}dx,\ 0\le\zeta\le1,\ \tau\ge0, \tag{35}$$

$$\Theta^*(\zeta,\tau) = \vartheta_0 - \frac{2}{\pi}\int_0^\infty F(x)G_2(\zeta,x)e^{-x^2\tau}dx,\ \zeta \le 0,\ \tau \ge 0, \tag{36}$$

where

$$F(x) = \frac{\Psi(x)}{x^2\left\{[\Phi(x)]^2 + [\varepsilon\,\Psi(x)]^2\right\}}, \tag{37}$$

$$G_1(\zeta,x) = \mathrm{J}(x)\mathrm{Y}_1\left(\frac{2}{\gamma^*}e^{-\frac{1}{2}\gamma^*\zeta}x\right) - \mathrm{Y}(x)\mathrm{J}_1\left(\frac{2}{\gamma^*}e^{-\frac{1}{2}\gamma^*\zeta}x\right), \tag{38}$$

$$G_2(\zeta,x) = \varepsilon\,\Psi(x)\cos\left(\frac{\zeta}{\sqrt{k^*}}x\right) - \Phi(x)\sin\left(\frac{\zeta}{\sqrt{k^*}}x\right), \tag{39}$$

$$\Phi(x) = \mathrm{J}(x)\mathrm{Y}_0\left(\frac{2}{\gamma^*}x\right) - \mathrm{Y}(x)\mathrm{J}_0\left(\frac{2}{\gamma^*}x\right), \tag{40}$$

$$\Psi(x) = \mathrm{J}(x)\mathrm{Y}_1\left(\frac{2}{\gamma^*}x\right) - \mathrm{Y}(x)\mathrm{J}_1\left(\frac{2}{\gamma^*}x\right), \tag{41}$$

$$\mathrm{J}(x) = \mathrm{J}_0\left(\frac{2}{\gamma^*}e^{-\frac{1}{2}\gamma^*}x\right) - Bi\,\frac{e^{-\frac{1}{2}\gamma^*}}{x}\mathrm{J}_1\left(\frac{2}{\gamma^*}e^{-\frac{1}{2}\gamma^*}x\right), \tag{42}$$

$$\mathrm{Y}(x) = \mathrm{Y}_0\left(\frac{2}{\gamma^*}e^{-\frac{1}{2}\gamma^*}x\right) - Bi\,\frac{e^{-\frac{1}{2}\gamma^*}}{x}\mathrm{Y}_1\left(\frac{2}{\gamma^*}e^{-\frac{1}{2}\gamma^*}x\right), \tag{43}$$

$$\vartheta(\zeta) = (\gamma^* Bi)^{-1}[\gamma^* - Bi(e^{-\gamma^*} - e^{-\gamma^*\zeta})],\ \vartheta_0 \equiv \vartheta(0) = (\gamma^* Bi)^{-1}[\gamma^* - Bi(e^{-\gamma^*} - 1)]. \tag{44}$$

On the contact surface $\zeta = 0$, from Equations (38) and (39) follows that $G_1(0,x) = \Psi(x)$, $G_2(0,x) = \varepsilon\,\Psi(x)$. Then, the solutions (35) and (36) have the form:

$$\Theta^*(0^+,\tau) = \Theta^*(0^-,\tau) \equiv \Theta^*(\tau) = \vartheta_0 - \frac{2}{\pi}\varepsilon\int_0^\infty F(x)\Psi(x)e^{-x^2\tau}dx,\ \tau \ge 0, \tag{45}$$

which confirms the fulfillment of the boundary condition (13). Differentiating solutions (35)–(44) with respect to the variable ζ, with account of the derivatives [40]:

$$\mathrm{J}_1'(x) = \mathrm{J}_0(x) - x^{-1}\mathrm{J}_1(x),\ \mathrm{Y}_1'(x) = \mathrm{Y}_0(x) - x^{-1}\mathrm{Y}_1(x), \tag{46}$$

the dimensionless intensities of heat fluxes absorbed by the strip and the semi-space were found:

$$\frac{\partial\Theta^*(\zeta,\tau)}{\partial\zeta} = -e^{-\gamma^*\zeta} + \frac{2}{\pi}\varepsilon\,e^{-\gamma^*\zeta}\int_0^\infty xF(x)\hat{G}_1(\zeta,x)e^{-x^2\tau}dx,\ 0 \le \zeta \le 1,\ \tau \ge 0, \tag{47}$$

$$K^*\frac{\partial\Theta^*(\zeta,\tau)}{\partial\zeta} = \frac{2}{\pi}\varepsilon\int_0^\infty xF(x)\hat{G}_2(\zeta,x)e^{-x^2\tau}dx,\ \zeta \le 0,\ \tau \ge 0, \tag{48}$$

where

$$\hat{G}_1(\zeta,x) = \mathrm{J}(x)\mathrm{Y}_0\left(\frac{2}{\gamma^*}e^{-\frac{1}{2}\gamma^*\zeta}x\right) - \mathrm{Y}(x)\mathrm{J}_0\left(\frac{2}{\gamma^*}e^{-\frac{1}{2}\gamma^*\zeta}x\right), \tag{49}$$

$$\hat{G}_2(\zeta,x) = \varepsilon\,\Psi(x)\sin\left(\frac{\zeta}{\sqrt{k^*}}x\right) + \Phi(x)\cos\left(\frac{\zeta}{\sqrt{k^*}}x\right), \tag{50}$$

and K^*, k^* are the dimensionless thermal conductivity and diffusivity of the system, respectively, as determined by Equation (9).

Substituting $\zeta = 0$ in Equations (49) and (50), we found $\hat{G}_1(0,x) = \hat{G}_2(0,x) = \Phi(x)$ and Formulas (47) and (48) take the form:

$$\left.\frac{\partial\Theta^*(\zeta,\tau)}{\partial\zeta}\right|_{\zeta=0^+} \equiv q_1^*(\tau) = -1 + \frac{2}{\pi}\varepsilon\int_0^\infty xF(x)\Phi(x)e^{-x^2\tau}dx,\ \tau \geq 0, \tag{51}$$

$$K^*\left.\frac{\partial\Theta^*(\zeta,\tau)}{\partial\zeta}\right|_{\zeta=0^-} \equiv q_2^*(\tau) = \frac{2}{\pi}\varepsilon\int_0^\infty xF(x)\Phi(x)e^{-x^2\tau}dx,\ \tau \geq 0. \tag{52}$$

From Equations (51) and (52), it is easy to obtain the confirmation of the fulfillment of the boundary condition (14). For the purpose of checking the boundary condition (15) on the free surface of the strip $\zeta = 1$, from relations (35), (38), (44), (47) and (49), it was found:

$$\Theta^*(1,\tau) = \frac{1}{Bi} - \frac{2}{\pi}\varepsilon e^{-\frac{1}{2}\gamma^*}\int_0^\infty F(x)G_1(1,x)e^{-x^2\tau}dx,\ \tau \geq 0, \tag{53}$$

$$\left.\frac{\partial\Theta^*(\zeta,\tau)}{\partial\zeta}\right|_{\zeta=1} = -e^{-\gamma^*} + \frac{2}{\pi}\varepsilon e^{-\gamma^*}\int_0^\infty xF(x)\hat{G}_1(1,x)e^{-x^2\tau}dx,\ \tau \geq 0, \tag{54}$$

where

$$G_1(1,x) = \mathrm{J}(x)\mathrm{Y}_1\left(\frac{2}{\gamma^*}e^{-\frac{1}{2}\gamma^*}x\right) - \mathrm{Y}(x)\mathrm{J}_1\left(\frac{2}{\gamma^*}e^{-\frac{1}{2}\gamma^*}x\right), \tag{55}$$

$$\hat{G}_1(1,x) = \mathrm{J}(x)\mathrm{Y}_0\left(\frac{2}{\gamma^*}e^{-\frac{1}{2}\gamma^*}x\right) - \mathrm{Y}(x)\mathrm{J}_0\left(\frac{2}{\gamma^*}e^{-\frac{1}{2}\gamma^*}x\right). \tag{56}$$

Bearing in mind Equations (53)–(56) and the left-hand side of the boundary condition (15), it was written in the form:

$$e^{\gamma^*}\left.\frac{\partial\Theta^*(\zeta,\tau)}{\partial\zeta}\right|_{\zeta=1} + Bi\,\Theta^*(1,\tau) = \frac{2}{\pi}\varepsilon e^{-\gamma^*}\int_0^\infty F(x)[x\hat{G}_1(1,x) - e^{-\frac{1}{2}\gamma^*}Bi\,G_1(1,x)]e^{-x^2\tau}dx,\ \tau > 0. \tag{57}$$

Substituting the functions $\mathrm{J}(x)$ (42) and $\mathrm{Y}(x)$ (43) to Equations (55) and (56), the relations were determined [40]:

$$G_1(1,x) = \mathrm{J}_0\left(\frac{2}{\gamma^*}e^{-\frac{1}{2}\gamma^*}x\right)\mathrm{Y}_1\left(\frac{2}{\gamma^*}e^{-\frac{1}{2}\gamma^*}x\right) - \mathrm{Y}_0\left(\frac{2}{\gamma^*}e^{-\frac{1}{2}\gamma^*}x\right)\mathrm{J}_1\left(\frac{2}{\gamma^*}e^{-\frac{1}{2}\gamma^*}x\right) \equiv -\frac{\gamma^* e^{\frac{1}{2}\gamma^*}}{\pi x}, \tag{58}$$

$$\hat{G}_1(1,x) = e^{-\frac{1}{2}\gamma^*}Bi\,x^{-1}G_1(1,x). \tag{59}$$

Proving the zero value of the integral in the right side of Equation (57), and thus, the boundary condition (15) is met.

The fulfillment of the boundary condition (16) and the initial condition (17) were verified numerically.

4. Some Special Cases of Solution

Usage of the exact solutions (35) and (36) requires the numerical integration over a semi-limited interval, which requires the application of appropriate, sometimes complex, software. In this section, the asymptotic solutions (for small and large values of the Fourier number τ) will be developed, which are devoid of this problem. In addition, the exact solutions for the temperature generated in the strip and semi-space in the process of frictional heating during braking with constant deceleration will be presented.

Small values of Fourier number τ (large values of the Laplace transform parameter p). Including Formulas (26)–(30), the asymptotes of the modified Bessel functions for large values of arguments [40]:

$$\mathrm{I}_n(x) \cong \frac{e^x}{\sqrt{2\pi x}},\ \mathrm{K}_n(x) \cong \sqrt{\frac{\pi}{2x}}e^{-x},\ n = 0,\ 1, \tag{60}$$

where the transformed solutions (25) were written in the form:

$$\overline{\Theta}^*(\zeta, p) \cong \frac{e^{-\frac{1}{4}\gamma^*\zeta}}{(1+\varepsilon)}\frac{e^{-\alpha\sqrt{p}}}{p\sqrt{p}},\ 0 \le \zeta < 1,\ \overline{\Theta}^*(\zeta, p) \cong \frac{1}{(1+\varepsilon)}\frac{e^{-\zeta^*\sqrt{p}}}{p\sqrt{p}},\ \zeta \le 0, \tag{61}$$

where

$$\alpha = \frac{2}{\gamma^*}(1 - e^{-\frac{1}{2}\gamma^*\zeta}),\ \zeta^* = \frac{|\zeta|}{\sqrt{k^*}}. \tag{62}$$

Proceeding in the transforms (61) to the originals by means of the relation [41]:

$$L^{-1}\left[\frac{e^{-\alpha\sqrt{p}}}{p\sqrt{p}}; \tau\right] = 2\sqrt{\tau}\,\mathrm{ierfc}\left(\frac{\alpha}{2\sqrt{\tau}}\right),\ \alpha \ge 0, \tag{63}$$

where $\mathrm{ierfc}(x) = \pi^{-0.5}e^{-x^2} - x\,\mathrm{erfc}(x)$, $\mathrm{erfc}(x) = 1 - \mathrm{erf}(x)$, and $\mathrm{erf}(x)$ is the Gauss error function [40]. Asymptotes of the dimensionless temperature rise at the initial moments of the heating process were found:

$$\Theta^*(\zeta, \tau) \cong \frac{2\sqrt{\tau}e^{-\frac{1}{4}\gamma^*\zeta}}{(1+\varepsilon)}\mathrm{ierfc}\left(\frac{\alpha}{2\sqrt{\tau}}\right),\ 0 \le \zeta \le 1,\ 0 \le \tau << 1, \tag{64}$$

$$\Theta^*(\zeta, \tau) \cong \frac{2\sqrt{\tau}}{(1+\varepsilon)}\mathrm{ierfc}\left(\frac{\zeta^*}{2\sqrt{\tau}}\right),\ \zeta \le 0,\ 0 \le \tau << 1. \tag{65}$$

On the contact surface $\zeta = 0$, with consideration of the parameter α definition in Equation (62), from Equations (64) and (65), the following was obtained:

$$\Theta^*(0^+, \tau) = \Theta^*(0^-, \tau) = \Theta^*(\tau) \cong \frac{2}{(1+\varepsilon)}\sqrt{\frac{\tau}{\pi}}. \tag{66}$$

In the case of homogeneous strip ($\gamma^* \to 0$) from Equation (62), we determined $\alpha \to \zeta$ and the solutions (64) and (65) take the form of a known solution of the problem for two homogeneous semi-spaces under uniform sliding [5]:

$$\Theta^*(\zeta, \tau) = \frac{2\sqrt{\tau}}{(1+\varepsilon)}\mathrm{ierfc}\left(\frac{\zeta}{2\sqrt{\tau}}\right),\ \zeta \ge 0,\ \tau \ge 0, \tag{67}$$

$$\Theta^*(\zeta, \tau) = \frac{2\sqrt{\tau}}{(1+\varepsilon)}\mathrm{ierfc}\left(-\frac{\zeta}{2\sqrt{k^*\tau}}\right),\ \zeta \ge 0,\ \tau \ge 0. \tag{68}$$

For $\zeta = 0$ from Equations (67) and (68), we received also the solution (66) to determine the temperature on the contact surface.

Analyzing the obtained asymptotic solutions (64) and (65), it can be noticed that at the beginning of the friction heating process, the effect of convective cooling on the free surface of the strip on the temperature of both elements is insignificant, and the gradient nature of the material affects the temperature only inside the layer.

Large values of Fourier number τ (small values of Laplace parameter p). For small values of arguments, the modified Bessel function behaves as follows [40]:

$$\mathrm{I}_0(x) \cong 1,\ \mathrm{I}_1(x) \cong 0.5x,\ \mathrm{K}_0(x) \cong -\ln(x),\ \mathrm{K}_1(x) \cong x^{-1}. \tag{69}$$

Considering the asymptotes (69) in Equations (26)–(30), the transform solutions (26) were presented in the form:

$$\overline{\Theta}^*(\zeta,p) \cong \frac{\phi^*(\zeta)}{p(\sqrt{p}+a)},\ 0 \le \zeta \le 1,\ \overline{\Theta}^*(\zeta,p) \cong \frac{e^{-\zeta^*\sqrt{p}}}{\varepsilon p(\sqrt{p}+a)},\ \zeta \le 0, \tag{70}$$

where

$$\phi^*(\zeta) = \frac{\phi(\zeta)}{\varepsilon\phi_0},\ a = \frac{\gamma^* Bi}{\varepsilon\phi_0}, \tag{71}$$

$$\phi(\zeta) = \gamma^* + Bi(e^{-\gamma^*\zeta} - e^{-\gamma^*}),\ \phi_0 \equiv \phi(0) = \gamma^* + Bi(1 - e^{-\gamma^*}). \tag{72}$$

Taking into consideration the relation [41]:

$$L^{-1}\left[\frac{a\,e^{-b\sqrt{p}}}{p(\sqrt{p}+a)};\tau\right] = \mathrm{erfc}\left(\frac{b}{2\sqrt{\tau}}\right) - e^{ab+a^2\tau}\mathrm{erfc}\left(\frac{b}{2\sqrt{\tau}} + a\sqrt{\tau}\right),\ a > 0,\ b \ge 0, \tag{73}$$

the following asymptotes of dimensionless temperature rise for large values of the Fourier number, and τ were found:

$$\Theta^*(\zeta,\tau) \cong a^{-1}\phi^*(\zeta)[1 - e^{a^2\tau}\mathrm{erfc}(a\sqrt{\tau})],\ 0 \le \zeta \le 1,\ \tau >> 1, \tag{74}$$

$$\Theta^*(\zeta,\tau) \cong \frac{1}{a\varepsilon}\left[\mathrm{erfc}\left(\frac{\zeta^*}{2\sqrt{\tau}}\right) - e^{a\zeta^*+a^2\tau}\mathrm{erfc}\left(\frac{\zeta^*}{2\sqrt{\tau}} + a\sqrt{\tau}\right)\right],\ \zeta \le 0,\ \tau >> 1. \tag{75}$$

On the friction surface $\zeta = 0$ from Equations (62) and (72) follows that $\zeta^* = 0$ and $\phi^*(0) = \varepsilon^{-1}$, and from the solutions (74) and (75), yields:

$$\Theta^*(0^+,\tau) = \Theta^*(0^-,\tau) \equiv \Theta^*(\tau) \cong (a\varepsilon)^{-1}[1 - e^{a^2\tau}\mathrm{erfc}(a\sqrt{\tau})],\ \tau >> 1, \tag{76}$$

where parameters ε and a were designated on the basis of Equations (31) and (71).

Bearing in mind the values of limits:

$$\lim_{\gamma^*\to 0} a = \frac{Bi}{\varepsilon}\lim_{\gamma^*\to 0}\frac{\gamma^*}{[\gamma^* + Bi(1 - e^{-\gamma^*})]} = \frac{Bi}{\varepsilon(1+Bi)}, \tag{77}$$

$$\lim_{\gamma^*\to 0} a^{-1}\beta(\zeta) = \frac{1}{Bi}\lim_{\gamma^*\to 0}\frac{[\gamma^* + Bi(e^{-\gamma^*\zeta} - e^{-\gamma^*})]}{\gamma^*} = \frac{1 + Bi(1-\zeta)}{Bi}, \tag{78}$$

from the Equations (74) and (75), the solution were obtained to determine the temperature in the considered system with a homogeneous strip [10].

Linearly decreasing the time profile of specific friction power. Demonstrated above, the exact solutions (35)–(44) were found for the invariable specific friction power over time, $q(t) = q_0, t \ge 0$ in the boundary condition (4). In the relevant thermal problems of friction concerning the modeling of the frictional heating process during braking with a constant deceleration, the temporal profile of specific friction power has the form [1]:

$$\hat{q}(t) = \hat{q}_0 q^*(t),\ \hat{q}_0 = 2q_0,\ q^*(t) = 1 - t\,t_s^{-1},\ 0 \le t \le t_s, \tag{79}$$

where t_s is the stop moment of the vehicle. Dimensionless temperature rise $\hat{\Theta}^*(\zeta,\tau)$, corresponding to the specific power of friction (79), was found based on the Duhamel's theorem [42]:

$$\hat{\Theta}^*(\zeta,\tau) = \frac{\partial}{\partial\tau}\int_0^{\tau} q^*(\tau - s)\Theta^*(\zeta,s)ds,\ \zeta \geq 0,\ 0 \leq \tau \leq \tau_s, \tag{80}$$

where $\Theta^*(\zeta,\tau)$ is the dimensionless temperature rise (35)–(44) and function $q^*(\tau)$:

$$q^*(\tau) = 1 - \tau\tau_s^{-1},\ 0 \leq \tau \leq \tau_s,\ \tau_s = k_1 t_s d^{-2}. \tag{81}$$

Substituting solutions (35) and (36) and function $q^*(\tau)$ (81) to the right side of Equation (80) yields:

$$\hat{\Theta}^*(\zeta,\tau) = \vartheta(\zeta)q^*(\tau) - \frac{2}{\pi}\varepsilon e^{-\frac{1}{2}\gamma^*\zeta}\int_0^{\infty} F(x)G_1(\zeta,x)P(\tau,x)dx,\ 0 \leq \zeta \leq 1,\ 0 \leq \tau \leq \tau_s, \tag{82}$$

$$\hat{\Theta}^*(\zeta,\tau) = \vartheta_0 q^*(\tau) - \frac{2}{\pi}\int_0^{\infty} F(x)G_2(\zeta,x)P(\tau,x)dx,\ \zeta \leq 0,\ 0 \leq \tau \leq \tau_s, \tag{83}$$

where

$$P(\tau,x) = e^{-x^2\tau} - \frac{1}{x^2\tau_s}(1 - e^{-x^2\tau}), \tag{84}$$

and the constant ϑ_0 and functions $\vartheta(\zeta)$, $F(x)$, $G_l(\zeta,x)$, $l = 1,2$ were determined from Equations (37)–(44).

5. Numerical Analysis

Approval of the developed calculation model was performed for a friction system, in which the strip is made of a two-component FGM, and the counterbody is homogeneous (cast iron ChNMKh). The base of the FGM is zirconium dioxide ZrO_2, and on the second component of material the titanium alloy Ti-6Al-4V was selected. Properties of these materials at the initial temperature $T_0 = 20\ ^\circ C$ are given in Table 1. For the same volume fraction of ZrO_2 and Ti-6Al-4V, effective specific heat and density of the strip material amounted to $c_1 = 495.55\ J\,kg^{-1}K^{-1}$ and $\rho_1 = 5266.98\ kg\,m^{-3}$, respectively. By means of Equation (8), the dimensionless gradient of selected FGM was also established $\gamma^* = 1.26$. The rest dimensionless input parameters for the calculations are spatial variable ζ, Fourier number τ and Biot number Bi (9).

Table 1. Materials' properties [36].

Material	Thermal Conductivity $Wm^{-1}K^{-1}$	Specific Heat $J\,kg^{-1}K^{-1}$	Density $kg\,m^{-3}$
ZrO_2	$K_{1,1} = 1.94$	$c_{1,1} = 452.83$	$\rho_{1,1} = 6102.16$
Ti-6Al-4V	$K_{1,2} = 6.87$	$c_{1,2} = 538.08$	$\rho_{1,2} = 4431.79$
ChNMKh	$K_2 = 52.17$	$c_2 = 444.6$	$\rho_2 = 7100$

The aim of this numerical analysis was to establish the qualitative effect of the intensity of the convective heat exchange with the environment (parameter Bi) on the free surface of the FGM strip $z = d$ ($\zeta = 1$) on the temperature of friction system. It should be noted that the case $Bi \to 0$ corresponds to the thermal isolation of the surface $\zeta = 1$; however, for $Bi \to \infty$ on this surface during the whole process of heating, the initial temperature is sustained $T(d,t) = T_0$ ($\Theta^*(1,\tau) = 0$). Results for dimensionless temperature rise $\Theta^*(\zeta,\tau)$ (35)–(44) and the intensity of the heat fluxes $q_l^*(\tau), l = 1,2$ (51) and (52) obtained by means of the numerical integral procedure QAGI from the package QUADPACK [43] were presented in Figures 2–6.

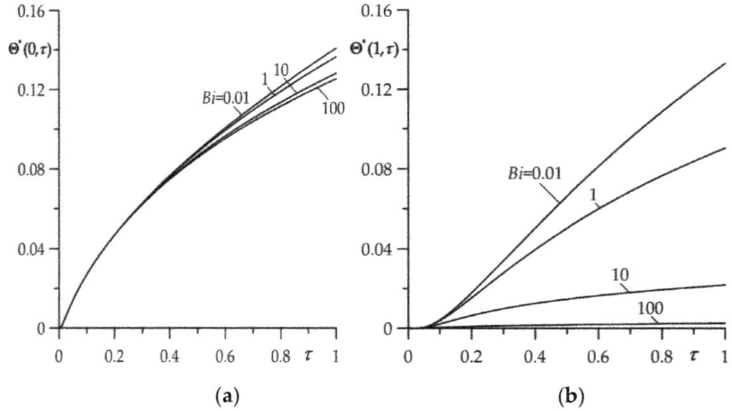

Figure 2. Evolutions of dimensionless temperature rise Θ^* for selected values of Biot number: (**a**) on the contact surface $\zeta = 0$; (**b**) on the free surface of the strip $\zeta = 1$.

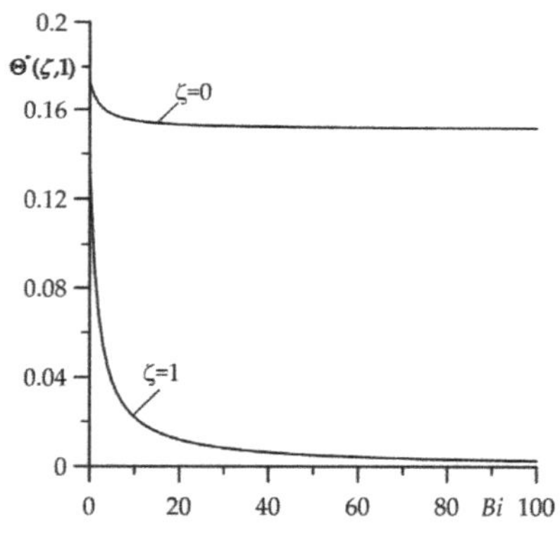

Figure 3. Dependence of dimensionless temperature rise Θ^* on the Biot number Bi on the surfaces $\zeta = 0$ and $\zeta = 1$ for $\tau = 1$.

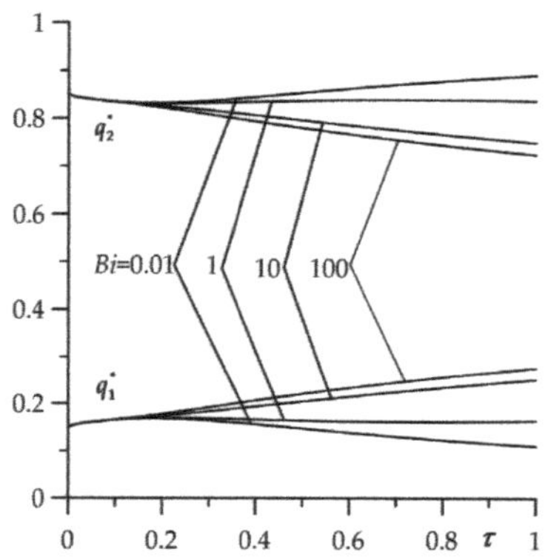

Figure 4. Evolutions of dimensionless intensities of a $q_l^*(\tau)$ heat fluxes absorbed by FGM strip ($l = 1$) and homogeneous half-space ($l = 2$) for selected values of the Biot number Bi.

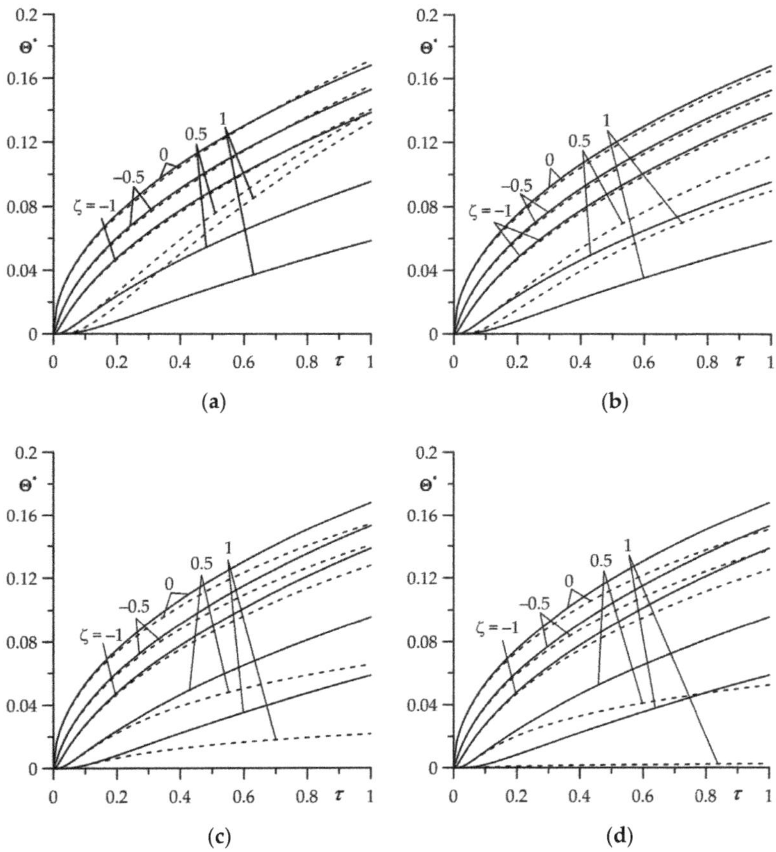

Figure 5. Evolutions of dimensionless temperature rise Θ^* for different values of dimensionless spatial variable ζ for (**a**) $Bi = 0.01$; (**b**) $Bi = 1$; (**c**) $Bi = 10$; and (**d**) $Bi = 100$. The solid curves are the exact solution (35)–(44), and the dashed curves are the asymptotic solution (64) and (65) for small values of the Fourier number τ.

Influence of the Biot number Bi value on the evolution of $\Theta^*(\zeta, \tau)$ is shown in Figure 2. A slight decrease in temperature on the friction surface $\zeta = 0$ with the increase in Bi becomes visible for $\tau \geq 0.6$ (Figure 2a). While the free surface $\zeta = 1$ of the FGM strip is more sensitive to changes in the Biot number. Growth of the convective cooling intensity causes a significant decrease in the temperature of this surface much earlier, at $\tau \geq 0.1$ (Figure 2b). For the fixed value of Fourier number τ, the temperature on both surfaces drops for higher values of parameter Bi. This effect is most noticeable on the free surface of the FGM strip $\zeta = 1$.

The drop in temperature on the contact surface $\zeta = 0$ and the free surface $\zeta = 1$ of the strip with the increase in the Biot number Bi for the fixed value $\tau = 1$ is demontrated in Figure 3. The highest temperature on both surfaces is achieved for the assumption of the adiabatic ($Bi = 0$) free strip surface. Next, growth of Bi causes cooling of the considered surfaces. An explicit decrease in temperature on the contact surface $\zeta = 0$ is visible for $0 \leq Bi \leq 10$. A further increase in Bi shows practically no effect on the temperature on this surface.

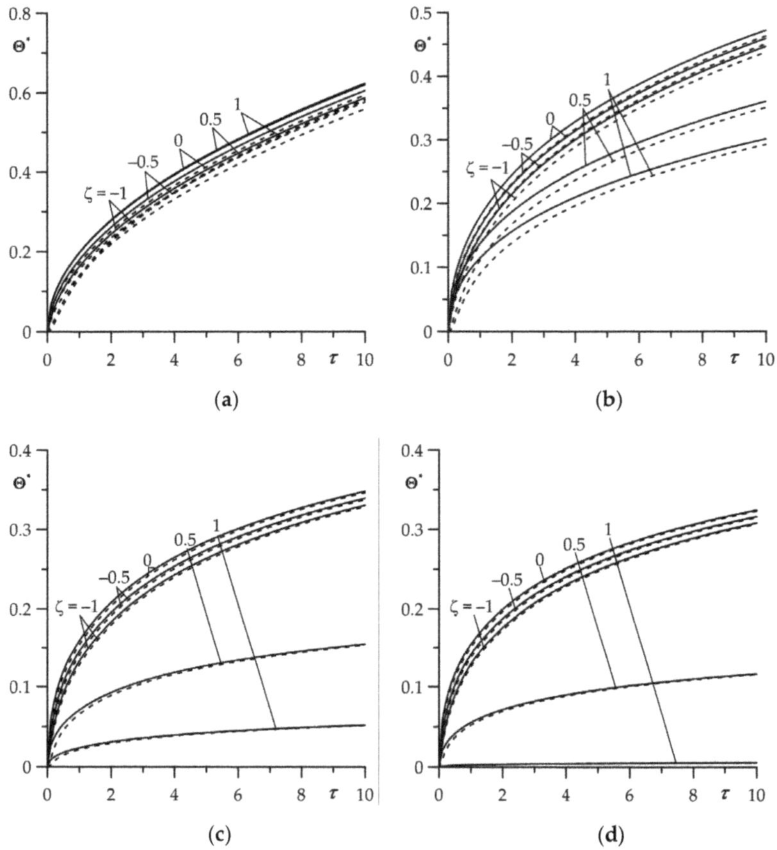

Figure 6. Evolutions of dimensionless temperature rise Θ^* for different values of dimensionless spatial variable ζ for (**a**) $Bi = 0.01$; (**b**) $Bi = 1$; (**c**) $Bi = 10$; and (**d**) $Bi = 100$. The solid curves are the exact solution(35)–(44), and the dashed curves are the asymptotic solution (74) and (75) for large values of the Fourier number τ.

Thus, when estimating the maximum (achievable at the contact surface $\zeta = 0$) temperature of the selected friction system, the cooling of the free strip surface should be taken into consideration for the values of the Biot number outside the specified range. If $Bi > 10$, then the boundary condition (5) in the formulation to the thermal problem of friction has to be replaced by its simplified variant $T(d, t) = T_0$, $t > 0$ ($\Theta^*(1, \tau) = 0$ in the condition (15)). A decrease in Bi causes the drop in temperature on the surface $\zeta = 1$ to the level of the initial temperature $\Theta^* = 0$, the most noticeable in the range $0 \le Bi \le 60$.

Most of the heat generated by friction on the contact surface is absorbed by the semi-space (Figure 4). This is due to the much better thermal conductivity of cast iron compared to zirconium dioxide (Table 1). In the initial heating period ($0 < \tau \le 0.2$), the strip absorbs $\approx 15\%$, and the semi-space the remaining $\approx 85\%$ of the heat. The proportion of heat distribution between the strip and the half-space changes with the sliding time depending on the value of Bi. For small values of Biot number, the amount of heat absorbed by the strip decreases slightly in time of heating, achieving for $\tau = 1$ the values $\approx 10\%$ and $\approx 13\%$ for $Bi = 0.01$ and $Bi = 1$, respectively. At the same time, the amount of heat directed to the half-space increases proportionally. The growth of the cooling intensity on the free strip surface increases the heat absorbed by it for $\tau = 1$ to 20% and 27% for $Bi = 10$ and $Bi = 100$, respectively.

The comparison of temperature values found by means of exact solution (35)–(44) (solid lines) and asymptotic solutions (64) and (65) for small values of the Fourier number τ (dashed lines) are illustrated in Figure 5. Good agreement of the results for $0 \leq \tau \leq 1$, determined based on the exact and asymptotic solutions, take place on the contact surface $\zeta = 0$ and inside the semi-space for $\zeta = -0.5$ and $\zeta = -1$, for all four selected values of Bi. Whereas inside the FGM strip, using the asymptotic solution should be restricted to the range $0 \leq \tau \leq 0.2$.

Asymptotic solutions (74) and (75) for large values of τ allow for estimation of the temperature in the range $0 \leq \tau \leq 10$ both in the strip and in the half-space (Figure 6). The accuracy of such an estimation rises with the increasing intensity of convective cooling on the free strip surface. Such a good agreement of the temperature time profiles, found with the use of exact and asymptotic solutions, allows for extensive use of the latter in engineering calculations of the temperature mode for the selected friction pair. The advantage of the asymptotic solution is the lack of numerical integration, which occurs when using the exact solution (35)–(44).

An influence of functionally graded material of the strip on the temperature field of the friction system is presented in Figure 7. Calculations were carried out based on the exact solutions for sliding under the constant (35)–(44) or linearly decreasing (82)–(84) temporal profile of specific friction power. Heating of the strip over the entire thickness is visible with low ($Bi = 1$) intensity of convective cooling on the free strip surface (Figure 7a,c). However, when the Biot number is increased to the value $Bi = 100$, the initial temperature is maintained on the free surface of the strip (Figure 7b,d).

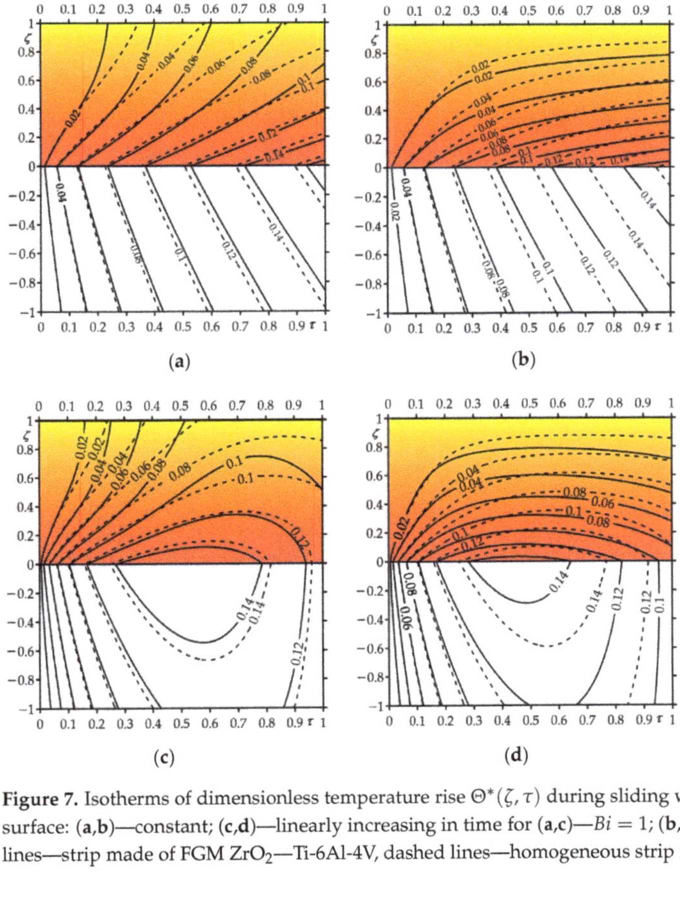

Figure 7. Isotherms of dimensionless temperature rise $\Theta^*(\zeta, \tau)$ during sliding with specific friction surface: (**a**,**b**)—constant; (**c**,**d**)—linearly increasing in time for (**a**,**c**)—$Bi = 1$; (**b**,**d**)—$Bi = 100$. Solid lines—strip made of FGM ZrO_2—Ti-6Al-4V, dashed lines—homogeneous strip made of ZrO_2.

Hence, we conclude that when determining such an important parameter as the effective depth of heat transfer during calculations of the temperature mode of the friction node [7], in the case $Bi = 1$, the entire thickness of the strip should be taken, and in the case $Bi = 100$, only that part of it should be taken that is determined by the distance from the contact surface at which the temperature is 5% of the maximum value.

The reduction in the temperature of the friction system as a result of the use of FGM is most noticeable near the contact surface during braking with a constant deceleration (Figure 7c,d). During uniform sliding with constant velocity, the temperature of the strip and the half-space at a fixed distance from the contact surface increase monotonically over the heating time (Figure 7a,b). By contrast, in the case of sliding with a linearly decreasing velocity, the temperature first quickly increases to the maximum value, after which the stage of slight cooling begins and lasts until the standstill (Figure 7c,d).

6. Conclusions

An analytical model was developed to simulate the processes of frictional heating on the contact surface and convection cooling on the free surface of the friction pair, in which one element was made of FGM and the other of a homogeneous material. Ignoring changes in the temperature gradient in directions parallel to the contact surface, the transient, one-dimensional temperature field in such a system was found from the exact solution of the thermal friction problem for the strip/half-space scheme at constant specific friction power. It was assumed that the two-component FGM strip has a thermal conductivity coefficient increasing exponentially along the thickness, and the material of the half-space is homogeneous. The friction thermal contact of the strip and the semi-space is perfect, and on the free surface of strip, the heat exchange with the surrounding environment takes place according to Newton's law. In addition, the exact, asymptotic solutions were also obtained for small and large values of the Fourier number. Using the Duhamel's formula and the solution at a constant specific friction power, appropriate solutions were determined with a linearly decreasing time profile of the specific friction power. This made it possible to simulate the frictional heating process during braking with a constant deceleration.

Numerical analysis was performed for a functionally graded strip (ZrO_2—Ti-6Al-4V), sliding against the cast iron half-space (ChNMKh). The following was established:

(1) Applying of FGM for one element of the friction couple (strip) allows for a decrease in the temperature on the contact surface in comparison to the case of the homogeneous strip (zirconium dioxide);
(2) A convective heat exchange with the environment on the free surface of the strip causes a decrease in the temperature on the contact surface at the values of the Biot number $0 \leq Bi \leq 10$. However, the greatest drop in temperature on the free surface of the strip occurs in the range of changes $0 \leq Bi \leq 60$;
(3) Most part of the frictional heat is absorbed by the cast iron semi-space (≈85%) in the initial stage of the heating process. With the elapse of the slipping time and the increase in the cooling intensity of the free surface of strip, the amount of heat absorbed by the half-space decreases to 73% for $\tau = 1$ and $Bi = 100$. The amount of heat directed to the FGM strip increases accordingly;
(4) Obtained asymptotic solutions for small and large values of the Fourier number τ can be used to quickly estimate the temperature of both elements of the system, with high accuracy. At the same time, the solution for large values of the Fourier number is useful for determining the temperature at any time during the friction heating process at $\tau > 0$;
(5) Convective cooling of the FGM strip allows for a reduction in the effective depth of heating, i.e., the distance from the contact surface at which the temperature of each element reaches significant values;
(6) The space–time distribution of isotherms in the strip and semi-space depends on the time profile of the specific friction power. With a constant friction power during sliding, the temperature monotonically increases with the increasing heating time (Fourier

number τ). However, in the case of braking with constant deceleration, the temperature of the friction surface reaches its maximum value around half the stopping time τ_s.

Summing up, the analysis carried out on the basis of the developed mathematical model of the frictional heating process showed that both the use of FGM and convection cooling in the process allow for an effective reduction in the temperature of the friction pair elements.

We would like to note that analytical solutions of one-dimensional friction thermal problems allow for estimating with good accuracy the maximum temperature of friction systems. This has been confirmed in many papers containing relevant experimental measurements [44,45]. These solutions are used to estimate the temperature on the nominal contact surface of the braking system (an average during one cycle of heating and cooling during disc rotation). It is one of the components to determine the maximum temperature in such a system, and the second component is the flash temperature. The maximum temperature is a design parameter for the development of a methodology for the initial selection of friction materials in various types of brake systems, including disc brake systems.

Such models regarding FGMs, proposed by us so far, assume the perfect thermal friction contact between the elements. This is fully justified when the friction surfaces of these elements are sufficiently smooth. However, in reality, these surfaces are rough, depending on the level of treatment and operating conditions. This causes the thermal resistance of the contact surface, and as a result, the appearance of a temperature jump on the friction surfaces. One of the approaches to solve this problem is to introduce into the formulation of the relevant problems, the conditions of imperfect thermal contact of friction. We plan to implement it at the next stage of our research, obtaining a solution to this problem in the case of friction pair elements made of FGM and investigating the effect of thermal resistance on temperature.

Author Contributions: Conceptualization and methodology, A.Y.; software, P.Z.; validation, A.Y. and K.T., formal analysis, A.Y., M.K. and P.Z.; investigation, A.Y., M.K. and K.T.; writing—original draft preparation A.Y. and K.T.; writing–review and editing, K.T.; visualization and figures preparation, P.Z.; supervision, A.Y. and M.K.; project administration, M.K. All authors have read and agreed to the published version of the manuscript.

Funding: This investigation was performed within the framework of research project No. 2017/27/B/ST8/01249, funded by the National Science Centre, Poland, and project financing through the program of the Ministry of Science and Higher Education, Poland, named "Regional Initiative of Excellence" in 2019–2022, project No. 011/RID/2018/19.

Institutional Review Board Statement: Not applicable.

Informed Consent Statement: Not applicable.

Data Availability Statement: No new data were created or analyzed in this study. Data sharing is not applicable to this article.

Conflicts of Interest: The authors declare no conflict of interest.

Nomenclature

Bi	Biot number
c	Specific heat (J kg^{-1}K^{-1})
d	Thickness of the strip (m)
f	Coefficient of friction
h	Coefficient of convective heat exchange (W m^{-2}K^{-1})
$\mathrm{I}_n(\cdot)$	Modified Bessel functions of the nth order of the first kind
$\mathrm{J}_n(\cdot)$	Bessel functions of the nth order of the first kind
$\mathrm{K}_n(\cdot)$	Modified Bessel functions of the nth order of the second kind
k	Thermal diffusivity (m^{2}s^{-1})
K	Thermal conductivity (W m^{-1}K^{-1})

p	Parameter of the Laplace integral transform
p_0	Nominal pressure on the contact surface (Pa)
q	Specific power of friction ($W\,m^{-2}$)
q_0	Nominal value of the specific friction power ($W\,m^{-2}$)
t	Time (s)
t_s	Stop moment of the process (s)
T	Temperature (°C)
T_0	Initial temperature (°C)
V_0	Sliding velocity ($m\,s^{-1}$)
$Y_n(\cdot)$	Bessel functions of the *n*th order of the second kind
z	Spatial coordinate in axial direction (m)
γ^*	Gradient parameter of FGM
Λ	Temperature rise scaling factor (°C)
ε	Dimensionless coefficient of thermal activity
Θ	Temperature rise (°C)
Θ^*	Dimensionless temperature rise
ρ	Density ($kg\,m^{-3}$)
τ	Dimensionless time
τ_s	Fourier number
ζ	Dimensionless spatial coordinate in axial direction

References

1. Newcomb, T.P. Temperatures reached in disc brakes. *J. Mech. Eng. Sci.* **1960**, *2*, 167–177. [CrossRef]
2. Day, A.J. *Braking of Road Vehicles*; Butterworth-Heinemann/Elsevier: Oxford, UK, 2014.
3. Bauzin, J.G.; Keruzore, N.; Laraqi, N.; Gapin, A.; Diebold, J.F. Identification of the heat flux generated by friction in an aircraft braking system. *Int. J. Therm. Sci.* **2018**, *130*, 449–456. [CrossRef]
4. Nosko, A.L.; Belyakov, N.S.; Nosko, A.P. Application of the generalized boundary condition to solving thermal friction problems. *J. Frict. Wear* **2009**, *30*, 455–462. [CrossRef]
5. Carlslaw, H.C.; Jaeger, J.J. *Conduction of Heat in Solids*, 2nd ed.; Clarendon Press: Oxford, UK, 1959.
6. Ling, F.F. *Surface Mechanics*; Wiley: New York, NY, USA, 1973.
7. Chichinadze, A.V. (Ed.) *Polymers in Friction Assemblies of Machines and Devices: A Handbook*; Allerton Press Inc.: New York, NY, USA, 1984.
8. Chichinadze, A.V.; Kozhemyakina, V.D.; Suvorov, A.V. Method of temperature-field calculation in model ring specimens during bilateral friction in multidisk aircraft brakes with the IM-58-T2 new multipurpose friction machine. *J. Frict. Wear* **2010**, *31*, 23–32. [CrossRef]
9. Yevtushenko, A.A.; Kuciej, M. Temperature in a frictionally-heated ceramic-metal patch and cast iron disc during braking. *Numer. Heat Transf. Part A Appl.* **2009**, *56*, 97–108. [CrossRef]
10. Yevtushenko, A.A.; Kuciej, M. Influence of convective cooling on the temperature in a frictionally heated strip and foundation. *Int. Comm. Heat Mass Trans.* **2009**, *36*, 129–136. [CrossRef]
11. Yevtushenko, A.A.; Kuciej, M.; Różniakowska, M. The contact heat transfer between the plane-parallel strip and the semi-infinite foundation. *Int. Comm. Heat Mass Trans.* **2009**, *36*, 787–793. [CrossRef]
12. Yevtushenko, A.A.; Kuciej, M. Influence of the convective cooling nd the thermal resistance on the temperature of the pad/disc tribosystem. *Int. Comm. Heat Mass Trans.* **2010**, *37*, 337–342. [CrossRef]
13. Barber, J.R.; Comninou, M. Thermoelastic contact problems. In *Thermal Stresses*, 3rd ed.; Hetnarsky, R.R., Ed.; Elsevier: Amsterdam, The Netherlands, 1989; pp. 1–105.
14. Nosko, A.L.; Nosko, A.P. Solution of contact heating problem with account for heat transfer between the friction members. *J. Frict. Wear* **2006**, *27*, 35–40.
15. Sathish, M.; Radhika, N.; Saleh, B.A. Critical review on functionally graded coatings: Methods, properties, and challenges. *Compos. Part B Eng.* **2021**, *225*, 109278. [CrossRef]
16. Polajnar, M.; Kalin, M.; Thorbjornsson, I.; Thorgrimsson, J.T.; Valle, N.; Botor-Probierz, A. Friction and wear performance of functionally graded ductile iron for brake pads. *Wear* **2017**, *382*, 85–94. [CrossRef]
17. Govindaraju, M.; Megalingam, A.; Murugasan, J.; Vignesh, V.R.; Kota, K.P.; Ram, S.A.; Lakshana, P.; Kumar, N.V. Investigations on the tribological behavior of functionally gradient iron-based brake pad material. *Proc. Inst. Mech. Eng. Part C J. Mech. Eng. Sci.* **2020**, *234*, 2474–2486. [CrossRef]
18. Genao, F.Y.; Kim, J.; Żur, K.K. Nonlinear finite element analysis of temperature-dependent functionally graded porous micro-plates under thermal and mechanical loads. *Compos. Struct.* **2021**, *256*, 112931. [CrossRef]
19. Mahamood, R.M.; Akinlabi, E.T. *Functionally Graded Materials*; Springer: Cham, Switzerland, 2017.

20. Yaghoobi, M.P.; Ghannad, M. An analytical solution for heat conduction of FGM cylinders with varying thickness subjected to non-uniform heat flux using a first-order temperature theory and perturbation technique. *Int. Commun. Heat Mass Transf.* **2020**, *116*, 104684. [CrossRef]
21. Hosseini, T.P.; Talebi, M. Stress and temperature distribution study in a functionally graded brake disk. *Int. J. Automot. Eng.* **2012**, *2*, 172–179.
22. Jin, Z.H. An asymptotic solution of temperature field in a strip of a functionally graded material. *Int. Commun. Heat Mass Transf.* **2002**, *29*, 887–895. [CrossRef]
23. Zhao, J.; Li, Y.; Ai, X. Analysis of transient thermal stress in sandwich plate with functionally graded coatings. *Thin Solid Film.* **2008**, *516*, 7581–7587. [CrossRef]
24. Delouei, A.A.; Emamian, A.; Karimnejad, S.; Sajjadi, H.; Jing, D. Two-dimensional temperature distribution in FGM sectors with the power-law variation in radial and circumferential directions. *J. Therm. Anal. Calorim.* **2021**, *144*, 611–621. [CrossRef]
25. Delouei, A.A.; Emamian, A.; Karimnejad, S.; Sajjadi, H.; Jing, D. Asymmetric Conduction in an Infinite Functionally Graded Cylinder: Two-Dimensional Exact Analytical Solution Under General Boundary Conditions. *ASME J. Heat Transfer.* **2020**, *142*, 044505. [CrossRef]
26. Zhou, J.; Shen, F.; Liu, J.; El-Borgi, S.; Ke, L. Thermoelastic rotating contact of an FGM coating with temperature-dependent and arbitrary varying properties. *Sci. China Technol. Sci.* **2023**, *66*, 1038–1049. [CrossRef]
27. Fu, P.; Zhao, J.; Zhang, X.; Kang, G.; Wang, P.; Kan, Q. Thermo-mechanically coupled sliding contact shakedown analysis of functionally graded coating-substrate structures. *Int. J. Mech. Sci.* **2022**, *222*, 107241. [CrossRef]
28. Balci, M.N.; Dag, S. Solution of the dynamic frictional contact problem between a functionally graded coating and a moving cylindrical punch. *Int. J. Solids Struct.* **2019**, *161*, 267–281. [CrossRef]
29. Shahzamanian, M.M.; Shahrjerdi, A.; Sahari, B.B.; Wu, P.D. Steady-State Thermal Analysis of Functionally Graded Rotating Disks Using Finite Element and Analytical Methods. *Materials* **2022**, *15*, 5548. [CrossRef] [PubMed]
30. Burlayenko, V.N.; Altenbach, H.; Sadowski, T.; Dimitrova, S.D.; Bhaskar, A. Modelling functionally graded materials in heat transfer and thermal stress analysis by means of graded finite elements. *Appl. Math. Model.* **2017**, *45*, 422–438. [CrossRef]
31. Liu, J.; Ke, L.L.; Wang, Y.S. Two-dimensional thermoelastic contact problem of functionally graded materials involving frictional heating. *Int. J. Solids Struct.* **2011**, *48*, 2536–2548. [CrossRef]
32. Liu, J.; Ke, L.; Zhang, C. Axisymmetric thermoelastic contact of an FGM-coated half-space under a rotating punch. *Acta Mech.* **2021**, *232*, 2361–2378. [CrossRef]
33. Tian, J.H.; Jiang, K. Heat conduction investigation of the functionally graded materials plates with variable gradient parameters under exponential heat source load. *Int. J. Heat Mass Trans.* **2018**, *122*, 22–30. [CrossRef]
34. Yevtushenko, A.; Topczewska, K.; Zamojski, P. The Effect of Functionally Graded Materials on Temperature during Frictional Heating: Under Uniform Sliding. *Materials* **2021**, *14*, 4285. [CrossRef]
35. Yevtushenko, A.; Topczewska, K.; Zamojski, P. Use of functionally graded material to decrease maximum temperature of a coating–substrate system. *Materials* **2023**, *16*, 2265. [CrossRef]
36. Yevtushenko, A.; Topczewska, K.; Zamojski, P. Influence of functionally graded protective coating on the temperature in braking system. *Materials* **2023**, *16*, 4308. [CrossRef]
37. Mao, J.J.; Ke, L.L.; Yang, J.; Kitipornchai, S.; Wang, Y.S. The coupled thermoelastic instability of FGM coatings with arbitrarily varying properties: In-plane sliding. *Acta Mech.* **2018**, *229*, 2979–2995. [CrossRef]
38. Alinia, Y.; Espo, M. Sliding contact problem of an FGM coating/substrate system with two-dimensional material property grading. *Acta Mech.* **2020**, *231*, 649–659. [CrossRef]
39. Sneddon, I.N. *The Use of Integral Transforms*; McGraw-Hill: New York, NY, USA, 1972.
40. Abramowitz, M.; Stegun, I. *Handbook of Mathematical Functions with Formulas, Graphs, and Mathematical Tables*; United States Department of Commerce; National Bureau of Standards (NBS): Washington, DC, USA, 1964.
41. Bateman, H.; Erdelyi, A. *Tables of Integrals Transforms*; McGraw-Hill: New York, NY, USA, 1954; Volume 1.
42. Özisik, N.M. *Heat Conduction*; John Wiley: New York, NY, USA, 1993.
43. Piessens, R.; De Doneker-Kapenga, E.; Überhuber, C.W.; Kahaner, D.K. *Quadpack: A Subroutine for Automatic Integration*; Springer Science & Business Media: Berlin/Heidelberg, Germany, 2012; Volume 1.
44. Yang, Y.; Wu, B.; Shen, Q.; Xiao, G. Numerical simulation of the frictional heat problem of subway brake discs considering variable friction coefficient and slope track. *Eng. Fail. Anal.* **2021**, *130*, 105794. [CrossRef]
45. Wang, Z.; Han, J.; Liu, X.; Li, Z.; Yang, Z.; Chen, E. Temperature evolution of the train brake disc during high-speed braking. *Adv. Mech. Eng.* **2019**, *11*, 1687814018819563. [CrossRef]

materials

MDPI

Article

The Mutual Influence of Thermal Contact Conductivity and Convective Cooling on the Temperature Field in a Tribosystem with a Functionally Graded Strip

Aleksander Yevtushenko, Katarzyna Topczewska * and Przemysław Zamojski

Faculty of Mechanical Engineering, Bialystok University of Technology (BUT), 45C Wiejska Street, 15-351 Bialystok, Poland; a.yevtushenko@pb.edu.pl (A.Y.); p.zamojski@pb.edu.pl (P.Z.)
* Correspondence: k.topczewska@pb.edu.pl

Abstract: An analytical model to find the temperature field that has been developed for friction systems consists of a strip and semi-space. The strip is made of a two-component functionally graded material (FGM) with an exponentially changing coefficient of thermal conductivity. In contrast, the material of the semi-space is homogeneous. An appropriate boundary-value problem of heat conduction with constant specific friction power was formulated and solved using the Laplace integral transform method. The model takes into consideration the imperfect thermal friction contact between the strip and the semi-space, and also the convective cooling on the exposed surface of the strip. The appropriate asymptotic solutions to this problem for low and high values of Fourier number were obtained. It is shown how the determined exact solution can be generalized using Duhamel's formula for the case of a linearly reduction in time-specific friction power (a braking process with constant deceleration). Numerical analysis for selected materials of the friction pair was carried out in terms of examining the mutual impact on the temperature of the two Biot numbers, characterizing the intensity of the thermal contact conductivity and convective heat exchange on the exposed surface of the strip. The obtained results can be used to predict the temperature of friction systems containing elements made of FGM. In particular, such systems include modern disc braking systems.

Keywords: functionally graded material; frictional heating; temperature; thermal contact conductivity; convective cooling

1. Introduction

In order to model the frictional heating process at the interface of elements in sliding contact, the thermal problems of friction are formulated in the form of classical heat conduction equations with the appropriate boundary and initial conditions. Solutions of such problems result in temperature distributions in both components of the friction pair. One of the various approaches used during the formulation of the heat conduction problems is the separated bodies concept [1,2], which considers the sliding elements of the friction pair to be uncoupled. Naturally, this simplification forces the introduction of a virtual heat source in order to take into account relevant thermal effects associated with frictional contact interactions. The heat flux absorbed on the friction surface by each component is then established using a suitable heat partition ratio [3,4]. This coefficient is mostly introduced to the models a priori, which might be determined based on experimental studies or analytical solutions [4–6].

Unlike the separation approach, the formulation for the thermal friction problems in accordance with the second variant assumes the presence of two bodies in contact. In coupled problems, it is necessary to formulate the boundary conditions at the interface between bodies [7]. Most often, it is assumed that the distribution of temperature and the intensity of heat flux on the contact surface are continuous, which allow one to determine the mean surface temperature and partition of generated frictional heat [8]. The assumption of these perfect thermal contact conditions is a classical approach, applied in most of the models proposed to date for the frictional heating process in sliding systems [9]. However, these conditions only describe the macroscopic geometry of coupled elements with ideally smooth rubbing surfaces, which does not account well for the tribological phenomena occurring at the sliding interface. This means that the perfect thermal contact conditions at the nominal friction surface do not allow one to model the heating process adequately. Considering the microscopic tribological effects from mechanical and thermal perspectives more closely, contact between two sliding solids is not perfect. Since the friction surfaces are rough rather than smooth, the real contact area is much smaller than the apparent (nominal) contact area [10]. When heat flows across the non-ideal contact zone, temperature discontinuity occurs at the interface. So, precise predictions of temperature fields in a friction zone require the consideration of imperfect thermal contact at the interface [11,12]. Experimental studies show a significant thermal gradient at the macroscopic level between rubbing surfaces of the friction pair elements [13,14]. For instance, in disc braking systems, typically, the values of temperature on the pad's contact surface are higher than the corresponding values related to the disc [1]. During braking, the pad surface is hotter, since it is in constant friction contact with the disc and it is made of softer material, so it can be more easily plastically deformed and worn than the harder disc [3,15]. Moreover, not only the temperature distribution in a friction couple element but also the heat partition between the sliding components is sensitive to the interfacial boundary conditions [16].

Non-ideal contact is more representative of the heat transfer phenomenon at the interface of coupled bodies sliding against each other. The imperfect thermal contact conditions state that the intensity of heat flux is continuous, according to the law of energy conservation, and there is a gap between temperature values achieved on the contacting surfaces of the sliding elements [9]. This temperature drop is said to be a result of thermal contact resistance existing at the interface [5,17], which is defined as the ratio between the temperature gradient at the contacting surfaces and the heat flux flowing across the interface. Study [16] shows that the value of thermal resistance at the interface of the coupled elements decreases with an increase in relative sliding velocity. The estimation of the thermal contact resistance for a tribosystem made of a semi-infinite foundation (the disc) sliding over the surface of a strip (the braking pad) has been performed in [18]. The inverse of thermal contact resistance is termed heat contact conductance, and it is dependent on numerous factors, such as load, sliding velocity, temperature, and the properties of friction materials [3,19–21]. The effect of the thermal contact conductance on the temperature and heat partition at the interface of the disc/pad system during braking has been investigated in [22]. A numerical model has been proposed in [23] to simultaneously determine the heat flux generated by friction, the thermal contact conductance, and the intrinsic heat partition coefficient for the problem of sliding contacts. The experimental validation of the thermal behavior of a system consisting of two hollow cylinders under imperfect thermal sliding contact has been presented in [24]. Godet introduced the third body concept [25,26], i.e., the presence of a third body between friction pair elements, as an explanation for the existence of thermal contact resistance. The formation of a third body in disc/pad contact in a material sense can be defined as an interface zone between contacting surfaces formed from the accumulation of wear debris and contaminants from the surrounding environment, e.g., dust or water [27,28]. A third body has been introduced to an FE model of a contact interface as a thin layer with uniform volumetric heat generation in article [29]. Based on this model, the temperature and the heat distributions in both components of a disc brake system have been evaluated. It was established that the results were primarily sensitive to the third body layer thickness and its conductivity. Another numerical simulation in [15] demonstrated that due to the thermal resistance of a third body, the contact temperatures of the pad were higher and that the increase in the third body's thickness resulted in a greater temperature gradient between the contacting surfaces of the disc and pad. In another study [1], the temperature fields in the disc and pin have been modeled using a finite element analysis. The authors developed three models using three different approaches, i.e., considering the heat partition ratio, the perfect thermal contact conditions between the sliding surfaces, and the third body concept. The model assuming perfect contact with thermal continuity at the interface was found to better fit the experimental temperature measurements and to be in good agreement with the observed wear occurring at the pin–disc interface. This pin-on-disc tribotest was performed under mild sliding conditions, so the thermal resistance at the interface was low, which was similar to the perfect heat contact. Likewise, by proceeding with the thermal contact conductance to the infinite value in the condition of imperfect thermal contact related to the temperature drop between sliding surfaces, the temperature continuity condition was obtained [9,25,30].

Most analytical studies dealing with thermal analysis during braking processes have been proposed based on the perfect contact assumptions, which is a less realistic but an easier development [9]. However, in some analytical models of transient temperature distributions in the braking systems, the conditions of imperfect thermal contact have been involved. Exact solutions to the non-stationary problems of heat conduction for two homogeneous semi-spaces under non-ideal thermal contact conditions have been obtained in papers [31,32]. A three-dimensional exact solution to determine the mean and flash temperature and the thermal resistance have been proposed in [33]. Problems of non-stationary heat conduction for two sliding semi-spaces under imperfect conditions at the interface have been considered in [34,35] to determine the contact temperatures and heat partition coefficient. The generalized thermal contact conditions have been involved in the problem formulated in [36] to establish the effect of time-dependent thermal contact heat conductance. The analytical solutions to a thermal problem of friction during braking considering heat transfer through the disc/pad interface have been obtained for a strip/semi-space system with the constant intensity of heat generation [37] for a three-element tribosystem (disc/pad/caliper) with time-dependent specific power of friction [38] and for a two-element system during braking with constant deceleration accounting for the convective cooling on a free surface of the pad [39].

The above-mentioned studies concerned tribosystems with friction pair elements made of homogeneous materials. Basically, modern friction systems are more likely to be made of the nonhomogeneous composites, such as functionally graded materials (FGMs) [40]. These materials are characterized by a smooth change in their properties along certain direction(s). The ability to design the continuous spatial distribution of material properties is a great advantage, which makes FGMs widely used in many varied applications [41,42]. The imperfect thermal contact of friction systems with FGM has been studied in some papers. Frictionally excited thermoelastic instability with consideration of the thermal contact resistance at the interface has been considered for a functionally graded layer and a homogeneous half-plane [43], for two sliding FGMs [44], and for a system consisting of an FGM semi-infinite body sliding against a homogeneous half-space [45]. Analytical solutions to the heating problems of thermoelastic semi-space with imperfectly bonded, functionally graded coating have been proposed in [46,47].

A functionally graded strip heated on the friction surface and cooled by convection on the free surface and sliding against the homogeneous semi-space under perfect thermal contact conditions has been considered in our previous study [48]. It was assumed that the strip was made of a two-component FGM with exponentially changing thermal conductivity along its thickness. This paper presents the exact solution to the boundary-value problem of heat conduction for the same friction system with consideration of the thermal contact conductivity at the strip/semi-space interface. Furthermore, asymptotic solutions were also identified for both low and high values of Fourier number. The developed mathematical model allowed for the investigation of the thermal conductance on the contact surface and convective heat exchange with the environment on the temperature distribution.

The results contained in this article are a continuation of our previous research on the use of FGM to reduce the temperature of friction elements [48,49]. In these previous papers, the relevance of this topic was justified, and a broad literature review was presented. In this article, we focused on obtaining an accurate solution to the problem of a friction pair, in which one of the elements is made of FGM (strip) under conditions of imperfect thermal contact of friction. The solutions known to date concern the contact of smooth friction surfaces (perfect thermal contact of friction). Solutions to problems with imperfect thermal contact of friction only concerned homogeneous materials [11,18]. In our work, we fill this research gap.

The proposed research methodology for developing a mathematical model of heat generation due to friction in the strip/semi-space system with imperfect thermal contact of friction consisted of the following stages:

(1) Formulating and obtaining an exact solution to the appropriate boundary-value thermal conductivity problem.
(2) Verification of the received solution.
(3) Obtaining asymptotic solutions for small and large values of the Fourier number.
(4) Numerical analysis for selected materials of the friction pair.
(5) Summary and conclusions.

2. Materials and Methods

Let us consider the strip/semi-space system heated as a result of the friction on their sliding contact surfaces. The strip $0 \leq z \leq d$ is made of a two-component functionally graded material (FGM) and the semi-space $z \leq 0$ of a homogeneous material (Figure 1).

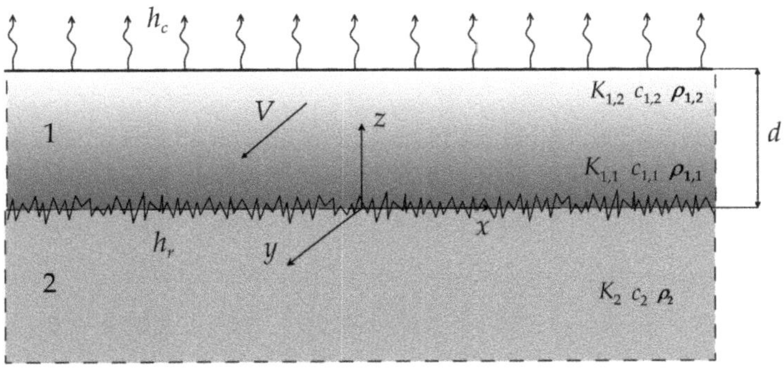

Figure 1. Scheme of the imperfect thermal contact of the strip and the semi-space.

The coefficient of thermal conductivity of FGM increases exponentially with the distance from the sliding surface of the body:

$$K_1(z) = K_{1,1}e^{\gamma^* z/d},\ \gamma^* = \ln(K_{1,2}K_{1,1}^{-1}),\ 0 \leq z \leq d, \tag{1}$$

where $K_{1,1}$, $K_{1,2}$, and K_2—coefficients of thermal conductivity, respectively, of two components of the FGM layer and the half-plane. The thermal friction contact of the strip with the semi-space is imperfect, and the specific friction power remains constant and equal q_0 during the entire heating process. Also, the coefficients of contact heat conductivity h_r through the surfaces $z = 0$ and convective heat exchange h_c on the free surface of the strip $z = d$ are constant. At the initial moment $t = 0$, the temperature of the strip and the semi-space is the same and equal T_0. Other simplifying assumptions can be found in article [50]. It should be emphasized that the transient temperature field $T(z,t)$, $-\infty < z \leq d$, $t > 0$ of the system will be found from the following boundary-value problem of heat conduction:

$$\frac{\partial^2 \Theta^*(\zeta,\tau)}{\partial \zeta^2} + \gamma^* \frac{\partial \Theta^*(\zeta,\tau)}{\partial \zeta} - e^{-\gamma^* \zeta} \frac{\partial \Theta^*(\zeta,\tau)}{\partial \tau} = 0,\ 0 < \zeta < 1,\ \tau > 0, \tag{2}$$

$$\frac{\partial^2\Theta^*(\zeta,\tau)}{\partial\zeta^2} - \frac{1}{k^*}\frac{\partial\Theta^*(\zeta,\tau)}{\partial\tau} = 0,\ \zeta < 0,\ \tau > 0, \tag{3}$$

$$K^* \left.\frac{\partial\Theta^*(\zeta,\tau)}{\partial\zeta}\right|_{\zeta=0^-} - \left.\frac{\partial\Theta^*(\zeta,\tau)}{\partial\zeta}\right|_{\zeta=0^+} = 1,\ \tau > 0, \tag{4}$$

$$K^* \left.\frac{\partial\Theta^*(\zeta,\tau)}{\partial\zeta}\right|_{\zeta=0^-} + \left.\frac{\partial\Theta^*(\zeta,\tau)}{\partial\zeta}\right|_{\zeta=0^+} + Bi_r[\Theta^*(0^-,\tau) - \Theta^*(0^+,\tau)] = 0,\ \tau > 0, \tag{5}$$

$$e^{\gamma^*} \left.\frac{\partial\Theta^*(\zeta,\tau)}{\partial\zeta}\right|_{\zeta=1} + Bi_c\,\Theta^*(1,\tau) = 0,\ \tau > 0, \tag{6}$$

$$\Theta^*(\zeta,\tau) \to 0\,,\ \zeta \to -\infty,\ \tau > 0, \tag{7}$$

$$\Theta^*(\zeta,0) = 0,\ |\zeta| < \infty, \tag{8}$$

where

$$\zeta = \frac{z}{d},\ \tau = \frac{k_1 t}{d^2},\ K^* = \frac{K_2}{K_{1,1}},\ k^* = \frac{k_2}{k_1},\ \Theta^* = \frac{\Theta}{\Theta_0},\ Bi_r = \frac{h_r d}{K_{1,1}},\ Bi_c = \frac{h_c d}{K_{1,1}}, \tag{9}$$

$$k_1 = \frac{K_{1,1}}{c_1\rho_1},\ k_2 = \frac{K_2}{c_2\rho_2},\ \Theta_0 = \frac{q_0 d}{K_{1,1}}, \tag{10}$$

$\Theta(z,t) = T(z,t) - T_0$—temperature rise; ρ_l, c_l—the density and specific heat of materials of the strip ($l = 1$) and the semi-space ($l = 2$).

Exact solution. By applying into the initial-boundary problem (2)–(8), the integral Laplace transform [51]:

$$\overline{\Theta}^*(\zeta,p) \equiv L[\Theta^*(\zeta,\tau);p] = \int_0^\infty \Theta^*(\zeta,\tau)e^{-p\tau}d\tau,\ \mathrm{Re}p \geq 0, \tag{11}$$

the following boundary problem was received for a system of two ordinary differential equations:

$$\frac{d^2\overline{\Theta}^*(\zeta,p)}{d\zeta^2} + \gamma^*\frac{d\overline{\Theta}^*(\zeta,\tau)}{d\zeta} - pe^{-\gamma^*\zeta}\overline{\Theta}^*(\zeta,p) = 0,\ 0 < \zeta < 1, \tag{12}$$

$$\frac{d^2\overline{\Theta}^*(\zeta,p)}{d\zeta^2} - \frac{p}{k^*}\overline{\Theta}^*(\zeta,p) = 0,\ \zeta < 0, \tag{13}$$

$$K^* \left.\frac{d\overline{\Theta}^*(\zeta,p)}{d\zeta}\right|_{\zeta=0^-} - \left.\frac{d\overline{\Theta}^*(\zeta,p)}{d\zeta}\right|_{\zeta=0^+} = \frac{1}{p}, \tag{14}$$

$$K^* \left.\frac{d\overline{\Theta}^*(\zeta,p)}{d\zeta}\right|_{\zeta=0^-} + \left.\frac{d\overline{\Theta}^*(\zeta,p)}{d\zeta}\right|_{\zeta=0^+} + Bi_r[\overline{\Theta}^*(0^-,\tau) - \overline{\Theta}^*(0^+,\tau)] = 0, \tag{15}$$

$$e^{\gamma^*} \left.\frac{d\overline{\Theta}^*(\zeta,p)}{d\zeta}\right|_{\zeta=1} + Bi_c\,\overline{\Theta}^*(1,p) = 0, \tag{16}$$

$$\overline{\Theta}^*(\zeta,p) \to 0\,,\ \zeta \to -\infty. \tag{17}$$

Using the methodology described in article [50], the solutions to the problem (12)–(17) have the following form:

$$\overline{\Theta}^*(\zeta, p) = \frac{e^{-\frac{1}{2}\gamma^*\zeta}}{p\sqrt{p}}\left(\varepsilon + \frac{Bi_r}{\sqrt{p}}\right)\frac{\Delta_1(\zeta, p)}{\Delta(p)},\ 0 \le \zeta \le 1, \tag{18}$$

$$\overline{\Theta}^*(\zeta, p) = \frac{\Delta_2(\zeta, p)}{p\sqrt{p}\,\Delta(p)},\ \zeta \le 0, \tag{19}$$

where

$$\begin{aligned}\Delta(p) = A_1(p)\left[\left(2\varepsilon + \tfrac{Bi_r}{\sqrt{p}}\right)\mathrm{I}_0\left(\tfrac{2}{\gamma^*}\sqrt{p}\right) + \varepsilon\tfrac{Bi_r}{\sqrt{p}}\,\mathrm{I}_1\left(\tfrac{2}{\gamma^*}\sqrt{p}\right)\right] - \\ -B_1(p)\left[\left(2\varepsilon + \tfrac{Bi_r}{\sqrt{p}}\right)\mathrm{K}_0\left(\tfrac{2}{\gamma^*}\sqrt{p}\right) - \varepsilon\tfrac{Bi_r}{\sqrt{p}}\,\mathrm{K}_1\left(\tfrac{2}{\gamma^*}\sqrt{p}\right)\right],\end{aligned} \tag{20}$$

$$\begin{aligned}\Delta_1(\zeta, p) = A_1(p)\mathrm{I}_1\left(\tfrac{2}{\gamma^*}e^{-\frac{1}{2}\gamma^*\zeta}\sqrt{p}\right) + B_1(p)\mathrm{K}_1\left(\tfrac{2}{\gamma^*}e^{-\frac{1}{2}\gamma^*\zeta}\sqrt{p}\right), \\ \Delta_2(\zeta, p) = A_2(p)e^{\sqrt{\frac{p}{k^*}}\,\zeta},\end{aligned} \tag{21}$$

$$A_1(p) = \mathrm{K}_0\left(\frac{2}{\gamma^*}e^{-\frac{1}{2}\gamma^*}\sqrt{p}\right) + Bi_c\frac{e^{-\frac{1}{2}\gamma^*}}{\sqrt{p}}\mathrm{K}_1\left(\frac{2}{\gamma^*}e^{-\frac{1}{2}\gamma^*}\sqrt{p}\right), \tag{22}$$

$$B_1(p) = \mathrm{I}_0\left(\frac{2}{\gamma^*}e^{-\frac{1}{2}\gamma^*}\sqrt{p}\right) - Bi_c\frac{e^{-\frac{1}{2}\gamma^*}}{\sqrt{p}}\mathrm{I}_1\left(\frac{2}{\gamma^*}e^{-\frac{1}{2}\gamma^*}\sqrt{p}\right), \tag{23}$$

$$\begin{aligned}A_2(p) = A_1(p)\left[\mathrm{I}_0\left(\tfrac{2}{\gamma^*}\sqrt{p}\right) + \tfrac{Bi_r}{\sqrt{p}}\,\mathrm{I}_1\left(\tfrac{2}{\gamma^*}\sqrt{p}\right)\right] - \\ -B_1(p)\left[\mathrm{K}_0\left(\tfrac{2}{\gamma^*}\sqrt{p}\right) - \varepsilon\tfrac{Bi_r}{\sqrt{p}}\,\mathrm{K}_1\left(\tfrac{2}{\gamma^*}\sqrt{p}\right)\right],\end{aligned} \tag{24}$$

$$\varepsilon = \frac{K^*}{\sqrt{k^*}}, \tag{25}$$

$\mathrm{I}_n(x), \mathrm{K}_n(x)$—modified Bessel functions of the nth order $n = 0, 1$, respectively, of the first and the second kind [52].

Applying to the transformed solution (18)–(25) the inverse Laplace transform [51]:

$$\Theta^*(\zeta, \tau) \equiv L^{-1}[\overline{\Theta}^*(\zeta, p);\ \tau] = \frac{1}{2\pi i}\int\limits_{\omega - i\,\infty}^{\omega + i\,\infty}\overline{\Theta}^*(\zeta, p)e^{p\tau}dp,\ \omega \equiv \mathrm{Re}p > 0,\ i \equiv \sqrt{-1}, \tag{26}$$

and performing the integration on the plane of the complex variable p in line with the methodology presented in the study [49] using the relations [52]:

$$\mathrm{I}_0(\pm ix) = \mathrm{J}_0(x),\ \mathrm{K}_0(\pm ix) = -0.5\pi[\mathrm{Y}_0(x) \pm i\mathrm{J}_0(x)], \tag{27}$$

$$\mathrm{I}_1(\pm ix) = \pm i\mathrm{J}_1(ix),\ \mathrm{K}_1(\pm ix) = 0.5\pi[\mathrm{J}_1(x) \pm i\mathrm{Y}_1(x)], \tag{28}$$

($\mathrm{J}_n(x)$ and $\mathrm{Y}_n(x)$—Bessel functions of the nth order $n = 0, 1$, respectively, of the first and the second kind), dimensionless temperature rises in the strip and semi-space were found in the following form:

$$\Theta^*(\zeta, \tau) = \vartheta_1(\zeta) - \frac{2}{\pi}\varepsilon\, e^{-\frac{1}{2}\gamma^*\zeta}\int\limits_0^\infty F(x)G_1(\zeta, x)e^{-x^2\tau}dx,\ 0 \le \zeta \le 1,\ \tau \ge 0, \tag{29}$$

$$\Theta^*(\zeta, \tau) = \vartheta_2 - \frac{2}{\pi}\int\limits_0^\infty F(x)G_2(\zeta, x)e^{-x^2\tau}dx,\ \zeta \le 0,\ \tau \ge 0, \tag{30}$$

where

$$\vartheta_1(\zeta) = \frac{1}{Bi_c} + \frac{1}{\gamma^*}(e^{-\gamma^*\zeta} - e^{-\gamma^*}),\ \vartheta_2 = \frac{1}{Bi_r} + \frac{1}{Bi_c} + \frac{1}{\gamma^*}(1 - e^{-\gamma^*}), \tag{31}$$

$$F(x) = \frac{[\Psi(x) - x\,Bi_r^{-1}\Phi(x)]}{x^2\left\{[\Phi(x)]^2 + [\varepsilon\,\Psi(x)]^2\right\}}, \tag{32}$$

$$G_1(\zeta, x) = \mathrm{J}(x)\mathrm{Y}_1\left(\frac{2}{\gamma^*}e^{-\frac{1}{2}\gamma^*\zeta}x\right) - \mathrm{Y}(x)\mathrm{J}_1\left(\frac{2}{\gamma^*}e^{-\frac{1}{2}\gamma^*\zeta}x\right), \tag{33}$$

$$G_2(\zeta, x) = \varepsilon\,\Psi(x)\cos\left(\frac{\zeta}{\sqrt{k^*}}x\right) - \Phi(x)\sin\left(\frac{\zeta}{\sqrt{k^*}}x\right), \tag{34}$$

$$\Phi(x) = \mathrm{J}(x)\mathrm{Y}_0\left(\frac{2}{\gamma^*}x\right) - \mathrm{Y}(x)\mathrm{J}_0\left(\frac{2}{\gamma^*}x\right), \tag{35}$$

$$\Psi(x) = \mathrm{J}(x)\left[\mathrm{Y}_1\left(\frac{2}{\gamma^*}x\right) + \frac{2x}{Bi_r}\mathrm{Y}_0\left(\frac{2}{\gamma^*}x\right)\right] - \mathrm{Y}(x)\left[\mathrm{J}_1\left(\frac{2}{\gamma^*}x\right) + \frac{2x}{Bi_r}\mathrm{J}_0\left(\frac{2}{\gamma^*}x\right)\right], \tag{36}$$

$$\mathrm{J}(x) = \mathrm{J}_0\left(\frac{2}{\gamma^*}e^{-\frac{1}{2}\gamma^*}x\right) - Bi_c\frac{e^{-\frac{1}{2}\gamma^*}}{x}\mathrm{J}_1\left(\frac{2}{\gamma^*}e^{-\frac{1}{2}\gamma^*}x\right), \tag{37}$$

$$\mathrm{Y}(x) = \mathrm{Y}_0\left(\frac{2}{\gamma^*}e^{-\frac{1}{2}\gamma^*}x\right) - Bi_c\frac{e^{-\frac{1}{2}\gamma^*}}{x}\mathrm{Y}_1\left(\frac{2}{\gamma^*}e^{-\frac{1}{2}\gamma^*}x\right). \tag{38}$$

Substituting $\zeta = 0$ in the solutions (29)–(38), the following expressions were used to estimate temperature on the sliding surfaces of the strip ($z = 0^+$) and of the semi-space ($z = 0^-$) for imperfect thermal contact of friction:

$$\Theta^*(0^+, \tau) = \vartheta_0 - \frac{2}{\pi}\varepsilon\int_0^\infty F(x)\Psi_0(x)e^{-x^2\tau}dx,\ \tau \geq 0, \tag{39}$$

$$\Theta^*(0^-, \tau) = (Bi_r^{-1} + \vartheta_0) - \frac{2}{\pi}\varepsilon\int_0^\infty F(x)\Psi(x)e^{-x^2\tau}dx,\ \tau \geq 0, \tag{40}$$

where

$$\vartheta_0 \equiv \vartheta_1(0) = \frac{1}{Bi_c} + \frac{1}{\gamma^*}(1 - e^{-\gamma^*}), \tag{41}$$

$$\Psi_0(x) \equiv G_1(0, x) = \mathrm{J}(x)\mathrm{Y}_1\left(\frac{2}{\gamma^*}x\right) - \mathrm{Y}(x)\mathrm{J}_1\left(\frac{2}{\gamma^*}x\right). \tag{42}$$

In the case of ideal heat contact of the strip and the semi-space ($Bi_r \to \infty$), Equations (36) and (42) give $\Psi(x) = \Psi_0(x)$, and from solutions (39) and (40), it follows that the temperature of the friction surfaces is the same and equal [50]:

$$\Theta^*(0^+, \tau) = \Theta^*(0^-, \tau) \equiv \Theta^*(\tau) = \vartheta_0 - \frac{2}{\pi}\varepsilon\int_0^\infty F_0(x)\Psi_0(x)e^{-x^2\tau}dx,\ \tau \geq 0, \tag{43}$$

where

$$F_0(x) = \frac{\Psi_0(x)}{x^2\left\{[\Phi(x)]^2 + [\varepsilon\,\Psi_0(x)]^2\right\}}. \tag{44}$$

Solutions (29)–(38) have been obtained for constant-value q_0 of the specific friction power. The form of these expressions allows for the usage of Duhamel's theorem in order to obtain corresponding solutions to this problem for time-dependent specific friction power. In this way, for the case of the specific friction power linearly decreasing from the nominal value $\hat{q}_0 = 2q_0$ at the initial moment $t = 0$ to zero at the final moment $t = t_s$, the dimensionless temperature rise $\hat{\Theta}^*(\zeta,\tau)$, based on Duhamel's formula, was sought as [53]:

$$\hat{\Theta}^*(\zeta,\tau) = \frac{\partial}{\partial\tau}\int_0^{\tau} q^*(\tau - s)\Theta^*(\zeta,s)ds,\ \zeta \geq 0,\ 0 \leq \tau \leq \tau_s, \tag{45}$$

where

$$q^*(\tau) = 1 - \tau\,\tau_s^{-1}, 0 \leq \tau \leq \tau_s,\ \tau_s = k_1 t_s d^{-2}, \tag{46}$$

and $\Theta^*(\zeta,\tau)$ is the dimensionless temperature rise (29)–(38) achieved for $q^*(\tau) = q_0,\ \tau \geq 0$.

After substituting under the integral sign in the right side of Formula (45) functions $\Theta^*(\zeta,\tau)$ (29), (30) and $q^*(\tau)$ (53), the following was found:

$$\hat{\Theta}^*(\zeta,\tau) = \vartheta_1(\zeta)q^*(\tau) - \tfrac{2}{\pi}\varepsilon\, e^{-\frac{1}{2}\gamma^*\zeta}\int_0^{\infty} F(x)G_1(\zeta,x)P(\tau,x)dx,$$
$$0 \leq \zeta \leq 1,\ 0 \leq \tau \leq \tau_s, \tag{47}$$

$$\hat{\Theta}^*(\zeta,\tau) = \vartheta_2 q^*(\tau) - \frac{2}{\pi}\int_0^{\infty} F(x)G_2(\zeta,x)P(\tau,x)dx,\ \zeta \leq 0,\ 0 \leq \tau \leq \tau_s{}' \tag{48}$$

where

$$P(\tau,x) = e^{-x^2\tau} - \frac{1}{x^2\tau_s}(1 - e^{-x^2\tau}), \tag{49}$$

and the rest of the functions have been determined in Equations (31)–(28).

Verification of the solution. Validation of the obtained exact solution (29)–(38) was carried out by checking the fulfillment of boundary conditions (4)–(6). For this purpose, on account of the derivatives [52]:

$$\mathrm{J}_1'(x) = \mathrm{J}_0(x) - x^{-1}\mathrm{J}_1(x),\ \mathrm{Y}_1'(x) = \mathrm{Y}_0(x) - x^{-1}\mathrm{Y}_1(x), \tag{50}$$

and from solutions (29) and (30), the dimensionless intensities of heat fluxes in the strip and in the semi-space were found:

$$\frac{\partial\Theta^*(\zeta,\tau)}{\partial\zeta} = -e^{-\gamma^*\zeta} + \frac{2}{\pi}\varepsilon\, e^{-\gamma^*\zeta}\int_0^{\infty} xF(x)\hat{G}_1(\zeta,x)e^{-x^2\tau}dx,\ 0 \leq \zeta \leq 1,\ \tau \geq 0, \tag{51}$$

$$K^*\frac{\partial\Theta^*(\zeta,\tau)}{\partial\zeta} = \frac{2}{\pi}\varepsilon\int_0^{\infty} xF(x)\hat{G}_2(\zeta,x)e^{-x^2\tau}dx,\ \zeta \leq 0,\ \tau \geq 0, \tag{52}$$

where

$$\hat{G}_1(\zeta,x) = \mathrm{J}(x)\mathrm{Y}_0\left(\frac{2}{\gamma^*}e^{-\frac{1}{2}\gamma^*\zeta}x\right) - \mathrm{Y}(x)\mathrm{J}_0\left(\frac{2}{\gamma^*}e^{-\frac{1}{2}\gamma^*\zeta}x\right), \tag{53}$$

$$\hat{G}_2(\zeta,x) = \varepsilon\,\Psi(x)\sin\left(\frac{\zeta}{\sqrt{k^*}}x\right) + \Phi(x)\cos\left(\frac{\zeta}{\sqrt{k^*}}x\right). \tag{54}$$

On the sliding surfaces $\zeta = 0^{\pm}$, Equations (51)–(54) yield:

$$q_1^*(\tau) \equiv -\left.\frac{\partial\Theta^*(\zeta,\tau)}{\partial\zeta}\right|_{\zeta=0^+} = 1 - \frac{2}{\pi}\varepsilon\int_0^{\infty} xF(x)\Phi(x)e^{-x^2\tau}dx,\ \tau \geq 0, \tag{55}$$

$$q_2^*(\tau) \equiv K^* \frac{\partial \Theta^*(\zeta,\tau)}{\partial \zeta}\bigg|_{\zeta=0^-} = \frac{2}{\pi}\varepsilon \int_0^\infty x F(x)\Phi(x) e^{-x^2\tau} dx,\ \tau \geq 0, \tag{56}$$

where functions $F(x)$ and $\Phi(x)$ have the forms, respectively, of (32) and (35). The addition of Formulas (55) and (56) confirms the fulfillment of boundary condition (4). However, after subtracting them, it was found that

$$K^* \frac{\partial \Theta^*(\zeta,\tau)}{\partial \zeta}\bigg|_{\zeta=0^-} + \frac{\partial \Theta^*(\zeta,\tau)}{\partial \zeta}\bigg|_{\zeta=0^+} = -1 + \frac{4}{\pi}\varepsilon \int_0^\infty x F(x)\Phi(x) e^{-x^2\tau} dx,\ \tau \geq 0. \tag{57}$$

On the other hand, subtracting Equations (39) and (40) results in

$$\Theta^*(0^-,\tau) - \Theta^*(0^+,\tau) = \frac{1}{Bi_r} - \frac{2}{\pi}\int_0^\infty F(x)[\Psi(x) - \varepsilon\, \Psi_0(x)] e^{-x^2\tau} dx,\ \tau \geq 0. \tag{58}$$

Considering the form of the functions $\Phi(x)$ (35), $\Psi(x)$ (36), and $\Psi_0(x)$ (42), it was established that

$$\Psi(x) - \varepsilon\, \Psi_0(x) = 2\varepsilon\, Bi_r^{-1}\, x\Phi(x). \tag{59}$$

Then, based on the expression (58) and (59), the following was received:

$$Bi_r[\Theta^*(0^+,\tau) - \Theta^*(0^-,\tau)] = 1 - \frac{4}{\pi}\varepsilon \int_0^\infty x F(x)\Phi(x) e^{-x^2\tau} dx,\ \tau \geq 0. \tag{60}$$

The result of the addition of Formulas (57) and (60) confirms that boundary condition (5) is met.

The dimensionless intensity of heat flux on the free surface of the strip $\zeta = 1$, from Equations (52) and (54), was determined:

$$\frac{\partial \Theta^*(\zeta,\tau)}{\partial \zeta}\bigg|_{\zeta=1} = -e^{-\gamma^*} - \varepsilon\, e^{-\gamma^*} \frac{2}{\pi}\int_0^\infty x F(x) \hat{G}_1(1,x) e^{-x^2\tau} dx,\ \tau \geq 0, \tag{61}$$

where

$$\hat{G}_1(1,x) = \mathrm{J}(x)\mathrm{Y}_0\left(\frac{2}{\gamma^*} e^{-\frac{1}{2}\gamma^*} x\right) - \mathrm{Y}(x)\mathrm{J}_0\left(\frac{2}{\gamma^*} e^{-\frac{1}{2}\gamma^*} x\right). \tag{62}$$

Then, from solution (29), a suitable dimensionless temperature rise was found:

$$\Theta^*(1,\tau) = \frac{1}{Bi_c} - \varepsilon\, e^{-\frac{1}{2}\gamma^*} \frac{2}{\pi}\int_0^\infty F(x) G_1(1,x) e^{-x^2\tau} dx,\ \tau \geq 0, \tag{63}$$

where, based on Equation (33), the following was obtained:

$$G_1(1,x) = \mathrm{J}(x)\mathrm{Y}_1\left(\frac{2}{\gamma^*} e^{-\frac{1}{2}\gamma^*} x\right) - \mathrm{Y}(x)\mathrm{J}_1\left(\frac{2}{\gamma^*} e^{-\frac{1}{2}\gamma^*} x\right). \tag{64}$$

From relations (61) and (63), it follows that

$$e^{\gamma^*} \frac{\partial \Theta^*(\zeta,\tau)}{\partial \zeta}\bigg|_{\zeta=1} + Bi_c \Theta^*(1,\tau) = \frac{2}{\pi}\varepsilon \int_0^\infty F(x)[x\hat{G}_1(1,x) - Bi_c e^{-\frac{1}{2}\gamma^*} G_1(1,x)] e^{-x^2\tau} dx,\ \tau \geq 0. \tag{65}$$

Taking into account forms of functions $\hat{G}_1(1,x)$ (62) and $G_1(1,x)$ (64), the following was achieved:

$$\begin{aligned} x\hat{G}_1(1,x) - Bi_c e^{-\frac{1}{2}\gamma^*} G_1(1,x) = \mathrm{J}(x)\left[x\mathrm{Y}_0\left(\tfrac{2}{\gamma^*}e^{-\frac{1}{2}\gamma^*}x\right) - Bi_c e^{-\frac{1}{2}\gamma^*}\mathrm{Y}_1\left(\tfrac{2}{\gamma^*}e^{-\frac{1}{2}\gamma^*}x\right)\right] - \\ -\mathrm{Y}(x)\left[x\mathrm{J}_0\left(\tfrac{2}{\gamma^*}e^{-\frac{1}{2}\gamma^*}x\right) - Bi_c e^{-\frac{1}{2}\gamma^*}\mathrm{J}_1\left(\tfrac{2}{\gamma^*}e^{-\frac{1}{2}\gamma^*}x\right)\right] = \\ = x[\mathrm{J}(x)\mathrm{Y}(x) - \mathrm{Y}(x)\mathrm{J}(x)] = 0, \end{aligned} \quad (66)$$

which confirms the fulfillment of boundary condition (6).

The fulfillment of the boundary condition (7) and the initial condition (8) were taken into account when obtaining the solution (18)–(25) in the space of the Laplace integral transform.

Asymptotic solutions. Small values of the Fourier number τ (large values of the Laplace transform parameter p). Bearing in mind Formulas (18)–(24) and the asymptotes of the modified Bessel functions for large values of the argument [52]:

$$\mathrm{I}_n(x) \cong \frac{e^x}{\sqrt{2\pi x}},\ \mathrm{K}_n(x) \cong \sqrt{\frac{\pi}{2x}}\,e^{-x},\ n = 0,\ 1, \quad (67)$$

it was established that the solution in the Laplace transform space has the following form:

$$\overline{\Theta}^*(\zeta,p) \cong \frac{1}{2}e^{-\frac{1}{4}\gamma^*\zeta}\left[\frac{e^{-\zeta_1\sqrt{p}}}{p(\alpha+\sqrt{p})} + \frac{Bi_r}{\varepsilon}\frac{e^{-\zeta_1\sqrt{p}}}{p\sqrt{p}(\alpha+\sqrt{p})}\right],\ 0 \le \zeta < 1, \quad (68)$$

$$\overline{\Theta}^*(\zeta,p) \cong \frac{1}{2\varepsilon}\left[\frac{e^{-\zeta_2\sqrt{p}}}{p(\alpha+\sqrt{p})} + Ei_r\frac{e^{-\zeta_2\sqrt{p}}}{p\sqrt{p}(\alpha+\sqrt{p})}\right],\ \zeta \le 0, \quad (69)$$

where

$$\alpha = \frac{(1+\varepsilon)}{2\varepsilon}Bi_r,\ \zeta_1 = \frac{2}{\gamma^*}(1-e^{-\frac{1}{2}\gamma^*\zeta}), \zeta_2 = \frac{|\zeta|}{\sqrt{k^*}}. \quad (70)$$

Using relations [54]:

$$L^{-1}\left[\frac{\alpha e^{-\zeta_l\sqrt{p}}}{p(\alpha+\sqrt{p})};\ \tau\right] = \mathrm{erfc}\left(\frac{\zeta_l}{2\sqrt{\tau}}\right) - e^{\alpha^2\tau+\alpha\zeta_l}\mathrm{erfc}\left(\frac{\zeta_l}{2\sqrt{\tau}} + \alpha\sqrt{\tau}\right),\ l = 1,2, \quad (71)$$

$$\begin{aligned} L^{-1}\left[\tfrac{\alpha e^{-\zeta_l\sqrt{p}}}{p\sqrt{p}(\alpha+\sqrt{p})};\ \tau\right] = 2\sqrt{\tfrac{\tau}{\pi}}\,e^{-\left(\frac{\zeta_l}{2\sqrt{\tau}}\right)^2} - \left(\zeta_l + \tfrac{1}{\alpha}\right)\mathrm{erfc}\left(\tfrac{\zeta_l}{2\sqrt{\tau}}\right) + \\ + \tfrac{1}{\alpha}e^{\alpha^2\tau+\alpha\zeta_l}\mathrm{erfc}\left(\tfrac{\zeta_l}{2\sqrt{\tau}} + \alpha\sqrt{\tau}\right),\ l = 1,2, \end{aligned} \quad (72)$$

where $\mathrm{erfc}(x) = 1 - \mathrm{erf}(x)$, $\mathrm{erf}(x)$—Gauss error function [52], from Equations (68)–(70), yields:

$$\begin{aligned} \Theta^*(\zeta,\tau) = \tfrac{e^{-\frac{1}{4}\gamma^*\zeta}}{(1+\varepsilon)}\left[2\sqrt{\tfrac{\tau}{\pi}}\,e^{-(\frac{\zeta_1}{2\sqrt{\tau}})^2} - \left(\zeta_1 + \tfrac{1}{\alpha} - \tfrac{\varepsilon}{Bi_r}\right)\mathrm{erfc}\left(\tfrac{\zeta_1}{2\sqrt{\tau}}\right) + \right. \\ \left. + \left(\tfrac{1}{\alpha} - \tfrac{\varepsilon}{Bi_r}\right)e^{\alpha^2\tau+\alpha\zeta_1}\mathrm{erfc}\left(\tfrac{\zeta_1}{2\sqrt{\tau}} + \alpha\sqrt{\tau}\right)\right],\ 0 \le \zeta \le 1,\ 0 \le \tau << 1, \end{aligned} \quad (73)$$

$$\begin{aligned} \Theta^*(\zeta,\tau) = \tfrac{1}{(1+\varepsilon)}\left[2\sqrt{\tfrac{\tau}{\pi}}\,e^{-(\frac{\zeta_2}{2\sqrt{\tau}})^2} - \left(\zeta_2 + \tfrac{1}{\alpha} - \tfrac{1}{Bi_r}\right)\mathrm{erfc}\left(\tfrac{\zeta_2}{2\sqrt{\tau}}\right) + \right. \\ \left. + \left(\tfrac{1}{\alpha} - \tfrac{1}{Bi_r}\right)e^{\alpha^2\tau+\alpha\zeta_2}\mathrm{erfc}\left(\tfrac{\zeta_2}{2\sqrt{\tau}} + \alpha\sqrt{\tau}\right)\right],\ \zeta \le 0,\ 0 \le \tau << 1. \end{aligned} \quad (74)$$

Proceeding with Equations (73) and (74) to the limit $Bi_r \to \infty$, the known asymptotic solution was obtained, related to the perfect thermal contact between the FGM strip and homogeneous semi-space [53]:

$$\Theta^*(\zeta,\tau) \cong \frac{2\sqrt{\tau}}{(1+\varepsilon)}e^{-\frac{1}{4}\gamma^*\zeta}\mathrm{ierfc}\left(\frac{\zeta_1}{2\sqrt{\tau}}\right),\ 0 \le \zeta \le 1,\ 0 \le \tau << 1, \quad (75)$$

$$\Theta^*(\zeta,\tau) \cong \frac{2\sqrt{\tau}}{(1+\varepsilon)}\mathrm{ierfc}\left(\frac{\zeta_2}{2\sqrt{\tau}}\right),\ \zeta \le 0,\ 0 \le \tau << 1, \tag{76}$$

where $\mathrm{ierfc}(x) = \pi^{-0.5}e^{-x^2} - x\,\mathrm{erfc}(x)$. From Equations (75) and (76), it follows that the temperature of the friction surfaces $\zeta = 0^{\pm}$ ($\zeta_1 = \zeta_2 = 0$) in this case is the same:

$$\Theta^*(0^+,\tau) = \Theta^*(0^-,\tau) = \Theta^*(\tau) \cong \frac{2}{(1+\varepsilon)}\sqrt{\frac{\tau}{\pi}}. \tag{77}$$

For the homogeneous material of the strip ($\gamma^* \to 0$) from the second formula from Equation (70), it follows that $\zeta_1 \to \zeta$ and solution (80) has the known form [37]:

$$\begin{aligned}\Theta^*(\zeta,\tau) = \tfrac{1}{(1+\varepsilon)}\Big[2\sqrt{\tfrac{\tau}{\pi}}\,e^{-(\frac{\zeta}{2\sqrt{\tau}})2} - \left(\zeta + \tfrac{1}{\alpha} - \tfrac{\varepsilon}{Bi_r}\right)\mathrm{erfc}\left(\tfrac{\zeta}{2\sqrt{\tau}}\right) + \\ + \left(\tfrac{1}{\alpha} - \tfrac{\varepsilon}{Bi_r}\right)e^{\alpha^2\tau+\alpha\zeta}\mathrm{erfc}\left(\tfrac{\zeta}{2\sqrt{\tau}} + \alpha\sqrt{\tau}\right)\Big],\ 0 \le \zeta \le 1,\ 0 \le \tau << 1,\end{aligned} \tag{78}$$

and solution (74) for semi-space remains without changes, as it is independent on parameter γ^*.

It should be noted that the form of asymptotic solutions (73) and (74) shows that at the beginning of the friction heating process, the effect of convection cooling of the exposed surface of the strip on the temperature of both bodies is negligible, and the gradient nature of the material only affects the temperature of the strip.

High values of Fourier number τ (small values of the Laplace parameter p). Considering the behavior of modified Bessel functions at small argument values [52]:

$$\mathrm{I}_0(x) \cong 1,\ \mathrm{I}_1(x) \cong 0.5x,\ \mathrm{K}_0(x) \cong -\ln(x),\ \mathrm{K}_1(x) \cong x^{-1}, \tag{79}$$

in Equations (18)–(25), the following was found:

$$\overline{\Theta}^*(\zeta,p) \cong \phi^*(\zeta)\left[\frac{1}{p(a+\sqrt{p})} + \frac{\varepsilon}{Bi_r}\frac{1}{\sqrt{p}(a+\sqrt{p})}\right],\ 0 \le \zeta \le 1, \tag{80}$$

$$\overline{\Theta}^*(\zeta,p) \cong \left(\frac{1}{\varepsilon} - \frac{a}{Bi_r}\right)\frac{e^{-\zeta_2\sqrt{p}}}{p(a+\sqrt{p})},\ \zeta \le 0, \tag{81}$$

where

$$a = \frac{\gamma^* Bi_r Bi_c}{\varepsilon(\phi_0 Bi_r + 2\gamma^* Bi_c)},\ \phi^*(\zeta) = \frac{\phi(\zeta)Bi_r}{\varepsilon(\phi_0 Bi_r + 2\gamma^* Bi_c)}, \tag{82}$$

$$\phi(\zeta) = \gamma^* + Bi_c(e^{-\gamma^*\zeta} - e^{-\gamma^*}),\ \phi_0 \equiv \phi(0) = \gamma^* + Bi_c(1 - e^{-\gamma^*}), \tag{83}$$

and parameter ζ_2 is defined in the last formula of Equation (70). Taking into account relation (71) and the following [54]:

$$L^{-1}\left[\frac{1}{\sqrt{p}(a+\sqrt{p})};\tau\right] = e^{a^2\tau}\mathrm{erfc}(a\sqrt{\tau}),\ L^{-1}\left[\frac{a}{p(a+\sqrt{p})};\tau\right] = 1 - e^{a^2\tau}\mathrm{erfc}(a\sqrt{\tau}), \tag{84}$$

asymptotes of dimensionless temperature rise for high values of Fourier number τ were found in the following form:

$$\Theta^*(\zeta,\tau) \cong \phi^*(\zeta)\left[\frac{1}{a} - \left(\frac{1}{a} - \frac{\varepsilon}{Bi_r}\right)\right]e^{a^2\tau}\mathrm{erfc}(a\sqrt{\tau}),\ 0 \le \zeta \le 1,\ \tau >> 1, \tag{85}$$

$$\Theta^*(\zeta,\tau) \cong \left(\frac{1}{a\varepsilon} - \frac{1}{Bi_r}\right)\left[\mathrm{erfc}\left(\frac{\zeta_2}{2\sqrt{\tau}}\right) - e^{a^2\tau+a\zeta_2}\mathrm{erfc}\left(\frac{\zeta_2}{2\sqrt{\tau}} + a\sqrt{\tau}\right)\right],\ \zeta \le 0,\ \tau >> 1. \tag{86}$$

In the case of perfect thermal contact of friction ($Bi_r \to \infty$) from Formula (82), it follows that

$$a = \frac{\gamma^* Bi_c}{\varepsilon\phi_0},\ \phi^*(\zeta) = \frac{\phi(\zeta)}{\varepsilon\phi_0}, \tag{87}$$

and solutions (85), (86) will take the form [50]

$$\Theta^*(\zeta,\tau) \cong \frac{\phi(\zeta)}{\gamma^* Bi_c}[1 - e^{a^2\tau}\operatorname{erfc}(a\sqrt{\tau})],\ 0 \le \zeta \le 1,\ \tau >> 1, \tag{88}$$

$$\Theta^*(\zeta,\tau) \cong \frac{1}{a\varepsilon}\left[\operatorname{erfc}\left(\frac{\zeta_2}{2\sqrt{\tau}}\right) - e^{a^2\tau + a\zeta_2}\operatorname{erfc}\left(\frac{\zeta_2}{2\sqrt{\tau}} + a\sqrt{\tau}\right)\right],\ \zeta \le 0,\ \tau >> 1. \tag{89}$$

Considering the limit

$$\lim_{\gamma^* \to 0}\frac{\gamma^*}{\phi_0} = \frac{Bi_c}{\varepsilon(1 + Bi_c)},\ \lim_{\gamma^* \to 0}\frac{\phi(\zeta)}{\gamma^*} = 1 + Bi_c(1 - \zeta), \tag{90}$$

from Equation (88), the dimensionless temperature rises for ideal heat contact between the layer and the half-plane were obtained:

$$\Theta^*(\zeta,\tau) \cong \left[\frac{1 + Bi_c(1 - \zeta)}{Bi_c}\right][1 - e^{a^2\tau}\operatorname{erfc}(a\sqrt{\tau})],\ 0 \le \zeta \le 1,\ \tau >> 1, \tag{91}$$

$$\Theta^*(\zeta,\tau) \cong \frac{(1 + Bi_c)}{Bi_c}\left[\operatorname{erfc}\left(\frac{\zeta_2}{2\sqrt{\tau}}\right) - e^{a^2\tau + a\zeta_2}\operatorname{erfc}\left(\frac{\zeta_2}{2\sqrt{\tau}} + a\sqrt{\tau}\right)\right],\ \zeta \le 0,\ \tau >> 1, \tag{92}$$

where

$$a = \frac{Bi_c}{\varepsilon(1 + Bi_c)}. \tag{93}$$

On the other hand, proceeding with Formulas (82) and (83) to the limit $\gamma^* \to 0$, the following was found:

$$a = \frac{Bi_r Bi_c}{\varepsilon[(1 + Bi_c)Bi_r + 2Bi_c]},\ \phi^*(\zeta) = \frac{[1 + (1 - \zeta)Bi_c]Bi_r}{\varepsilon[(1 + Bi_c)Bi_r + 2Bi_c]}, \tag{94}$$

and from solutions (85) and (86), the following was obtained:

$$\Theta^*(\zeta,\tau) \cong \left[\frac{1 + Bi_c(1 - \zeta)}{Bi_c}\right]\left\{1 - \left[\frac{(1 + Bi_c)Bi_r + Bi_c}{(1 + Bi_c)Bi_r + 2Bi_c}\right]e^{a^2\tau}\operatorname{erfc}(a\sqrt{\tau})\right\},\quad 0 \le \zeta \le 1,\ \tau >> 1, \tag{95}$$

$$\Theta^*(\zeta,\tau) \cong \frac{[(1 + Bi_c)Bi_r + Bi_c]}{Bi_r Bi_c}\left[\operatorname{erfc}\left(\frac{\zeta_2}{2\sqrt{\tau}}\right) - e^{a^2\tau + a\zeta_2}\operatorname{erfc}\left(\frac{\zeta_2}{2\sqrt{\tau}} + a\sqrt{\tau}\right)\right],\quad \zeta \le 0,\ \tau >> 1. \tag{96}$$

Additionally, accepting $Bi_r \to \infty$, solutions (94)–(96) become equal to solutions (90)–(92).

It should be noted that asymptotic solutions (91)–(93) for homogeneous materials of the layer and half-space, taking into account convective cooling on the free surface of the layer, are also new. So far, a suitable solution for maintaining the initial temperature on this surface has been known [37]. This solution is easy to obtain from Formulas (91)–(93), proceeding with them to the limit $Bi_c \to \infty$. Then,

$$\Theta^*(\zeta,\tau) \cong (1 - \zeta)[1 - e^{\varepsilon^{-2}\tau}\operatorname{erfc}(\varepsilon^{-1}\sqrt{\tau})],\ 0 \le \zeta \le 1,\ \tau >> 1, \tag{97}$$

$$\Theta^*(\zeta,\tau) \cong \operatorname{erfc}\left(\frac{\zeta_2}{2\sqrt{\tau}}\right) - e^{\varepsilon^{-2}\tau + \varepsilon^{-1}\zeta_2}\operatorname{erfc}\left(\frac{\zeta_2}{2\sqrt{\tau}} + \varepsilon^{-1}\sqrt{\tau}\right),\ \zeta \le 0,\ \tau >> 1. \tag{98}$$

3. Results and Discussion

Numerical analysis was carried out in the case of a two-component FGM strip, in which the friction surface was zirconium dioxide ZrO_2 and the titanium alloy $Ti-6Al-4V$ was chosen as the core. The gray cast iron ChNMKh served as the material of a homogeneous semi-space. The properties of these materials at the initial temperature $T_0 = 20\ ^\circ C$, necessary to perform the calculations, were taken from article [47]. The value $\gamma^* = 1.26$ of the dimensionless gradient of the selected FGM was determined from Formula (1). The remaining dimensionless input parameters were defined by Formulas (9): the spatial variable ζ, the Fourier number τ, and the Biot numbers Bi_r and Bi_c. Numerical integration in the obtained exact solutions was carried out using the QAGI procedure from the QUADPACK package [55]. QAGI is the standard Fortran subroutine for integration over infinite intervals. To launch this procedure, it was necessary to write the real function subprogram definiting the integrand function. Another input parameter was the relative accuracy requested. For our calculations, this parameter was equal to 10^{-4}.

The aim of the numerical analysis was to Investigate the mutual impact of dimensionless parameters Bi_r and Bi_c on the temperature of the friction system. It should be noted that with a fixed value of the Biot number Bi_r in the limit case $Bi_c \to 0$, the free surface $z = d$ ($\zeta = 1$) of the strip was adiabatic (the thermally insulated), and at $Bi_c \to \infty$, the initial temperature on this surface was maintained during the entire process of friction heating. On the other hand, with a fixed value of the Biot number Bi_c, the case $Bi_r \to 0$ corresponded to such a heating mode of the considered system, at which the sliding surfaces of the strip and the semi-space were heated separately by heat fluxes with the same intensity equal to the half of the specific power of friction, i.e., 0.5 q_0. If $Bi_r \to \infty$, then the temperature of the friction surfaces of the strip and the semi-space at any time of the heating process should be equal.

The results of the calculations are presented in Figures 2–9, where the solid curves correspond to the case of an FGM strip ($ZrO_2/Ti-6Al-4V$), and the dashed curves—to the case of the strip made entirely of ZrO_2. The results, shown by continuous and dashed curves, were obtained using exact solutions, while the dotted curves represent the results obtained on the basis of asymptotic solutions.

The influence of the thermal contact conductivity (the Biot number Bi_r) on the dimensionless temperature rises Θ^* (39) and (40) of the surfaces of friction of the strip ($\zeta = 0^+$) and the semi-space ($\zeta = 0^-$) for the four values of the Biot number Bi_c at the fixed value of the Fourier number $\tau = 1$ are shown in Figure 2. As expected, the greatest jump in the temperature on the surfaces of friction for both types of strip material took place at small ($Bi_r \to 0^+$) thermal contact conductivity (high thermal resistance). Increasing the Biot number Bi_r equalized the temperature of the friction surfaces of the strip and the semi-space. At $Bi_r \approx 100$, their temperature could be considered the same. During the entire heating process, the temperature of the friction system with the FGM strip (the solid curves) was lower compared to the temperature when the homogeneous material of the strip was used (the dashed curves). This effect increased with the growth of the convective cooling intensity (the Biot number Bi_c) on the free surface $\zeta = 1$ of the strip.

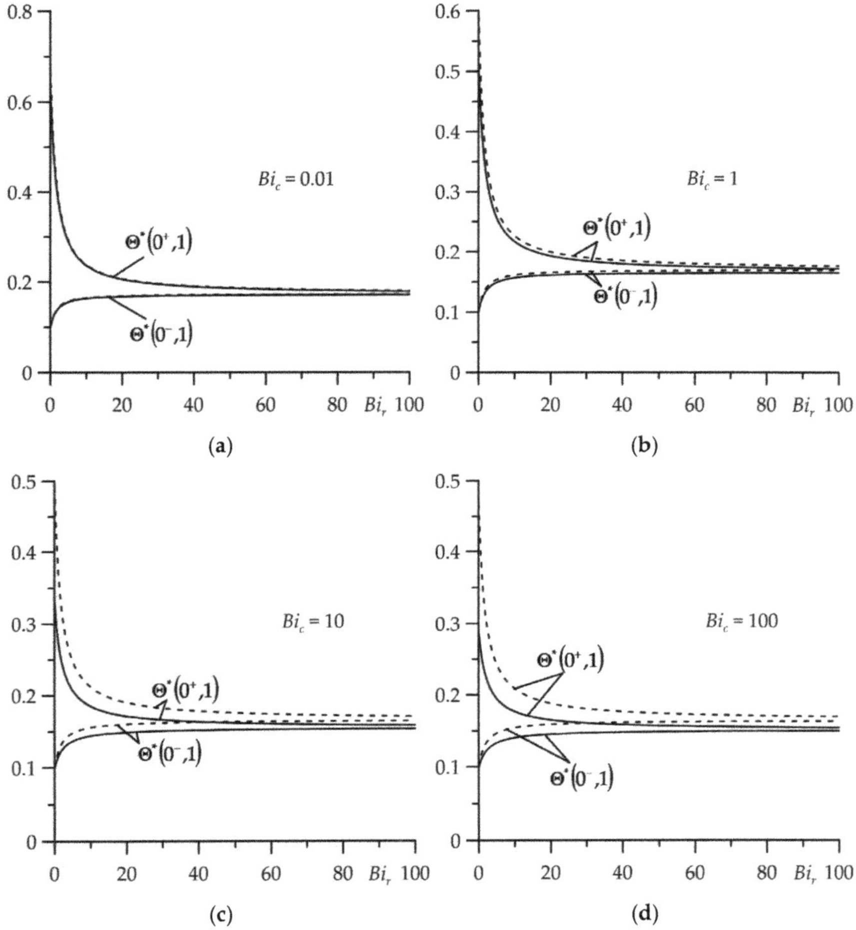

Figure 2. Dependences of dimensionless temperature rise Θ^* on the friction surface of the strip ($\zeta = 0^+$) and the semi-space ($\zeta = 0^-$) from the Biot number Bi_r for $\tau = 1$ at four values of the Biot number Bi_c: **(a)** 0.01; **(b)** 1; **(c)** 10; **(d)** 100. The continuous curves—the FGM strip; the dashed curves—the strip made of zirconium dioxide.

The appropriate results for dimensionless intensities of the heat fluxes q_i^*, $i = 1,2$ (54) and (55) are shown in Figure 3. It was noticed that the greatest changes in the intensity of heat fluxes, directed along the normal from the contact surface to the inside of the strip (q_1^*) and the semi-space (q_2^*), occurred in the range $0 \leq Bi_r \leq 10$. Increasing the thermal contact conductivity (reducing the thermal resistance) resulted in the fact that the amount of heat absorbed by the FGM strip decreased and the heat absorbed by cast iron semi-space increased. With an almost adiabatic process ($Bi_c = 0.01$), the free surface of the strip, the influence of the gradient of the strip material on q_i^*, $i = 1,2$ was negligible (Figure 3a). With the increase in the intensity of heat transfer (the Biot number Bi_c) between the free surface of the strip and the surrounding environment, the effect of the FGM of the strip on temperature became more noticeable (Figure 3b–d).

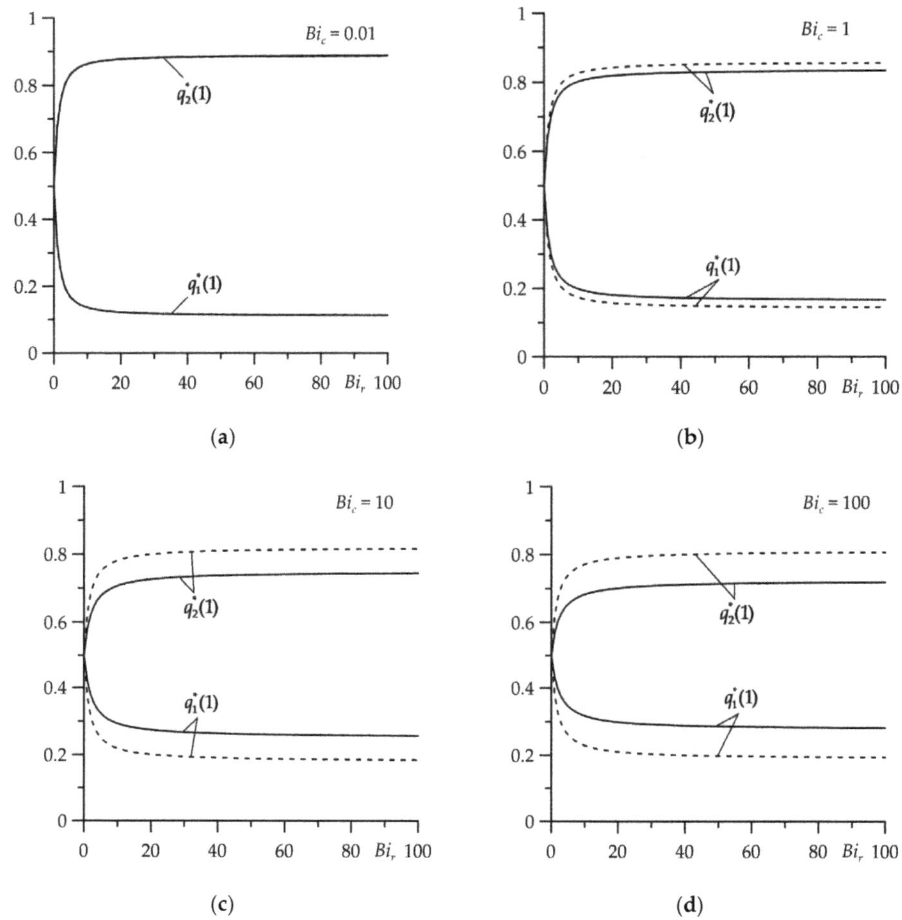

Figure 3. Dependences of dimensionless heat fluxes intensities q_i^*, $i = 1,2$ on the Biot number Bi_r for $\tau = 1$, at four values of the Biot number Bi_c: (**a**) 0.01; (**b**) 1; (**c**) 10; (**d**) 100. The continuous curves—the FGM strip, the dashed curves—the strip made of zirconium dioxide.

The evolutions of the dimensionless temperature rise Θ^* (39), (40) on the friction surfaces of the strip ($\zeta = 0^+$) and the semi-space ($\zeta = 0^-$) for different values of the Biot numbers Bi_r and Bi_c are shown in Figure 4. It can be seen that with the same input parameters, the use of an FGM strip caused a decrease in the temperature of both elements in comparison to the temperature found in the case in which the strip has a homogeneous material. This tendency is most clearly visible on the friction surface of the strip ($\zeta = 0^+$) with a small value of thermal contact conductivity ($Bi_r = 0.01$) and the intensive convective cooling ($Bi_c = 100$) of the strip. The influence of the gradient nature of the strip material on the temperature on the friction surface of the semi-space ($\zeta = 0^-$) is much smaller. On this surface, the noticeable temperature differences when using the FGM of the strip and the homogeneous strip material occur in the case of perfect thermal contact of friction ($Bi_r = 100$) with intensive ($Bi_c = 100$) convective cooling of the free surface the strip.

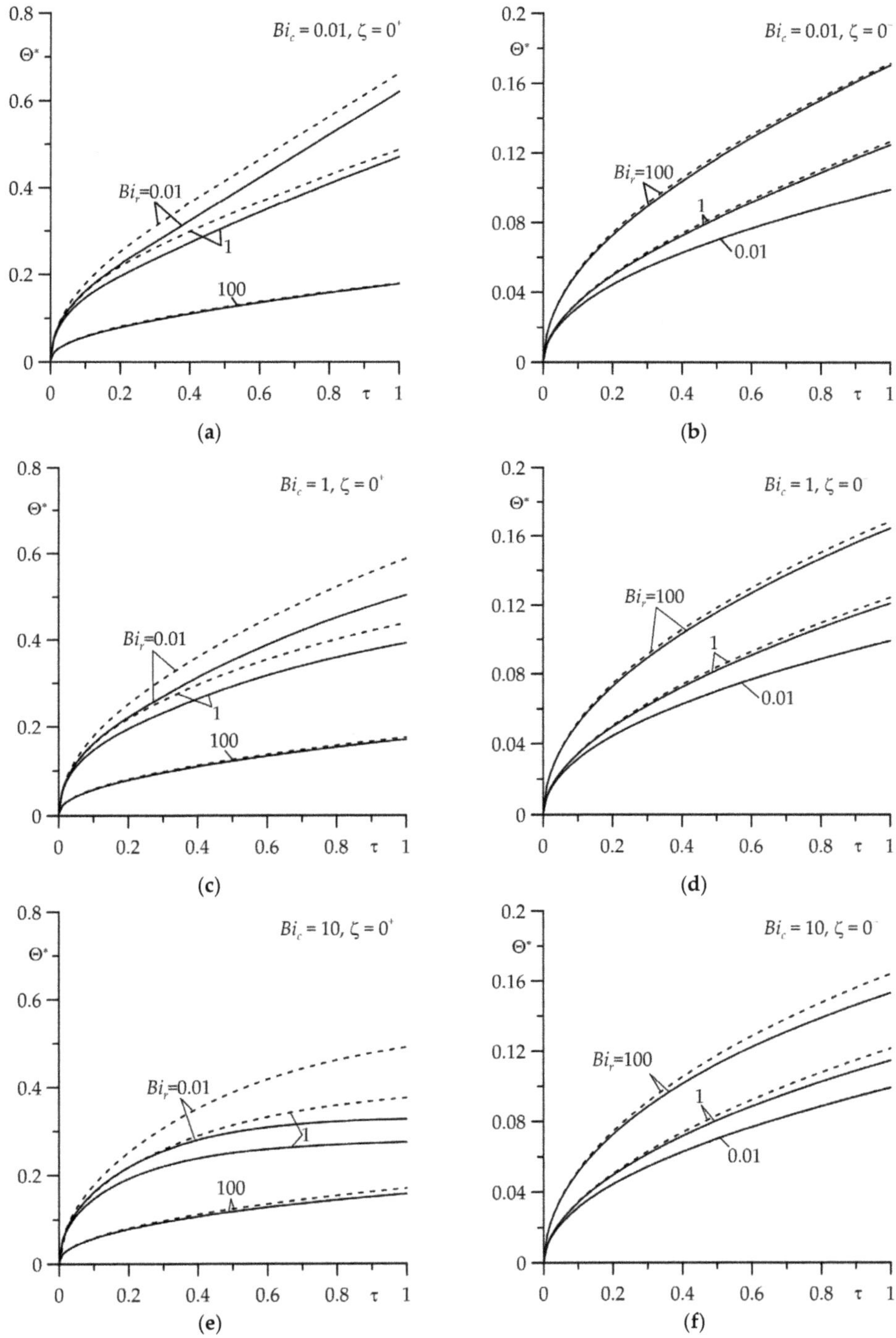

Figure 4. *Cont.*

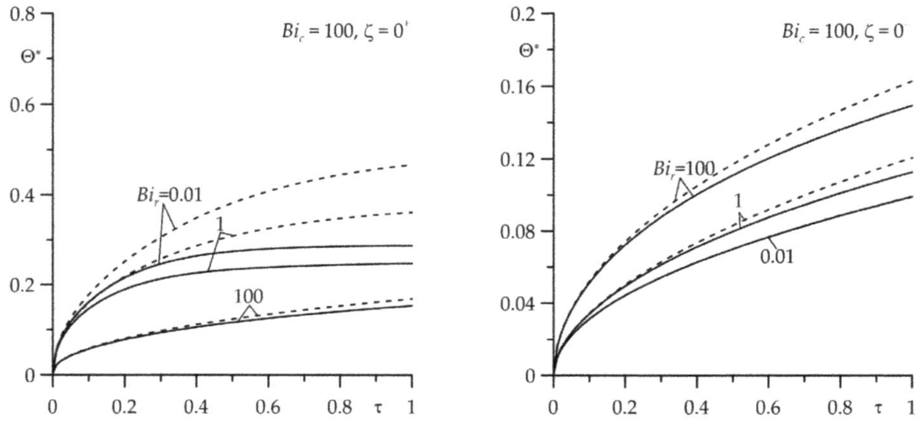

Figure 4. Evolutions of dimensionless temperature rise Θ^* on the friction surface of the strip ($\zeta = 0^+$) and the semi-space ($\zeta = 0^-$) for different values of the Biot number Bi_r at four values of the Biot number Bi_c: (**a**,**b**) 0.01; (**c**,**d**) 1; (**e**,**f**) 10; (**g**,**h**) 100. The continuous curves—the FGM strip; the dashed curves—the strip made of zirconium dioxide.

Corresponding changes with time (the Fourier number τ) of dimensionless heat flux intensities q_i^*, $i = 1,2$ (54), (55) are illustrated in Figure 5. The much higher thermal conductivity of cast iron ChNMKh compared to ZrO_2 and $Ti-6Al-4V$ resulted in the fact that the semi-space absorbed most of the heat generated during friction. At $Bi_r = 0.01$, when the conditions for the separate heating of friction elements were implemented, as expected, the intensities persisted at a constant level of 0.5 during the entire heating process, regardless of the intensity of cooling the exposed strip surface. But already at $Bi_r = 1$ with the time, the amount of heat directed to the strip monotonically decreased and to the semi-space decreased. The evolutions of the heat flux intensities looked slightly different, with perfect thermal contact of friction ($Bi_r = 100$). First of all, it was clearly seen that for a fixed value τ the difference between q_1^* and q_2^* in this case was the largest compared to the appropriate results obtained at $Bi_r = 0.01$ or $Bi_r = 1$. Secondly, the sensitivity q_i^*, $i = 1,2$ to the intensity of the convective cooling on the free surface of the strip increased significantly. If the initial temperature was maintained on this surface ($Bi_c = 0.01$), a rapid achievement of the minimum value by q_1^* and the maximum value by q_2^* was established. Then, with increasing the dimensionless time τ, q_1^* decreased slightly and q_2^* accordingly increased. When $Bi_c = 1$ and the above-mentioned extreme values q_i^*, $i = 1,2$ were reached, further heating practically did not cause their changes. A further increase in the intensity of convective heat transfer on the free surface of the strip ($Bi_c = 10$, and in particular $Bi_c = 100$) caused the amount of heat absorbed by the strip to increase after quickly reaching the minimum q_1^* and maximum q_2^* values, and that by the semi-space decreased accordingly over the time. The most noticeable such effect was in the case of an FGM strip.

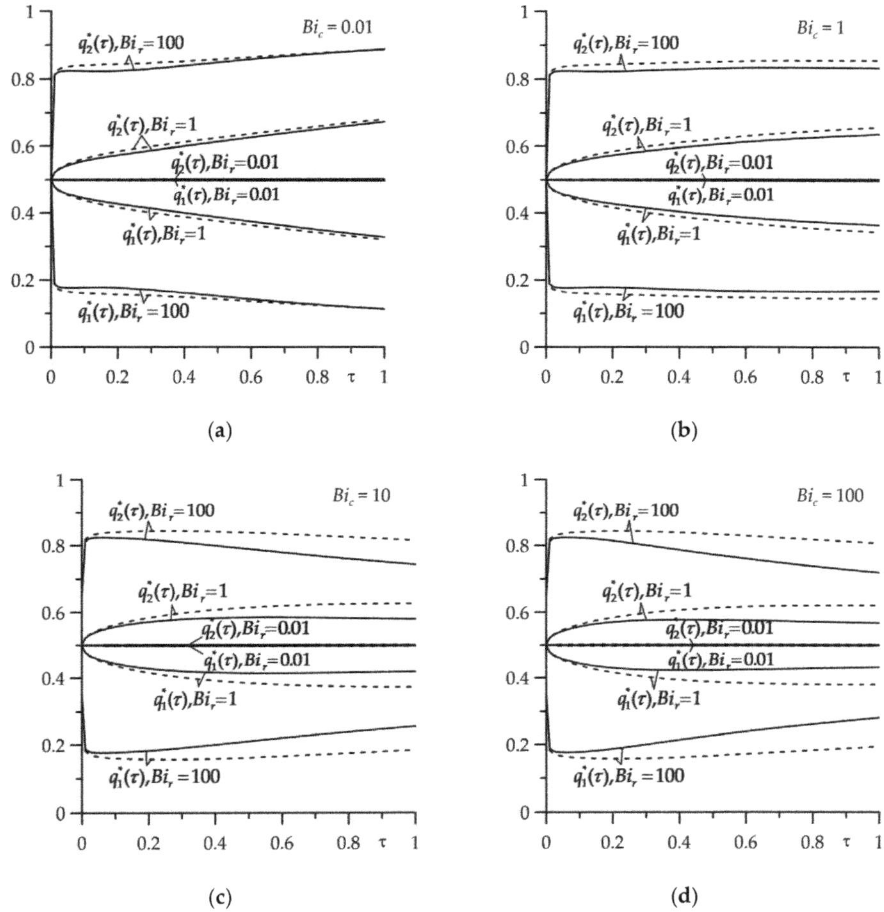

Figure 5. Evolutions of dimensionless intensities of the heat fluxes q_i^*, $i = 1, 2$ for different values of the Biot number Bi_r at four values of the Biot number Bi_c: (**a**) 0.01; (**b**) 1; (**c**) 10; (**d**) 100. The continuous curves—the FGM strip; the dashed curves—the strip made of zirconium dioxide.

Distributions of the dimensionless temperature rises Θ^* (29) and (30) along the thickness of the strip ($0 \leq \zeta \leq 1$) and semi-space ($-1 \leq \zeta \leq 0$) at $\tau = 1$ for different values of the Biot numbers Bi_r and Bi_c are presented in Figure 6. At a fixed-value Bi_c in the case of $Bi_r = 0.01$, the temperature jump was clearly visible on the contact surface $\zeta = 0$. At $Bi_r = 1$ this jump decreased, and at $Bi_r = 100$ it became practically imperceptible. Comparing to the case of a homogeneous strip, the FGM strip did not show a significant impact on the temperature of the semi-space. It could be observed that the temperature of the semi-space decreased linearly with distance from the contact surface, regardless of the choice of strip material. A different, non-linear nature of the temperature distribution was observed in the strip. At the same time, the use of FGM reduced the temperature of the strip in comparison to the temperature of the strip made of zirconium dioxide. This effect was most noticeable with intensive ($Bi_c = 10$ or $Bi_c = 100$) cooling on the free surface ($\zeta = 1$) of the strip.

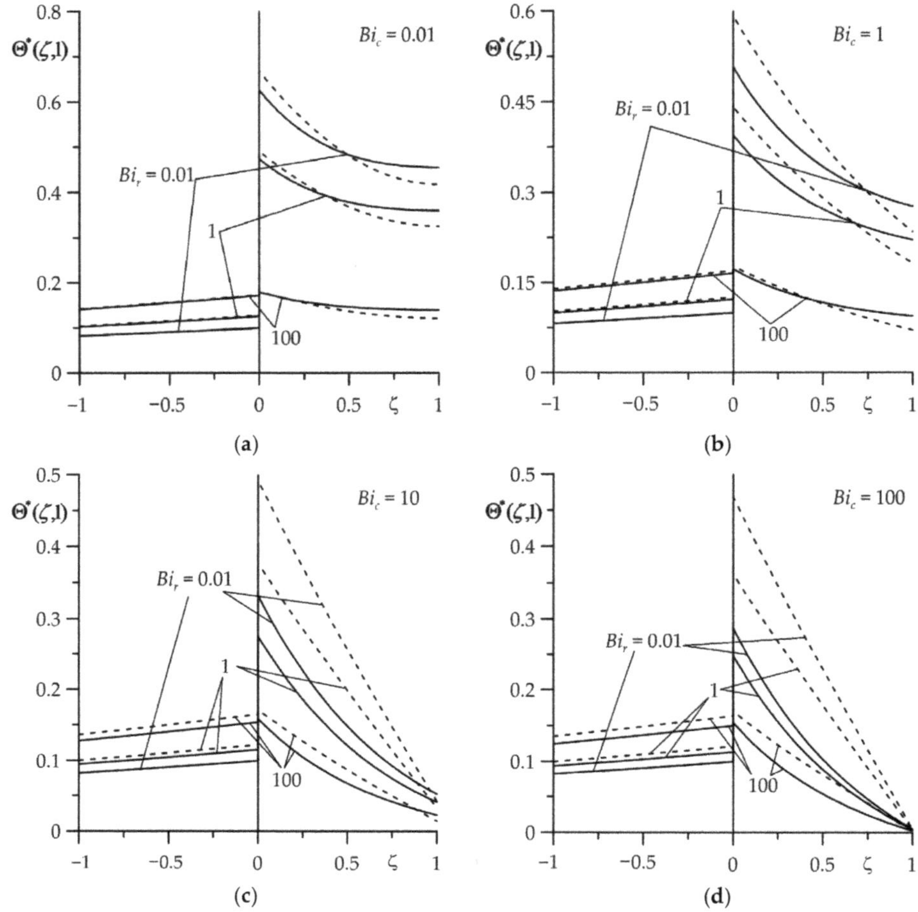

Figure 6. Change in dimensionless temperature rise Θ^* with dimensionless distance ζ from the contact surface $\zeta = 0$ at the final moment of time $\tau = 1$ of the heating process for selected values of Biot number Bi_r at four values of the Biot number Bi_c: (**a**) 0.01; (**b**) 1; (**c**) 10; (**d**) 100. The continuous curves—the FGM strip; the dashed curves—the strip made of zirconium dioxide.

The next part of the numerical analysis was devoted to determining the time intervals in which the asymptotic solutions could be used to estimate the temperature of the friction surfaces of the FGM strip ($\zeta = 0^+$) and the homogeneous semi-space ($\zeta = 0^-$) at small (73), (74) (Figure 7) and large (85), (86) (Figure 8) values of the Fourier number τ (9). The temperature values obtained by these solutions (the dotted curves) were compared with the corresponding results found using the exact solutions (29), (30) (the solid curves). It was established that the usefulness of the asymptotic solution (73) to determine the temperature of the friction surface of the FGM strip largely depended on the value of the Biot number Bi_r: results obtained from the exact (29) and asymptotic (73) solutions, which differed slightly for $Bi_r = 0.01$ or $Bi_r = 1$ at $0 \le \tau \le 0.1$ and for $Bi_r = 10$ or $Bi_r = 100$—if $0 \le \tau \le 1$ (Figure 7a). On the other hand, an asymptotic solution (74) for the semi-space could be used to estimate the temperature of the surface of friction at $0 \le \tau \le 1$ for four selected values of Bi_r (Figure 7b).

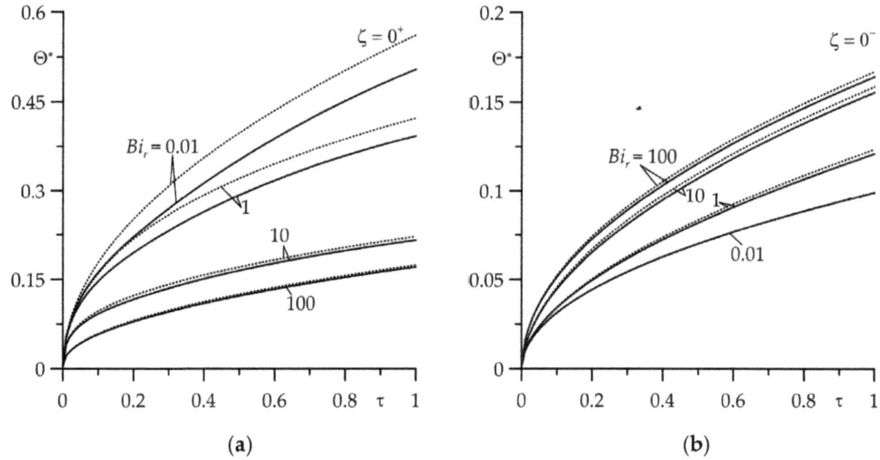

Figure 7. Evolutions of the dimensionless temperature rise Θ^* on the surfaces of friction: (**a**) FGM strip ($\zeta = 0^+$); (**b**) homogeneous semi-space ($\zeta = 0^-$) for $Bi_c = 1$ at four values of the Biot number $Bi_r = 0.01;\ 1;\ 10;\ 100$. The solid curves—the exact solutions (36) and (37); the dotted curves—the asymptotic solutions (80) and (81) at small values of the Fourier number τ.

Similar sensitivity to the value of the Biot number Bi_r of the results obtained by means of an asymptotic solution (85) for the FGM strip at high values τ is visible in Figure 8a. The difference in the results obtained by means of exact (30) and asymptotic (86) solutions for $Bi_r = 0.01;\ 1$ was insignificant at $\tau \geq 5$, and for $Bi_r = 10;\ 100$, it was permissible not only at large but also at small values τ. The asymptotic solution (86) for semi-space at high values of the Fourier number could be used at $\tau \geq 0$ (Figure 8b).

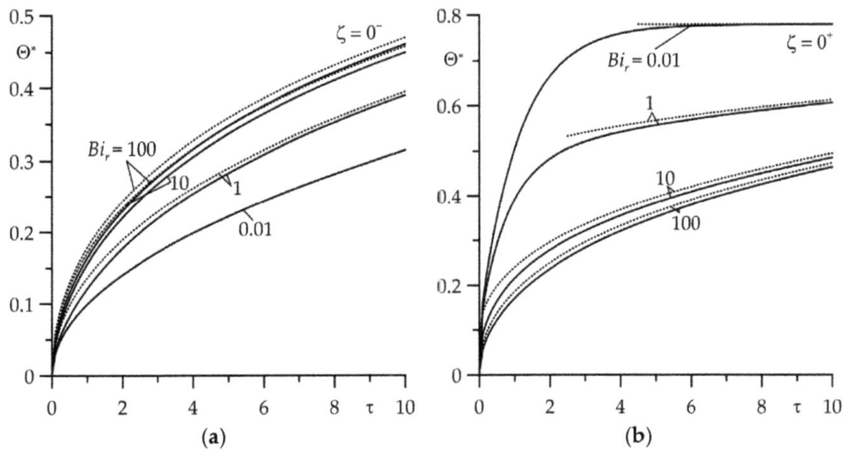

Figure 8. Evolutions of the dimensionless temperature rise Θ^* on the friction surfaces: (**a**) FGM strip ($\zeta = 0^+$); (**b**) homogeneous semi-space ($\zeta = 0^-$) for $Bi_c = 1$ at four values of the Biot number $Bi_r = 0.01;\ 1;\ 10;\ 100$. The solid curves—the exact solutions (29) and (30); the dotted curves—the asymptotic solutions (85) and (86) at large values of the Fourier number τ.

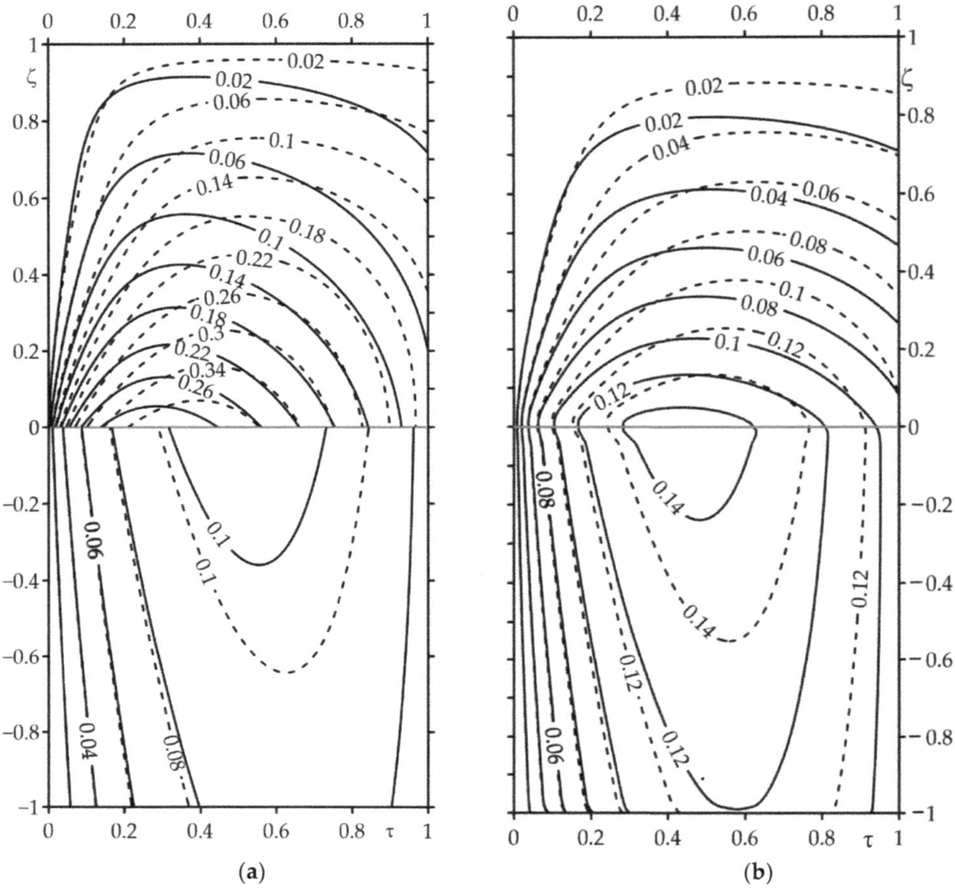

Figure 9. Isolines of the dimensionless temperature rise $\hat{\Theta}^*(\zeta, \tau)$ during braking for $Bi_c = 100$, $\tau_s = 1$ at two values of the Biot number Bi_r: (**a**) 1; (**b**) 100. The continuous curves—the FGM strip; the dashed curves—the strip made of zirconium dioxide.

Isolines of the dimensionless temperature rise $\hat{\Theta}^*(\zeta, \tau)$ (47)–(49) during single braking with constant deceleration with the forced ($Bi_c = 100$) convective cooling of the free surface of the strip for two values of the Biot number Bi_r is demonstrated in Figure 9. First of all, it should be noted that the temperature of the strip made of FGM was lower compared to the case of applying a layer of zirconium dioxide to the material. Secondly, just like during sliding with constant specific friction power, the jumps of isolines were also visible when passing through the contact surface in the case of $Bi_r = 1$ (Figure 9a). However, for $Bi_r = 100$, the temperature of the friction surfaces of the strip and semi-space were the same (Figure 9b). At low contact thermal conductivity ($Bi_r = 1$), the temperature of the strip was higher than the semi-space temperature (Figure 9a). The increase in parameter Bi_r equalized the temperature of both elements of the friction pair (Figure 9b). The effect of the linearly decreasing time profile of the specific friction power was visible primarily in the achievement of the maximum temperature of both elements at a fixed distance from the contact surface, not at the stop moment $\tau = \tau_s = 1$ (as it was in the case of sliding with constant specific friction power), but within the time interval $0 < \tau < \tau_s$.

4. Conclusions

A mathematical model to determine the temperature field resulting from heating during the sliding of a strip made of FGM on a homogeneous semi-space surface was developed. The influence of two dimensionless input parameters, namely the Biot numbers Bi_r and Bi_c, defined by Formula (9), on the temperature of such a friction system was examined. The first of them (Bi_r) was directly proportional to the thermal contact conductivity h_r, which in turn was inversely proportional to the thermal resistance of the friction surface of the strip and semi-space. Thus, at a fixed value of the contact pressure, the higher the roughness of these surfaces, the greater their thermal resistance and the lower the thermal contact conductivity.

Consideration of thermal contact conductivity in the modeling of the frictional heat generation process was carried out by means of two boundary conditions on the contact surface. The first of them assumed that the sum of the intensities of heat fluxes absorbed on the contact surface along the normal to this surfaceinside each element of the friction pair was equal to the specific power of friction, i.e., the product of the coefficient of friction, the contact pressure, and the sliding speed. The second condition stated that the difference in the above-mentioned intensity of heat fluxes was proportional to the difference in the temperature of friction surfaces. The ratio of proportionality here was the thermal contact conductivity h_r or in the dimensionless form of the Biot number Bi_r. Both of these conditions created the so-called conditions of imperfect thermal contact of friction. A characteristic of the solutions of appropriate heat problems of friction, obtained under such boundary conditions, is the temperature jump on the contact surfaces of the sliding bodies. It should be noted that during the sliding of smooth surfaces, the thermal resistance became negligible and the value of the parameter Bi_r was large. In the limit case, $Bi_r \to \infty$, the temperature difference in the friction surfaces disappeared. So, it can be said the generation of heat takes place in conditions of perfect thermal friction contact.

The Biot number Bi_c characterized the intensity of convective cooling on the exposed surface of the strip by means of the heat transfer coefficient h_c. The influence of the parameter Bi_c on the temperature of the system operating in conditions of perfect thermal contact of friction was examined in article [48].

Numerical analysis was performed for the FGM ($ZrO_2/Ti - 6Al - 4V$) strip and homogeneous (gray cast iron ChNMKh) semi-space. At the same time, the temperature of the system, consisting of a homogeneous (ZrO_2) strip and a cast iron (ChNMKh) semi-space, was analyzed. It was established that:

1. Regardless of the value Bi_c, the perfect thermal contact of friction occurs when $Bi_r \cong 100$.
2. The use of the functionally graded material on a strip reduces its temperature compared to the case of homogeneous material. The greater the decrease in temperature, the smaller the parameter Bi_r and the higher the parameter Bi_c is. The effect of the gradient nature of the strip material on lowering the temperature of the semi-space is insignificant.
3. Most of the heat generated during friction is absorbed by the cast iron semi-space. With a fixed value of the Biot number Bi_c, the greatest changes in the intensity of heat fluxes, directed along the normal to the contact surface between the strip and the semi-space, occur in the range of $0 \leq Bi_r \leq 10$. As the parameter Bi_c increases, the impact of the gradient of the strip material on the intensity of heat fluxes becomes more noticeable.
4. The obtained asymptotic solutions at small and large values of the Fourier number τ can be used with sufficient accuracies to estimate the temperature of the considered friction pair. Increasing the parameter Bi_r causes the widening of the interval of the parameter τ, in which the temperature of the strip obtained from asymptotic solutions slightly differs from the temperature determined by the exact solution. Such an effect was not noticed when determining the temperature on the friction surface of the semi-space: the appropriate asymptotic solutions allow one to obtain satisfactory results in the entire range of the Fourier number changes.

5. With a linearly decreasing time profile of the specific friction power, the spatial–temporal temperature distribution in the strip and semi-space is non-uniform. At each set distance from the contact surface, the temperature reaches its maximum value within the heating time interval. In a strip made of FGM, the moment of reaching the maximum temperature approaches the stop moment as the distance from the contact surface increases. However, when using a homogeneous layer and half-space materials, the maximum temperature is reached in approximately half the heating time.

In the future, the authors intend to develop a mathematical model to determine the temperature of the friction system of two FGM strips with imperfect thermal contact of friction and convective heat exchange with the surrounding environment on their free surfaces.

Author Contributions: Conceptualization and methodology, A.Y.; software, P.Z.; validation, A.Y. and K.T., formal analysis, A.Y. and P.Z.; investigation, A.Y., K.T. and P.Z.; writing—original draft preparation A.Y. and K.T.; Writing—Review and editing, K.T.; visualization and figures preparation, P.Z.; supervision, A.Y.; project administration, K.T. All authors have read and agreed to the published version of the manuscript.

Funding: This research was funded by the research project No. WZ/WM-IIM/4/2023 at the Bialystok University of Technology (BUT) in Poland.

Institutional Review Board Statement: Not applicable.

Informed Consent Statement: Not applicable.

Data Availability Statement: Data are contained within the article.

Conflicts of Interest: The authors declare no conflict of interest.

Nomenclature

Bi_c	Biot number on the free surface of the strip
Bi_r	Biot number through the contact surface
c	Specific heat (J $kg^{-1}K^{-1}$)
d	Thickness of the strip (m)
h_c	Coefficient of convective heat exchange on the free surface of the strip
h_r	Coefficient of contact heat conductivity through the surfaces
$I_n(\cdot)$	Modified Bessel functions of the nth order of the first kind
$J_n(\cdot)$	Bessel functions of the nth order of the first kind
$K_n(\cdot)$	Modified Bessel functions of the nth order of the second kind
k	Thermal diffusivity (m^2s^{-1})
K	Thermal conductivity (W $m^{-1}K^{-1}$)
q	Specific power of friction (W m^{-2})
q_0	Nominal value of the specific friction power (W m^{-2})
t	Time (s)
t_s	Stop moment of the process (s)
T	Temperature (°C)
T_0	Initial temperature (°C)
$Y_n(\cdot)$	Bessel functions of the nth order of the second kind
z	Spatial coordinate in axial direction (m)
γ^*	Gradient parameter of FGM
Λ	Temperature rise scaling factor (°C)
ε	Dimensionless coefficient of thermal activity
Θ	Temperature rise (°C)
Θ^*	Dimensionless temperature rise
$\hat{\Theta}^*$	Dimensionless temperature rise during sliding
ρ	Density (kg m^{-3})
τ	Dimensionless time
τ_s	Dimensionless stop time
ζ	Dimensionless spatial coordinate in axial direction

References

1. Straffelini, G.; Verlinski, S.; Verma, P.C.; Valota, G.; Gialanella, S. Wear and Contact Temperature Evolution in Pin-on-Disc Tribotesting of Low-Metallic Friction Material Sliding Against Pearlitic Cast Iron. *Tribol. Lett.* **2016**, *62*, 36. [CrossRef]
2. Yevtushenko, A.; Kuciej, M.; Topczewska, K. Frictional Heating during Braking of the C/C Composite Disc. *Materials* **2020**, *13*, 2691. [CrossRef] [PubMed]
3. Qiu, L.; Qi, H.-S.; Wood, A. Two-dimensional finite element analysis investigation of the heat partition ratio of a friction brake. *Proc. Inst. Mech. Eng. J. Eng. Tribol.* **2018**, *232*, 1489–1501. [CrossRef]
4. Grześ, P. Partition of heat in 2D finite element model of a disc brake. *Acta Mech. Autom.* **2011**, *5*, 35–41.
5. Vernersson, T. Temperatures at railway tread braking. Part 1: Modelling. *Proc. Inst. Mech. Eng. J. Rail Rapid Transit* **2007**, *221*, 167–182. [CrossRef]
6. Topczewska, K.; Gerlici, J.; Yevtushenko, A.; Kuciej, M.; Kravchenko, K. Analytical Model of the Frictional Heating in a Railway Brake Disc at Single Braking with Experimental Verification. *Materials* **2022**, *15*, 6821. [CrossRef]
7. Kreith, F.; Manglik, R.M. Principles of heat transfer. In *Cengage Learning*; Cengage Learning: Boston, MA, USA, 2016.
8. Ling, F.F. A quasi-iterative method for computing interface temperature distributions. *ZAMM-Z Angew. Math. Me. (ZAMP)* **1959**, *10*, 461–474. [CrossRef]
9. Yevtushenko, A.A.; Kuciej, M. One-dimensional thermal problem of friction during braking: The history of development and actual state. *Int. J. Heat Mass Transfer* **2012**, *55*, 4148–4153. [CrossRef]
10. Eriksson, M.; Bergman, F.; Jacobson, S. On the nature of tribological contact in automotive brakes. *Wear* **2002**, *252*, 26–36. [CrossRef]
11. Gopal, V.; Whiting, M.; Chew, J.; Mills, S. Thermal contact conductance and its dependence on load cycling. *Int. J. Heat Mass Transf.* **2013**, *66*, 444–450. [CrossRef]
12. Bauzin, J.-G.; Keruzore, N.; Laraqi, N.; Gapin, A.; Diebold, J.-F. Identification of the heat flux generated by friction in an aircraft braking system. *Int. J. Therm. Sci.* **2018**, *130*, 449–456. [CrossRef]
13. Ling, F.F.; Simkins, T.E. Measurement of Pointwise Juncture Condition of Temperature at the Interface of Two Bodies in Sliding Contact. *J. Basic Eng.* **1963**, *85*, 481–485. [CrossRef]
14. Majcherczak, D.; Dufrenoy, P.; Berthier, Y. Tribological, thermal and mechanical coupling aspects of the dry sliding contact. *Tribol. Int.* **2007**, *40*, 834–843. [CrossRef]
15. Loizou, A.; Qi, H.S.; Day, A.J. A Fundamental Study on the Heat Partition Ratio of Vehicle Disk Brakes. *J. Heat Transf.* **2013**, *135*, 121302. [CrossRef]
16. Berry, G.A.; Barber, J.R.; Berry, G.A.; Barber, J.R.; Berry, G.A.; Barber, J.R. The Division of Frictional Heat—A Guide to the Nature of Sliding Contact. *J. Tribol.* **1984**, *106*, 405–415. [CrossRef]
17. Laraqi, N. Velocity and Relative Contact Size Effects on the Thermal Constriction Resistance in Sliding Solids. *J. Heat Transf.* **1997**, *119*, 173–177. [CrossRef]
18. Yang, Y.-C.; Lee, H.-L.; Chen, W.-L.; Salazar, J.L. Estimation of thermal contact resistance and temperature distributions in the pad/disc tribosystem. *Int. Commun. Heat Mass Transf.* **2011**, *38*, 298–303. [CrossRef]
19. Tariq, A.; Asif, M. Experimental investigation of thermal contact conductance for nominally flat metallic contact. *Heat Mass Transf.* **2015**, *52*, 291–307. [CrossRef]
20. Rosochowska, M.; Balendra, R.; Chodnikiewicz, K. Measurements of thermal contact conductance. *J. Mater. Process. Technol.* **2003**, *135*, 204–210. [CrossRef]
21. Rosochowska, M.; Chodnikiewicz, K.; Balendra, R. A new method of measuring thermal contact conductance. *J. Mater. Process. Technol.* **2004**, *145*, 207–214. [CrossRef]
22. Qui, L.; Qi, H.S.; Wood, A.S. A new paradigm for disc-pad interface models in friction brake system. Paper No. EB2015-TEF-012. Presented at EuroBrake 2015, Conference Proceedings, FISITA, Dresden, Germany, 4–6 May 2015.
23. Bauzin, J.-G.; Laraqi, N. Simultaneous Estimation Of Frictional Heat Flux And Two Thermal Contact Parameters For Sliding Contacts. *Numer. Heat Transfer Appl.* **2004**, *45*, 313–328. [CrossRef]
24. Bauzin, J.G.; Nguyen, M.N.; Laraqi, N.; Cherikh, M.B. Thermal characterization of frictional interfaces using experiments and inverse heat conduction methods. *Int. J. Ther. Sci.* **2019**, *137*, 431–437. [CrossRef]
25. Godet, M. The third-body approach: A mechanical view of wear. *Wear* **1984**, *100*, 437–452. [CrossRef]
26. Denape, J. Third Body Concept and Wear Particle Behavior in Dry Friction Sliding Conditions. *Key Eng. Mater.* **2015**, *640*, 1–12. [CrossRef]
27. Abbasi, S.; Teimourimanesh, S.; Vernersson, T.; Sellgren, U.; Olofsson, U.; Lundén, R. Temperature and thermoelastic instability at tread braking using cast iron friction material. *Wear* **2014**, *314*, 171–180. [CrossRef]
28. Godet, M. Third-bodies in tribology. *Wear* **1990**, *136*, 29–45. [CrossRef]
29. Majcherczak, D.; Dufrénoy, P.; Naït-Abdelaziz, M. Third Body Influence on Thermal Friction Contact Problems: Application to Braking. *J. Tribol.* **2005**, *127*, 89–95. [CrossRef]
30. Belyakov, N.S.; Nosko, A.P. Mathematical simulation of thermal friction processes under conditions of nonideal contact. *High Temp.* **2009**, *47*, 123–130. [CrossRef]
31. Sazonov, V.S. Exact solution of the problem of nonstationary heat conduction for two semi-spaces in nonideal contact. *J. Eng. Phys. Thermophys.* **2006**, *79*, 928–930. [CrossRef]

32. Sazonov, V.S. Nonideal contact problem of nonstationary heat conduction for two half-spaces. *J. Eng. Phys. Thermophys.* **2008**, *81*, 397–408. [CrossRef]
33. Laraqi, N.; Baïri, A.; Ségui, L. Temperature and thermal resistance in frictional devices. *Appl. Therm. Eng.* **2004**, *24*, 2567–2581. [CrossRef]
34. Nosko, O. Partition of friction heat between sliding semispaces due to adhesion-deformational heat generation. *Int. J. Heat Mass Transf.* **2013**, *64*, 1189–1195. [CrossRef]
35. Nosko, O. Thermal boundary conditions to simulate friction layers and coatings at sliding contacts. *Int. J. Heat Mass Transf.* **2018**, *127*, 1128–1137. [CrossRef]
36. Belyakov, N.; Nosko, O. Analytical solution of non-stationary heat conduction problem for two sliding layers with time-dependent friction conditions. *Int. J. Heat Mass Transf.* **2016**, *98*, 624–630. [CrossRef]
37. Yevtushenko, A.; Kuciej, M.; Rozniakowska, M. The contact heat transfer between the plane-parallel strip and the semi-infinite foundation. *Int. Commun. Heat Mass Transf.* **2009**, *36*, 787–793. [CrossRef]
38. Yevtushenko, A.; Kuciej, M.; Yevtushenko, O. Three-element model of frictional heating during braking with contact thermal resistance and time-dependent pressure. *Int. J. Therm. Sci.* **2011**, *50*, 1116–1124. [CrossRef]
39. Yevtushenko, A.; Kuciej, M.; Yevtushenko, O. Modelling of the frictional heating in brake system with thermal resistance on a contact surface and convective cooling on a free surface of a pad. *Int. J. Heat Mass Transf.* **2014**, *81*, 915–923. [CrossRef]
40. Ilie, F.; Cristescu, A.-C. Tribological Behavior of Friction Materials of a Disk-Brake Pad Braking System Affected by Structural Changes—A Review. *Materials* **2022**, *15*, 4745. [CrossRef]
41. Bayat, Y.; Ghannad, M.; Torabi, H. Analytical and numerical analysis for the FGM thick sphere under combined pressure and temperature loading. *Arch. Appl. Mech.* **2011**, *82*, 229–242. [CrossRef]
42. Zhou, W.; Ai, S.; Chen, M.; Zhang, R.; He, R.; Pei, Y.; Fang, D. Preparation and thermodynamic analysis of the porous ZrO2/(ZrO2+ Ni) functionally graded bolted joint. *Compos. Eng.* **2015**, *82*, 13–22. [CrossRef]
43. Mao, J.-J.; Ke, L.-L.; Wang, Y.-S. Thermoelastic contact instability of a functionally graded layer and a homogeneous half-plane. *Int. J. Solids Struct.* **2014**, *51*, 3962–3972. [CrossRef]
44. Mao, J.-J.; Ke, L.-L.; Wang, Y.-S.; Liu, J. Frictionally Excited Thermoelastic Instability of Functionally Graded Materials Sliding Out-of-Plane With Contact Resistance. *J. Appl. Mech.* **2015**, *83*, 021010. [CrossRef]
45. Mao, J.-J.; Ke, L.-L.; Yang, J.; Kitipornchai, S.; Wang, Y.-S. Thermoelastic instability of functionally graded materials with interaction of frictional heat and contact resistance. *Mech. Based Des. Struct. Mach.* **2017**, *46*, 139–156. [CrossRef]
46. Volkov, S.S. Displacements of a surface of FGM-coated half-space heated in a circular area taking into account imperfect coating-substrate interface. *IOP Conf. Ser. Mater. Sci. Eng.* **2021**, *1029*, 012062. [CrossRef]
47. Volkov, S.S.; Vasiliev, A.S.; Aizikovich, S.M.; Sadyrin, E.V. Compliance functions for a thermoelastic FGM coated half-plane with incomplete adhesion between the coating and substrate. *J. Phys. Conf. Ser.* **2020**, *1474*, 012034. [CrossRef]
48. Yevtushenko, A.; Kuciej, M.; Topczewska, K.; Zamojski, P. Effect of convective cooling on the temperature in a friction system with functionally graded strip. *Materials* **2023**, *16*, 5228. [CrossRef]
49. Yevtushenko, A.; Topczewska, K.; Zamojski, P. Use of Functionally Graded Material to Decrease Maximum Temperature of a Coating–Substrate System. *Materials* **2023**, *16*, 2265. [CrossRef]
50. Yevtushenko, A.; Topczewska, K.; Zamojski, P. The Effect of Functionally Graded Materials on Temperature during Frictional Heating at Single Braking. *Materials* **2021**, *14*, 6241. [CrossRef]
51. Sneddon, I.N. *The Use of Integral Transforms*; McGraw-Hill: New York, NY, USA, 1972.
52. Abramowitz, M.; Stegun, I. *Handbook of Mathematical Functions with Formulas, Graphs, and Mathematical Tables*; United States Department of Commerce, National Bureau of Standards (NBS): Washington, DC, USA, 1964.
53. Özişik, N.M. *Heat Conduction*; John Wiley: New York, NY, USA, 1993.
54. Bateman, H.; Erdelyi, A. *Tables of Integrals Transforms, Version 1*; McGraw-Hill: New York, NY, USA, 1954.
55. Piessens, R.; De Doneker-Kapenga, E.; Überhuber, C.W.; Kahaner, D.K. *Quadpack: A Subroutine for Automatic Integration*; Springer Science & Business Media: Berlin/Heidelberg, Germany, 2012; Volume 1.

materials

MDPI

Article

Stress Concentration Factors for Non-Load-Carrying Welded Cruciform Joints Subjected to Tension, Bending, and Shear

Krzysztof L. Molski [1] and Piotr Tarasiuk [2,*]

[1] SaMASZ Sp. z o.o., ul. Trawiasta 1, 16-060 Zabludow, Poland; krzysztof.molski@samasz.pl
[2] Faculty of Mechanical Engineering, Bialystok University of Technology, Wiejska 45C, 15-351 Bialystok, Poland
* Correspondence: p.tarasiuk@pb.edu.pl; Tel.: +48-504-130-045

Abstract: This paper deals with the problem of stress concentration at the weld toe of non-load-carrying-type plate cruciform joints under tension, bending, and shear. Theoretical stress concentration factors were derived using the finite element method. Five of the most important geometrical parameters: the thickness of the main plate and the attachments, the weld throat thickness, the weld toe radius, and the weld face inclination angle were treated as independent variables. For each loading mode—tension, bending, and shear—parametric expression of high accuracy was obtained, covering the range used in real structures for cruciform connections. The maximum percentage error was lower than 2.5% as compared to numerical values. The presented solutions proved to be valid for the toe radius ρ tending to zero.

Keywords: welded plate cruciform connections; stress concentration factor; weld toe; finite element analysis; tension; bending; shear

1. Introduction

Welded cruciform joints are commonly used in real welded structures. Two types of such connections are generally distinguished: a load-carrying one and a non-load-carrying one, depending on the way the external load is transmitted through the joint. In the former case, the loaded plates are connected to the transversal one, and the whole load passes through the welds. In such a case, so-called lack of penetration defects may occur in the form of crack-like transversal flaws [1–4]. In the latter case, the main plate can sustain the whole load while additional attachments are welded to both sides of the main plate.

It is well known that the fatigue fracture of welded connections subjected to variable loads is the most common type of damage [5–8], which usually begins at critical points related to the maximum stress concentration. During the last few decades, a great effort has been made in analyzing fatigue phenomena. However, the problem of efficient design procedures has not been definitely solved and many research programs are currently being developed to further address this issue.

Various approaches are used to assess the strength and durability of weldments. Some are based on local concepts including hot spot, local stress [9–12]. and fracture mechanics [13–15].

It is generally accepted that the geometry of a joint has a significant influence on the fatigue strength and durability of the welded structure. In particular, the weld toe radius ρ, the weld angle θ, and the bead profile are of primary importance [2,8,13,16–19]. Unfortunately, all of these parameters change in a random way along the weld seam. Some examples of statistical data may be found in [16,20]. In spite of all of these difficulties, various standards and recommendations, presented in References [21–26], have been developed, which have proven to be significantly helpful in the fatigue design of welded structures, including cruciform connections.

The mechanisms of fatigue damage are very specific and governed by the stress range $\Delta\sigma$ and the stress gradient at the critical point of a cruciform joint, which depends mainly on the weld toe radius or the size and shape of additional imperfections that may appear in a joint. The geometry of the joint is the key quantity because it determines the SCF values. There are several important arguments confirming this conclusion:

- In EN ISO 5817:2014 [27], quality levels for imperfections are described. Numerous geometrical cases of imperfections are depicted in the form of tables. All geometrical cases do not depend on the material properties and are valid for all specified materials.
- Many recommended mechanical methods applied to improve the fatigue strength of weldments, such as grinding, re-melting, weld reinforcement, various types of hammer-peening, etc., consist of changing the weld toe geometry by increasing the radius to reduce the stress concentration, i.e., to diminish SCFs (see [27] Welding and allied processes. Quality levels for imperfections and the report of IIW Commission XIII, Recommendations on Post Weld Improvements of Steel and Aluminium Structure, Section 1: 'Modification of weld toe geometry').
- There are many papers relating the FAT approach to the particular types of welded joints. 'Particular joints' means 'joints of different geometries' and the 'FAT approach' means 'fatigue life characteristics based on the variable stress range'. The conclusions are obvious. All remaining properties and factors are less important than the geometrical and loading ones.

The theoretical stress concentration factor (SCF) is one of the crucial parameters used in the design of weldments, as it enables relating the remote stresses to the local critical points. Some methods, applications, particular values, and formulas used for calculating SCFs are given in References [17,28–35].

From the analysis of the data available in the literature, one may conclude that there is a lack of information about the stress concentration in welded joints. Such data would be very helpful for designers in predicting the fatigue strength and life of welded structures including cruciform joints. As shown in [32], some existing formulas have generally undetermined accuracy and a relatively narrow range of application. Moreover, there are no solutions dealing with shearing loads.

The present work deals with the determination of stress concentration factors for non-load-carrying welded plate cruciform joints subjected to tension, bending, and shear, using the finite element method (FEM) and represented by appropriate parametric equations. Some SCF solutions have already been obtained earlier for welded cruciform connections [31] subjected to axial and bending loads, for $\theta = 45°$. In the present work, some extended solutions are given for the weld angle θ changing from 30° to 60° and for the shearing load.

2. Methodology

2.1. General Assumptions

The shape and geometrical dimensions of a cruciform joint and the three loading modes considered in the present work are depicted schematically in Figure 1. The location of maximum stresses is indicated by small circles.

Five geometrical parameters—ρ, a, θ, t, and T—influencing the SCF value and shown in Figure 1 were selected. Leg lengths, h and h_p, characterizing the weld size, are also used in the literature.

In the present analysis, the following was assumed:

1. The material of the joint is linear elastic, homogeneous, and isotropic;
2. The elastic properties of the main plate, attachment plates, and welds are the same;
3. The welded joint is free from residual stresses, structural irregularities, and imperfections, including a lack of penetration defects;
4. Both attached plates are co-linear and perpendicular to the main plate;
5. All four welds satisfy the double symmetry of the connection;
6. The weld faces are plane, and the contour of the weldment is smooth with a transition toe radius $\rho > 0$;

7. External load—axial, bending, and shearing—is applied far enough from the welds in order to satisfy the principle of Saint-Venant;
8. Small deformations occur in the whole body;
9. Five geometrical parameters: ρ, a, θ, t, and T vary in the following ranges: $0 < \rho/a \leq 1.3$, $0 < a/t \leq 1.3$, $1 \leq T/a \leq 4$, and $30° \leq \theta \leq 60°$;
10. Particular stress concentration factors for different loading modes are defined as follows: $K_{tt} = \sigma_{1\max}/\sigma_t$, $K_{tb} = \sigma_{1\max}/\sigma_b$, and $K_{ts} = \tau_{\max}/\tau_s$.

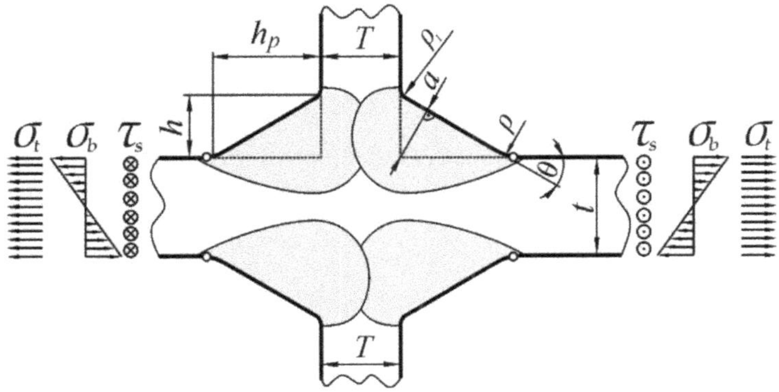

Figure 1. A cruciform welded joint, its geometrical parameters, and loading conditions.

Any deviation from these assumptions, including the material properties, the shape of the body, and the loading conditions may cause significant changes in the stress distribution, including the maximum stress value at the critical point.

The aim of this work was to derive SCF parametric equations covering all possible values of five geometrical parameters characterising the cruciform joint. For this reason, a wide margin for the normalized parameters was used.

2.2. General Approach

The procedure of deriving three parametric formulae consists of several steps described in detail in Refs. [34,35]. The difference between a modeling welded T-joint and a cruciform connection lays only in the boundary conditions considered in the numerical model. Therefore, only some general steps of the procedure are mentioned in the present study.

First, conveniently defined new geometrical parameters are specified as follows:

$$X = \rho/(\rho + a) = \frac{\rho/a}{\rho/a + 1} \tag{1}$$

$$Y = a/(a + t) = \frac{a/t}{a/t + 1} \tag{2}$$

$$Z = T/a \tag{3}$$

In this way, five independent parameters are converted into four variables X, Y, Z, and θ.

The second step consists of numerical modeling of the cruciform joint using the finite element method, where particular values of the four variables X, Y, Z, and θ systematically change. In this way, several thousand numerical solutions of SCF values are obtained. The third step of the approach required choosing a general form of the mathematical representation of the SCFs' approximating functions. The general form of expressions and procedures used in the present article is similar to that presented in Reference [35] dealing with welded T-joints.

Hence, a general form of the approximating function is given by Equation (4):

$$K_t = X^n P(X, Y, \theta, Z_0) \kappa(X, Y, \theta, Z_0) \tag{4}$$

where a singular term X^n accounts for the stress concentration effects when $\rho \to 0$. The function $P(X, Y, \theta, Z_0)$ is represented by polynomials and can be derived from numerical SCF solutions normalized with respect to the singular term, K_t/X^n for different X, Y, and θ, while $Z = Z_0$. From Equation (4), one may conclude that the P function cannot directly relate the SCF to Z because it is derived for a particular arbitrarily chosen value of $Z = Z_0$. Therefore, the correction function $\kappa(X, Y, Z, \theta, Z_0)$ must satisfy Equation (5):

$$\kappa(X, Y, \theta, Z_0) = 1 \tag{5}$$

As shown in [34,35], the exponent n, valid for tension and bending, and denoted as n_s for shearing load, is necessary to describe the stress concentration effects when the weld toe radius tends to zero. The numerical values of the exponent n, for tension and bending, are given by Equation (6):

$$n = \frac{-0.63662\theta - 0.09330\theta^2}{1 + 0.77635\theta + 0.040075\theta^{1.5} - 0.00499\theta^2 + 0.133656\theta^{2.5}} \tag{6}$$

which is valid in the range of $0 \leq \theta \leq \pi/2$ with an accuracy of five significant digits.

In the case of anti-plane deformation, produced by the shearing load, the exact value of the exponent n_s equals:

$$n_s = \frac{-\theta}{\theta + \pi} \tag{7}$$

where θ is in radians.

The next step of the procedure consists f approximating the normalized numerical SCF data with polynomial functions $P(X,Y,Z_0,\theta)$ using the least-squares method. Next, the approximation accuracy of the P function is verified.

The last step of the present procedure consists of deriving the unknown coefficients and exponents of the correction function $\kappa(X,Y,Z,\theta,Z_0)$. After performing the second validation of the full solution including κ, the close form approximating functions for calculating the SCFs were determined. The procedure was consecutively repeated for each loading mode. Approximating functions, in the form of parametric equations, are given in Appendix A.

3. Numerical FEM Modeling and Some SCF Results

3.1. Tensile and Bending Load

The ANSYS 19 Multiphysics FEM program was used for the systematic numerical modeling of cruciform joints. Four-node PLANE182 finite element, having 2 degrees of freedom at each node, was used in modeling. The elastic material constants: E = 210 GPa and ν = 0.3 were used.

The loading and displacement boundary conditions imposed on one quarter of the joint are shown in Figure 2. Considering the principle of Saint-Venant, the minimum length of the main plate measured from the weld toe was taken as 4*t*. This value was chosen by conducting preliminary numerical tests.

Before solving any FEM model, the finite element mesh density has to be found. This is usually carried out by solving the same problem using different numbers of elements and analyzing the subsequent results. This procedure was applied in the present case considering an increase in the mesh density in particular zones where the stress concentration occurs. For this reason, the dimensional ratio of neighboring finite elements was about 1.2, and approximately 40 to 70 finite elements were used along the weld toe circular arc. Approximately 730,000 finite elements were used for each geometrical case.

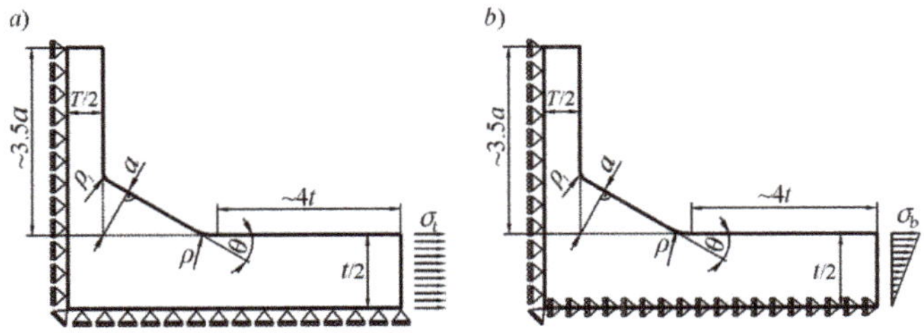

Figure 2. Boundary conditions used in the FEM model for tensile (**a**) and bending (**b**) loads.

Considering the fact that the geometry of the joint changed significantly and a great number of FEM solutions had to be obtained, automation in generating finite element meshes was necessary. Therefore, some unique meshing algorithms were applied. One example of a finite element mesh is shown in Figure 3.

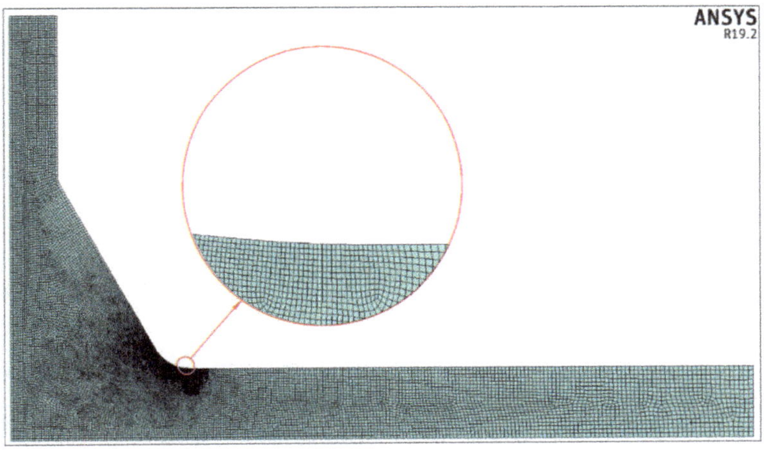

Figure 3. Example of a finite element mesh for $X = 0.3$, $Y = 0.4$, $\theta = 60°$, and $T/a = 1$.

The first principal stresses σ_1 are the most convenient values in deriving SCFs, because they are directly accessible, while nominal loads equal one.

Two examples of the solution obtained for tension and bending for the nominal stress $\sigma_t = 1$ MPa and $\sigma_b = 1$ MPa, respectively, are shown in Figures 4 and 5.

In the two cases presented here, the maximum principal stress σ_{1max} for axial load equals 1.78796 MPa, and for bending load, it equals 1.42917. These values are directly seen on both scales below the corresponding pictures and represent the stress concentration factors K_{tt} and K_{tb}.

3.2. Shearing Load

As described in refs. [34,35], an anti-plane state of deformation may be treated as a boundary value problem governed by Laplace's equation, represented in Cartesian coordinates by Equation (8):

$$\frac{\partial^2 \Psi}{\partial x^2} + \frac{\partial^2 \Psi}{\partial y^2} = 0 \quad (8)$$

The fact that the same relationship is also valid for in-plane steady-state heat conduction problems leads to the conclusion that thermal analogy may be applied to derive stress concentration factors for an anti-plane shear. This approach has been successfully applied in solving anti-plane problems of shearing stress fields in butt and T-joints [34,35].

In the present analysis, the ANSYS19 Multiphysics program with a thermal module and PLANE55 finite element was used. A PLANE55 finite element is defined by four nodes with a single degree of freedom corresponding to the temperature at each node. The finite element mesh of the modeled joint was the same as in the previous cases for tensile and bending loads. The shape of the modeled body and mixed boundary conditions are shown in Figure 6.

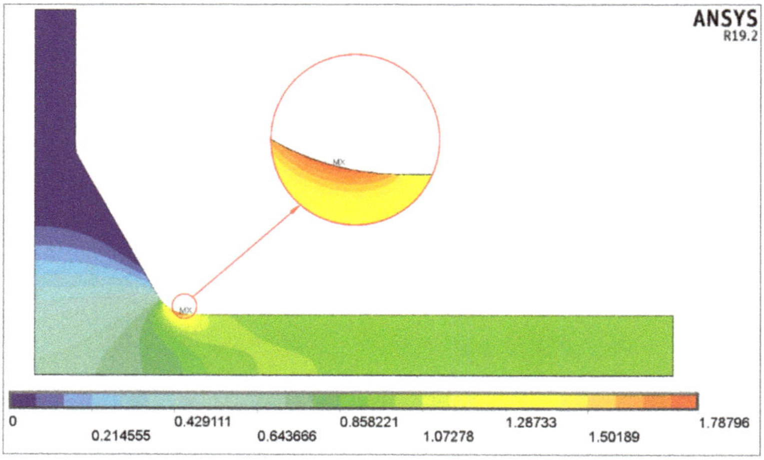

Figure 4. Distribution of the first principal stress σ_1. Tensile load σ_t = 1 MPa, θ = 60°, X = 0.3, Y = 0.4, and T/a = 1.

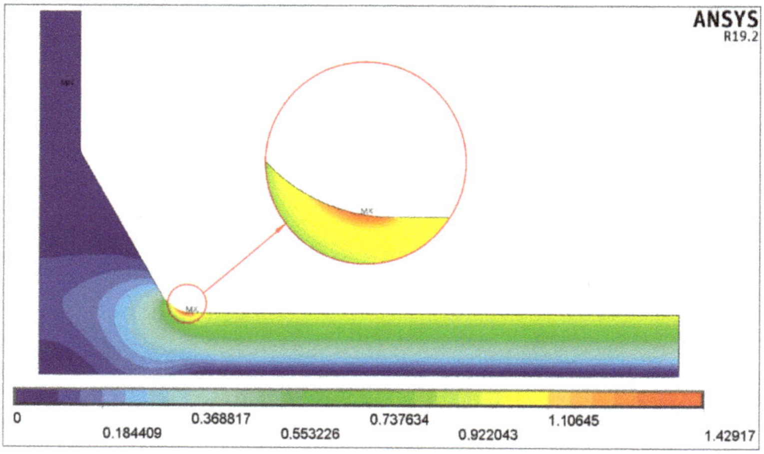

Figure 5. Distribution of the first principal stress σ_1. Bending load σ_b = 1 MPa, θ = 60°, X = 0.3, Y = 0.4, and T/a = 1.

It was convenient to apply nominal uniform heat flux q_{nom} = 1 W/m^2 along the right end of the body and zero temperature to the left end. Since all remaining faces of the joint are free from external shearing loads, they have to be insulated in the thermal model.

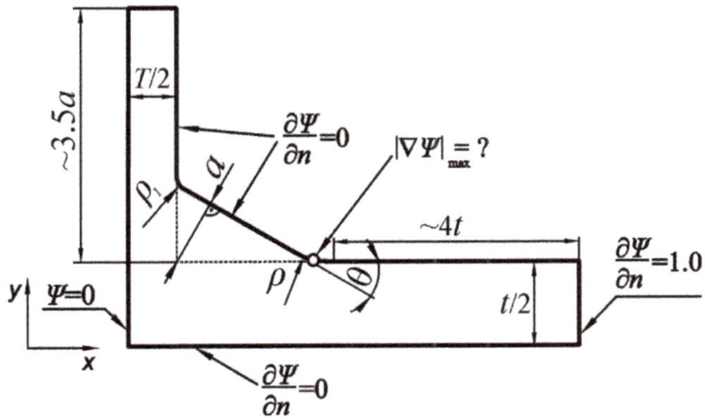

Figure 6. Boundary conditions used in the FEM model for shearing load.

The shear stress components, τ_{xz} and τ_{yz}, given by Equation (9):

$$\tau_{xz} = G\frac{\partial W}{\partial x};\ \tau_{yz} = G\frac{\partial W}{\partial y} \tag{9}$$

and related to the partial derivatives of the potential function $W(x,y)$ in particular directions are proportional to the corresponding heat flux components, q_x and q_y, represented by Equation (10):

$$q_x = -k\frac{\partial T_{temp}}{\partial x};\ q_y = -k\frac{\partial T_{temp}}{\partial y}. \tag{10}$$

After solving the particular boundary-value problem using the finite element method, numerical SCF values were calculated as a ratio of the maximum magnitude of the temperature gradient (or the maximum heat flux $q_{\max}$) at the weld toe zone to the magnitude of the nominal temperature gradient (or the nominal heat flux q_{nom}).

One example of a steady-state heat conduction solution is shown in Figure 7. Magnitudes of the flux q obtained at each point of the body correspond to the shear stresses τ. The maximum $q_{\max}$ value equals 1.34989 and is shown on the scale below the picture. This value directly represents the K_{ts} for this particular shape of the joint.

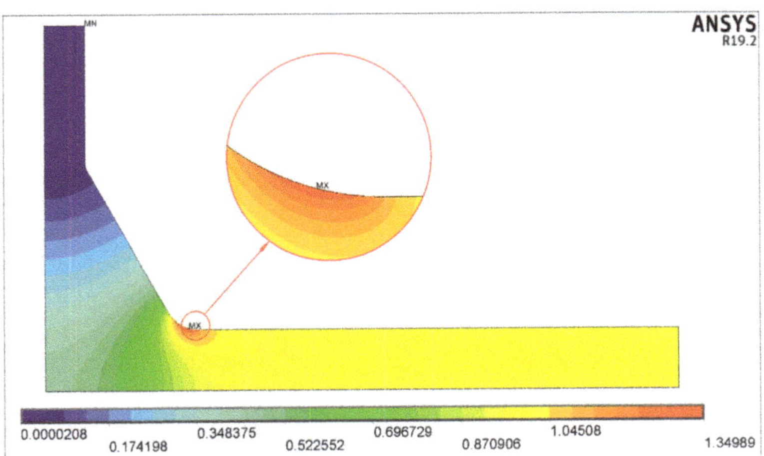

Figure 7. Distribution of the shearing stress τ. Shearing load τ_s = 1 MPa, θ = 60°, X = 0.3, Y = 0.4, and T/a = 1.

3.3. Numerical SCF Results

In subsequent models of cruciform joints, particular values of X, Y, and θ were systematically changed, while the relative attachment thickness T/a was constant and equal to one. Some examples of such a set of SCF results for a cruciform joint for $\theta = 30°, 40°, 50°$ and $60°$ are presented in Tables A1–A12 given in Appendix B.

A similar 12 sets of SCF data were obtained for other weld angles θ in the range of 30–60° with a step of 2.5°. In this way, several thousand SCF numerical solutions were obtained for each loading mode.

4. SCF Approximation Formulas

4.1. Representation of P Functions

It is well known that the SCF value becomes infinite when the notch root radius approaches zero. In such a case, any SCF approximating function must be singular, where the X^n term is responsible for this singular behavior. After normalizing each K_t value with respect to the properly chosen X^n term, three regular functions, P_t, P_b, and P_s, may be obtained. Their mathematical description by polynomials is shown in the brackets of Equations (A1)–(A3) given in Appendix A. Graphical representations of these functions, corresponding to the weld angle $\theta = 30°, 40°, 50°$ and $60°$ for three loading modes, are shown in Figures 8–10.

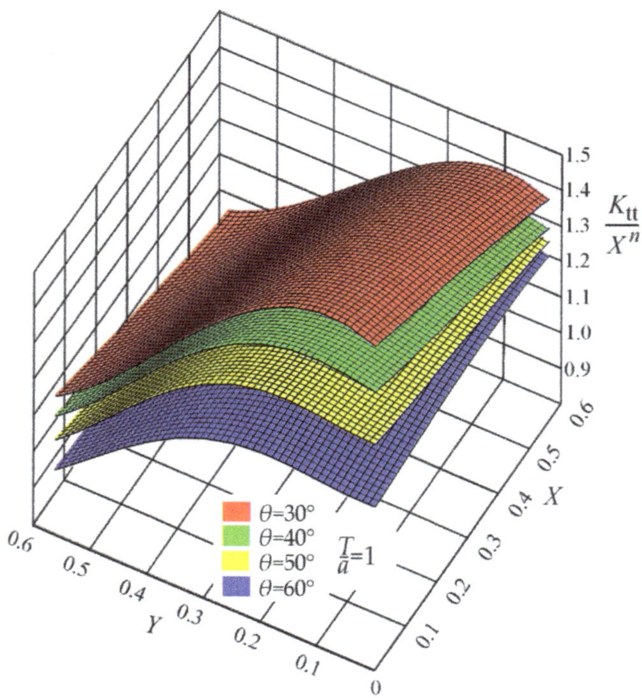

Figure 8. Values of the function $P_t = K_{tt}/X^n$ for $Z_0 = 1$, while $\theta = 30°, 40°, 50°$, and $60°$. Tensile load.

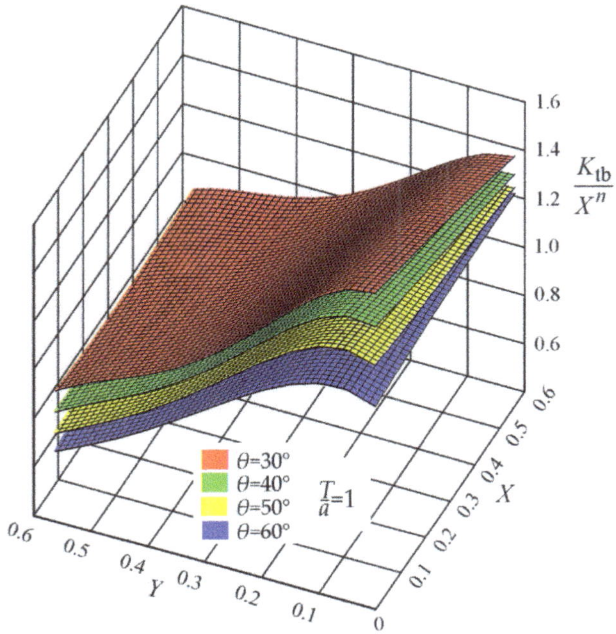

Figure 9. Values of the function $P_b = K_{tb}/X^n$ for $Z_0 = 1$, while $\theta = 30°, 40°, 50°$, and $60°$. Bending load.

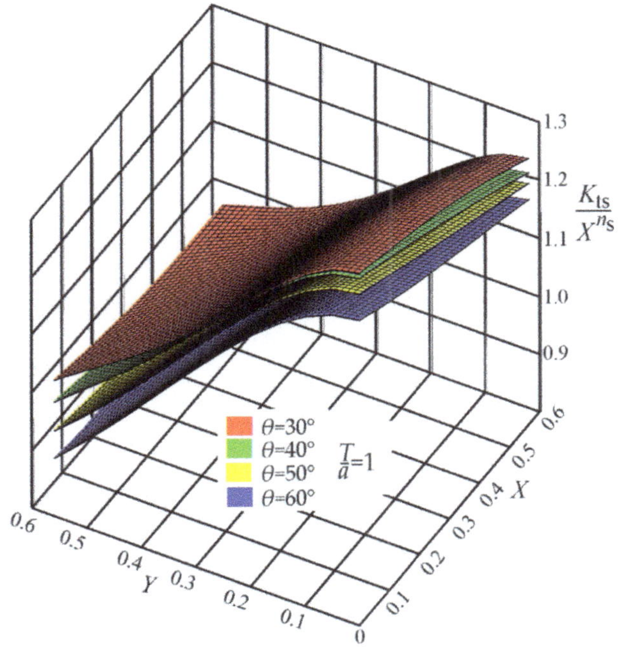

Figure 10. Values of the function $P_s = K_{ts}/X^{n_s}$ for $Z_0 = 1$, while $\theta = 30°, 40°, 50°$, and $60°$. Shearing load.

4.2. Validation of the P Functions

The accuracy of the approximating P functions, represented by polynomials shown in Appendix A, was verified for all loading modes, while the weld angle θ varied in the range of 30°–60°. Particular SCF values (*) calculated from Equation (11)

$$K_t = X^n P(X, Y, \theta, Z_0), \tag{11}$$

were compared to their numerical equivalencies obtained using the finite element method. Equation (11) is represented by the parametric formulas (A1)–(A3), when $Z = Z_0 = 1$ and $\kappa = 1$. Three examples of such comparisons are presented in Tables 1–3 for tension, bending, and shear.

Table 1. K_{tt} values obtained using the FEM and calculated from Equation (A1) (*). Tensile load, $\theta = 30°$, and $T/a = 1$.

θ = 30° *Y*	**0.050**	**0.100**	**0.200**	***X*** **0.300**	**0.400**	**0.500**	**0.562**
0.050	3.530→3.563 *	2.933→2.960 *	2.404→2.424 *	2.112→2.128 *	1.904→1.920 *	1.739→1.755 *	1.649→1.667 *
0.100	3.545→3.570 *	2.946→2.966 *	2.414→2.429 *	2.120→2.132 *	1.912→1.922 *	1.745→1.756 *	1.654→1.667 *
0.200	3.444→3.453 *	2.862→2.869 *	2.343→2.348 *	2.056→2.059 *	1.850→1.854 *	1.685→1.690 *	1.593→1.600 *
0.300	3.166→3.202 *	2.630→2.660 *	2.153→2.176 *	1.887→1.907 *	1.697→1.715 *	1.543→1.561 *	1.458→1.477 *
0.400	2.859→2.882 *	2.375→2.393 *	1.944→1.958 *	1.705→1.716 *	1.534→1.544 *	1.398→1.408 *	1.324→1.335 *
0.500	2.582→2.576 *	2.145→2.139 *	1.757→1.750 *	1.543→1.536 *	1.393→1.386 *	1.278→1.272 *	1.218→1.215 *
0.562	2.423→2.437 *	2.014→2.023 *	1.652→1.656 *	1.454→1.455 *	1.318→1.317 *	1.217→1.216 *	1.168→1.168 *

Table 2. K_{tb} values obtained using the FEM and calculated from Equation (A2) (*). Bending load, $\theta = 60°$, and $T/a = 1$.

θ = 60° *Y*	**0.050**	**0.100**	**0.200**	***X*** **0.300**	**0.400**	**0.500**	**0.562**
0.050	4.047→4.043 *	3.107→3.105 *	2.387→2.387 *	2.044→2.045 *	1.827→1.828 *	1.671→1.671 *	1.592→1.591 *
0.100	4.030→4.012 *	3.085→3.073 *	2.360→2.351 *	2.011→2.005 *	1.790→1.785 *	1.629→1.624 *	1.546→1.542 *
0.200	3.721→3.701 *	2.838→2.824 *	2.156→2.148 *	1.827→1.823 *	1.617→1.616 *	1.464→1.465 *	1.385→1.389 *
0.300	3.244→3.245 *	2.472→2.477 *	1.886→1.887 *	1.606→1.607 *	1.432→1.432 *	1.308→1.308 *	1.250→1.247 *
0.400	2.793→2.797 *	2.143→2.147 *	1.654→1.656 *	1.429→1.431 *	1.297→1.296 *	1.207→1.205 *	1.164→1.161 *
0.500	2.418→2.423 *	1.874→1.880 *	1.477→1.483 *	1.304→1.311 *	1.205→1.212 *	1.141→1.148 *	1.111→1.118 *
0.562	2.212→2.223 *	1.730→1.739 *	1.387→1.393 *	1.242→1.245 *	1.161→1.162 *	1.109→1.108 *	1.085→1.082 *

Table 3. K_{ts} values obtained using the FEM and calculated from Equation (A3) (*). Shearing load, $\theta = 45°$, and $T/a = 1$.

θ = 45° *Y*	**0.050**	**0.100**	**0.200**	***X*** **0.300**	**0.400**	**0.500**	**0.562**
0.0256	2.497→2.500 *	2.152→2.155 *	1.835→1.836 *	1.655→1.656 *	1.527→1.528 *	1.425→1.428 *	1.370→1.373 *
0.0526	2.487→2.487 *	2.143→2.144 *	1.827→1.827 *	1.648→1.647 *	1.520→1.520 *	1.419→1.420 *	1.364→1.366 *
0.1111	2.440→2.436 *	2.102→2.099 *	1.792→1.789 *	1.616→1.613 *	1.491→1.489 *	1.391→1.391 *	1.337→1.337 *
0.1765	2.353→2.351 *	2.028→2.026 *	1.729→1.727 *	1.559→1.558 *	1.438→1.439 *	1.343→1.345 *	1.291→1.293 *
0.2500	2.234→2.236 *	1.925→1.927 *	1.642→1.643 *	1.483→1.484 *	1.369→1.372 *	1.281→1.285 *	1.233→1.238 *
0.3333	2.095→2.098 *	1.807→1.808 *	1.543→1.543 *	1.396→1.397 *	1.294→1.296 *	1.216→1.219 *	1.175→1.177 *
0.3908	2.003→2.004 *	1.728→1.728 *	1.478→1.477 *	1.341→1.340 *	1.247→1.248 *	1.177→1.178 *	1.142→1.140 *

Similar comparisons were made for various weld angles θ in the range of 30°–60°. The maximum percentage error of the approximation was lower than 2.0%. Some examples of graphical representations of SCFs vs. X for different values of θ and Y are shown in Appendix C.

4.3. Determination of the Correction Functions κ

The influence of the relative attachment plate thickness $Z = T/a$ on the SCF was determined by performing additional FEM modeling. Several thousand new cases of a cruciform joint were modeled and solved for each loading mode. Particular values of T/a were changed in the range of $1 \leq T/a \leq 4$. Additional correction function κ was introduced to express these effects.

The general mathematical form of the correction functions κ, given by Equation (12),

$$\kappa(X, Y, Z, \theta, Z_0) = 1 + \left(\sqrt{Z} - \sqrt{Z_0}\right)\left[1 - \left(B_1 + B_2 Y^2\right)X^m\right] Exp\left[-(B_3 Y)^p - B_4\right] \quad (12)$$

is the same as in Reference [35] developed by the authors for a welded T-joint. The particular values of the coefficients B_1–B_4 and the exponents m and p were derived for each loading mode using the least-squares method. The best fits of the functions κ_t, κ_b and κ_s to the numerical FEM SCF results were obtained as follows: $m = 1$ for tensile and bending load and $m = 2$ for shearing load. The particular p values were equal to 2.7, 2.0, and 2.0 for tensile, bending, and shearing loads, respectively.

Since in the present case the coefficients B_i depend on the weld angle θ, additional functions $B_i(\theta)$ had to be developed. Parametric formulas of the correction functions κ_t, κ_b, and κ_s are given in Appendix A. Some geometrical representations of the correction functions κ_t, κ_b, and κ_s, for $T/a = 4$, are depicted in Figures 11–13.

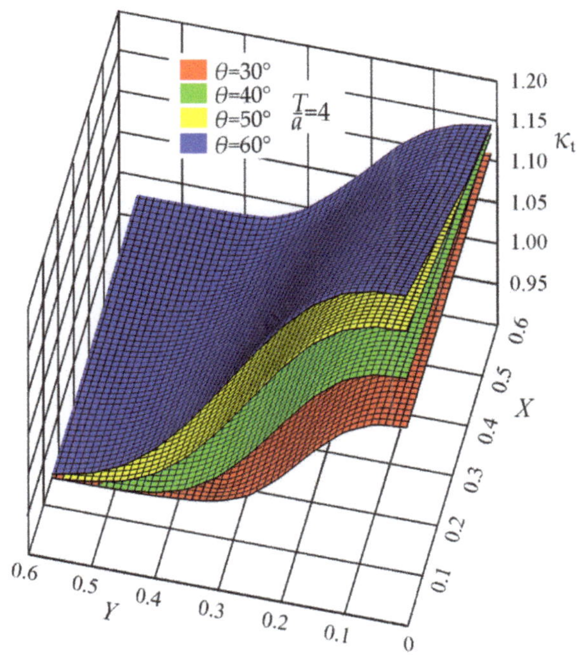

Figure 11. Correction functions κ_t for $\theta = 30°, 40°, 50°$, and $60°$. Tensile load.

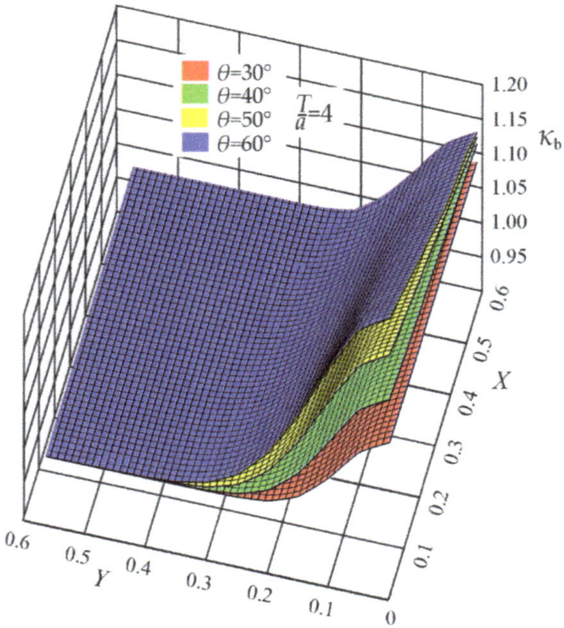

Figure 12. Correction functions κ_b for θ = 30°, 40°, 50°, and 60°. Bending load.

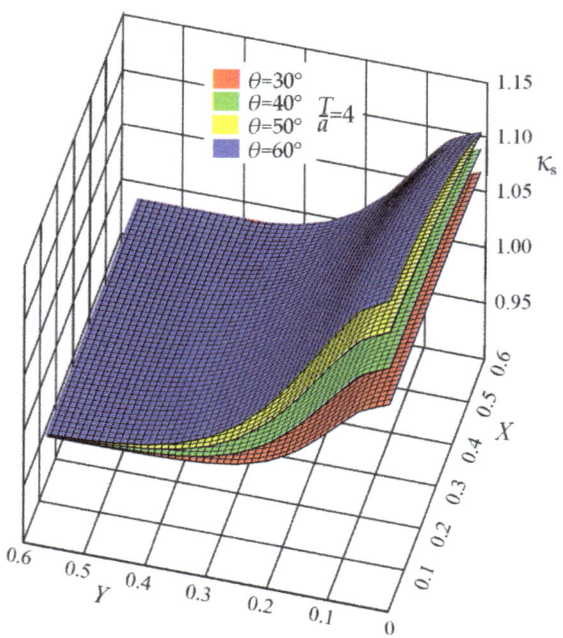

Figure 13. Correction functions κ_s for θ = 30°, 40°, 50°, and 60°. Shearing load.

4.4. Validation of SCF Parametric Equations

The accuracy of the SCF parametric equations, given in Appendix A, was verified for all loading modes, while the weld angle θ varied in the range of 30°–60°. Examples of such a comparison are shown in Tables 4–6 for the angle θ equal to 30°, 45°, and 60°.

Table 4. Comparison of SCF values obtained from Equations (A1)–(A3) (*) to the corresponding FEM results for $\theta = 30°$.

$\theta = 30°$		K_{tt}				K_{tb}				K_{ts}			
ρ/a	t/a	$T/a = 1$	$T/a = 2$	$T/a = 3$	$T/a = 4$	$T/a = 1$	$T/a = 2$	$T/a = 3$	$T/a = 4$	$T/a = 1$	$T/a = 2$	$T/a = 3$	$T/a = 4$
0.05	10	3.599 3.618 *	3.769 3.807 *	3.918 3.951 *	4.047 4.074 *	3.455 3.452 *	3.553 3.545 *	3.624 3.616 *	3.676 3.677 *	2.113 2.114 *	2.158 2.163 *	2.196 2.201 *	2.229 2.233 *
	7	3.598 3.602 *	3.753 3.761 *	3.878 3.884 *	3.977 3.987 *	3.318 3.330 *	3.375 3.379 *	3.410 3.417 *	3.430 3.448 *	2.077 2.078 *	2.111 2.113 *	2.137 2.139 *	2.158 2.162 *
	4	3.495 3.497 *	3.580 3.573 *	3.633 3.631 *	3.668 3.679 *	2.986 3.013 *	2.996 3.019 *	3.000 3.023 *	3.001 3.027 *	1.983 1.985 *	1.997 1.997 *	2.006 2.007 *	2.014 2.015 *
0.25	10	2.415 2.430 *	2.526 2.553 *	2.623 2.648 *	2.707 2.727 *	2.318 2.318 *	2.382 2.377 *	2.428 2.423 *	2.461 2.461 *	1.680 1.682 *	1.715 1.721 *	1.746 1.751 *	1.771 1.776 *
	7	2.412 2.419 *	2.513 2.522 *	2.594 2.601 *	2.659 2.668 *	2.227 2.237 *	2.263 2.267 *	2.285 2.290 *	2.299 2.310 *	1.651 1.654 *	1.678 1.681 *	1.698 1.702 *	1.715 1.719 *
	4	2.343 2.348 *	2.399 2.395 *	2.434 2.431 *	2.457 2.462 *	2.008 2.026 *	2.014 2.029 *	2.016 2.032 *	2.017 2.034 *	1.576 1.579 *	1.588 1.589 *	1.595 1.597 *	1.600 1.603 *
0.5	10	2.045 2.057 *	2.137 2.158 *	2.218 2.236 *	2.289 2.301 *	1.961 1.959 *	2.012 2.006 *	2.051 2.043 *	2.078 2.073 *	1.523 1.526 *	1.555 1.560 *	1.582 1.587 *	1.606 1.609 *
	7	2.043 2.047 *	2.126 2.131 *	2.193 2.195 *	2.248 2.249 *	1.883 1.890 *	1.912 1.914 *	1.930 1.932 *	1.941 1.948 *	1.497 1.500 *	1.521 1.524 *	1.539 1.543 *	1.554 1.558 *
	4	1.983 1.984 *	2.027 2.021 *	2.055 2.050 *	2.074 2.074 *	1.689 1.715 *	1.702 1.717 *	1.704 1.719 *	1.704 1.721 *	1.430 1.433 *	1.439 1.441 *	1.446 1.448 *	1.451 1.453 *
1	10	1.746 1.758 *	1.819 1.842 *	1.883 1.906 *	1.940 1.960 *	1.670 1.666 *	1.710 1.704 *	1.739 1.733 *	1.760 1.758 *	1.385 1.384 *	1.412 1.415 *	1.436 1.438 *	1.456 1.458 *
	7	1.743 1.747 *	1.807 1.816 *	1.860 1.868 *	1.903 1.913 *	1.604 1.608 *	1.625 1.627 *	1.638 1.641 *	1.647 1.653 *	1.361 1.361 *	1.381 1.382 *	1.397 1.398 *	1.410 1.412 *
	4	1.685 1.690 *	1.718 1.718 *	1.739 1.741 *	1.753 1.759 *	1.454 1.465 *	1.454 1.467 *	1.454 1.468 *	1.455 1.469 *	1.301 1.301 *	1.309 1.308 *	1.314 1.314 *	1.318 1.318 *

Table 5. Comparison of SCF values obtained from Equations (A1)–(A3) (*) to the corresponding FEM results for $\theta = 45°$.

$\theta = 45°$		K_{tt}				K_{tb}				K_{ts}			
ρ/a	t/a	$T/a = 1$	$T/a = 2$	$T/a = 3$	$T/a = 4$	$T/a = 1$	$T/a = 2$	$T/a = 3$	$T/a = 4$	$T/a = 1$	$T/a = 2$	$T/a = 3$	$T/a = 4$
0.05	10	4.036 4.034 *	4.361 4.394 *	4.645 4.671 *	4.893 4.904 *	3.996 3.982 *	4.249 4.223 *	4.438 4.408 *	4.579 4.564 *	2.485 2.482 *	2.585 2.585 *	2.666 2.663 *	2.732 2.730 *
	7	4.051 4.042 *	4.372 4.378 *	4.640 4.635 *	4.860 4.853 *	3.897 3.881 *	4.084 4.045 *	4.206 4.171 *	4.284 4.278 *	2.449 2.445 *	2.530 2.525 *	2.592 2.586 *	2.639 2.637 *
	4	4.037 4.011 *	4.295 4.253 *	4.478 4.439 *	4.609 4.596 *	3.562 3.562 *	3.630 3.608 *	3.661 3.643 *	3.677 3.673 *	2.340 2.340 *	2.385 2.379 *	2.413 2.408 *	2.433 2.434 *
0.25	10	2.439 2.445 *	2.620 2.642 *	2.778 2.794 *	2.918 2.921 *	2.403 2.400 *	2.541 2.527 *	2.646 2.625 *	2.724 2.708 *	1.807 1.804 *	1.878 1.877 *	1.935 1.932 *	1.982 1.979 *
	7	2.446 2.447 *	2.624 2.629 *	2.773 2.769 *	2.898 2.886 *	2.341 2.336 *	2.441 2.420 *	2.507 2.485 *	2.550 2.540 *	1.780 1.777 *	1.831 1.833 *	1.882 1.876 *	1.918 1.913 *
	4	2.430 2.422 *	2.568 2.547 *	2.666 2.644 *	2.737 2.725 *	2.140 2.142 *	2.174 2.164 *	2.190 2.180 *	2.197 2.194 *	1.702 1.701 *	1.733 1.728 *	1.753 1.749 *	1.767 1.766 *
0.5	10	2.009 2.015 *	2.143 2.163 *	2.263 2.276 *	2.369 2.372 *	1.969 1.966 *	2.068 2.058 *	2.143 2.128 *	2.201 2.188 *	1.584 1.581 *	1.642 1.642 *	1.691 1.688 *	1.730 1.727 *
	7	2.015 2.015 *	2.146 2.150 *	2.257 2.253 *	2.351 2.339 *	1.913 1.911 *	1.983 1.970 *	2.029 2.015 *	2.059 2.053 *	1.560 1.558 *	1.608 1.604 *	1.644 1.639 *	1.672 1.669 *
	4	1.993 1.987 *	2.091 2.076 *	2.161 2.144 *	2.211 2.201 *	1.746 1.753 *	1.767 1.765 *	1.776 1.775 *	1.780 1.784 *	1.492 1.492 *	1.517 1.514 *	1.533 1.530 *	1.544 1.544 *
1	10	1.698 1.703 *	1.793 1.812 *	1.878 1.895 *	1.955 1.966 *	1.651 1.648 *	1.716 1.712 *	1.766 1.761 *	1.803 1.802 *	1.403 1.402 *	1.448 1.450 *	1.486 1.486 *	1.517 1.517 *
	7	1.701 1.699 *	1.791 1.796 *	1.868 1.870 *	1.933 1.933 *	1.600 1.599 *	1.642 1.638 *	1.670 1.667 *	1.688 1.692 *	1.382 1.382 *	1.417 1.417 *	1.445 1.445 *	1.467 1.467 *
	4	1.672 1.665 *	1.732 1.723 *	1.775 1.768 *	1.806 1.806 *	1.462 1.469 *	1.471 1.475 *	1.474 1.480 *	1.476 1.484 *	1.323 1.326 *	1.340 1.341 *	1.351 1.353 *	1.359 1.363 *

Table 6. Comparison of SCF values obtained from Equations (A1)–(A3) (*) to the corresponding FEM results for θ = 60°.

θ = 60°		K_{tt}				K_{tb}				K_{ts}			
ρ/a	t/a	T/a = 1	T/a = 2	T/a = 3	T/a = 4	T/a = 1	T/a = 2	T/a = 3	T/a = 4	T/a = 1	T/a = 2	T/a = 3	T/a = 4
0.05	10	4.087	4.514	4.892	5.228	4.125	4.507	4.808	5.041	2.773	2.935	3.061	3.163
		4.061 *	4.535 *	4.899 *	5.206 *	4.102 *	4.474 *	4.760 *	5.000 *	2.763 *	2.921 *	3.043 *	3.145 *
	7	4.107	4.540	4.911	5.233	4.066	4.382	4.600	4.752	2.736	2.874	2.976	3.051
		4.076 *	4.531 *	4.881 *	5.175 *	4.035 *	4.314 *	4.529 *	4.709 *	2.725 *	2.853 *	2.951 *	3.035 *
	4	4.127	4.517	4.816	5.045	3.795	3.947	4.026	4.070	2.618	2.703	2.756	2.791
		4.081 *	4.457 *	4.746 *	4.990 *	3.772 *	3.879 *	3.961 *	4.030 *	2.612 *	2.682 *	2.736 *	2.782 *
0.25	10	2.378	2.585	2.769	2.935	2.368	2.546	2.686	2.797	1.876	1.978	2.059	2.124
		2.373 *	2.608 *	2.789 *	2.941 *	2.362 *	2.537 *	2.671 *	2.784 *	1.870 *	1.972 *	2.050 *	2.116 *
	7	2.386	2.593	2.773	2.928	2.324	2.465	2.564	2.632	1.850	1.937	2.002	2.050
		2.378 *	2.601 *	2.772 *	2.916 *	2.313 *	2.440 *	2.537 *	2.619 *	1.844 *	1.926 *	1.989 *	2.042 *
	4	2.383	2.563	2.699	2.802	2.156	2.216	2.246	2.262	1.771	1.823	1.855	1.877
		2.368 *	2.544 *	2.678 *	2.792 *	2.148 *	2.191 *	2.224 *	2.252 *	1.768 *	1.812 *	1.846 *	1.875 *
0.5	10	1.963	2.109	2.240	2.359	1.937	2.055	2.148	2.221	1.609	1.687	1.750	1.801
		1.965 *	2.129 *	2.255 *	2.361 *	1.934 *	2.048 *	2.137 *	2.211 *	1.602 *	1.681 *	1.741 *	1.793 *
	7	1.969	2.114	2.239	2.347	1.893	1.981	2.043	2.085	1.586	1.652	1.701	1.738
		1.967 *	2.120 *	2.237 *	2.336 *	1.887 *	1.967 *	2.028 *	2.080 *	1.579 *	1.642 *	1.690 *	1.730 *
	4	1.958	2.075	2.162	2.228	1.744	1.766	1.791	1.799	1.518	1.557	1.579	1.595
		1.948 *	2.062 *	2.149 *	2.223 *	1.745 *	1.768 *	1.785 *	1.800 *	1.515 *	1.548 *	1.572 *	1.593 *
1	10	1.680	1.778	1.867	1.947	1.639	1.709	1.764	1.807	1.410	1.464	1.508	1.545
		1.678 *	1.786 *	1.869 *	1.939 *	1.635 *	1.703 *	1.754 *	1.798 *	1.401 *	1.455 *	1.497 *	1.533 *
	7	1.683	1.777	1.859	1.928	1.593	1.640	1.672	1.694	1.390	1.433	1.466	1.492
		1.676 *	1.773 *	1.848 *	1.911 *	1.590 *	1.632 *	1.664 *	1.692 *	1.381 *	1.422 *	1.454 *	1.481 *
	4	1.657	1.725	1.774	1.809	1.464	1.471	1.476	1.478	1.330	1.352	1.366	1.376
		1.645 *	1.709 *	1.758 *	1.799 *	1.465 *	1.471 *	1.476 *	1.480 *	1.327 *	1.346 *	1.360 *	1.372 *

The same validation was carried out for weld angles θ changing with a step of 2.5°. The maximum error, calculated with respect to the SCF values obtained using the FEM was much lower than 2.5%.

5. Conclusions

Several thousand numerical FEM models of cruciform connections subjected to tensile, bending, and shearing loads were performed. The particular values of SCFs obtained from numerical solutions enabled the derivation of three approximating parametric formulas, covering the dimensions of welded cruciform connections used in real structures. Five characteristic geometrical parameters: t, T, a, ρ, and θ describing the shape of the joint were considered as independent variables. The accuracy of the formulas is higher than 97.5% and covers very wide ranges of application: $0 < \rho/a \leq 1.3$, $0 < a/t \leq 1.3$, $1 \leq T/a \leq 4$, and $30° \leq \theta \leq 60°$, including two limiting cases: for $\rho \to 0$ and when $t \to \infty$. In this way, all possible values of all geometrical parameters used in real cruciform joints are satisfied.

It is also important to note that the approximating equations presented herein agree with the results and formulas published in [32] and are valid for θ = 45°.

The use of such SCFs' solutions is recommended in EN 1993-1-9: 2005 [21] (Design of steel structures, Part 1–9: Fatigue, Sections 5 and 6. See Table A1 case 4 in addition). In such cases, the recommendations in Section 6.3 of ref. [21] state that… "SCF values may be taken from handbooks or from appropriate FE calculations".

The proposed formulas can be easily applied in computer-aided assessment in the fatigue design of cruciform joints, particularly in:

- Comparative studies of the stress concentration in cruciform joints of different dimensions;
- Hot spot approaches;
- The weight function method used for cracks initiated at the weld toe;
- Dealing with the possible mechanical improvements in the weld toe region.

It is important to note that the accuracy of the parametric formulas depends on the similarity of the real joint shape to the one assumed in the present models used in the numerical analysis. Wang et al. [17] have shown that the improper identification or interpretation of geometrical parameters and the shape of the joint may lead to significant errors in SCF estimation. The same may occur when the geometrical parameters' values lay beyond the specified range of validity.

Future work should be focused on local effects appearing in the critical zone where damage processes occur. It is usually assumed that small cracks initiate due to very high local stress ranges and there are at least two possible classes of methods to attack this problem. First, methods based on fracture mechanics may be used. Theoretical, regular, or singular stress fields (when ρ tends to zero) around the apex are averaged in small volumes or areas that are material constants. Second, there are methods of analyzing fatigue crack growth using weight functions, assuming that the fatigue life depends on the propagating crack. However, the aim of the present paper did not cover such problems.

Author Contributions: Conceptualization, K.L.M. Methodology, K.L.M. Software, P.T. Validation, K.L.M. and P.T. Formal analysis, K.L.M. and P.T. Investigation, K.L.M. and P.T. Resources, K.L.M. and P.T. Data curation, P.T. Writing—original draft preparation, K.L.M. Writing—review and editing, K.L.M. and P.T. Visualization, K.L.M. and P.T. Supervision, K.L.M. Project administration, P.T. Funding acquisition, P.T. All authors have read and agreed to the published version of the manuscript.

Funding: This research was financed by the Ministry of Science and Higher Education of Poland with allocation to the Faculty of Mechanical Engineering Bialystok University of Technology for the WZ/WM-IIM/5/2023 academic project in the mechanical engineering discipline.

Institutional Review Board Statement: Not applicable.

Informed Consent Statement: Not applicable.

Data Availability Statement: Data are contained within the article.

Conflicts of Interest: Author Krzysztof L. Molski was employed by the company SaMASZ sp. z o.o. The remaining authors declare that the research was conducted in the absence of any commercial or financial relationships that could be construed as a potential conflict of interest.

Nomenclature

a	weld throat thickness
G	shear modulus
h, h_p	leg lengths
k	thermal conductivity
K_t	stress concentration factor (SCF)
K_{tt}	stress concentration factor for tensile (axial) load
K_{tb}	stress concentration factor for bending load
K_{ts}	stress concentration factor for shearing load
m, p	exponents dependent on the loading mode
n	stress field exponent for a sharp corner for axial and bending load
n_s	stress field exponent for a sharp corner subjected to shearing load
P	regular function represented by polynomials
P_t, P_b, P_s	P functions corresponding to tensile, bending, and shearing load, respectively
Q	magnitude of the heat flux
$q_{\max}$	magnitude of the maximum heat flux
q_{nom}	magnitude of the nominal heat flux at the end of the body
t	thickness of the main plate
T	thickness of the attachment plates
T_{temp}	temperature
W	displacement component corresponding to the anti-plane shear
$X = \rho/(\rho + a)$	normalized toe radius parameter
$Y = a/(a + t)$	normalized weld thickness parameter

$Z = T/a$	normalized attachment plate thickness parameter
Z_0	reference T/a value
$\mid\nabla\Psi\mid$	magnitude of the displacement (temperature) gradient
$\mid\nabla\Psi\mid_{max}$	magnitude of the maximum displacement (temperature) gradient
$\mid\nabla\Psi\mid_{nom}$	magnitude of the nominal displacement (temperature) gradient
θ	weld angle
κ	correction function for the attachment plate thickness T/a
κ_t	correction function κ for tensile (axial) load
κ_b	correction function κ for bending load
κ_s	correction function κ for shearing load
$\Psi(x,y)$	potential function representing temperature or anti-plane displacement
$\partial\Psi/\partial n$	normal derivative at the bounding contour
ρ	weld toe radius
σ_1	first principal stress for tensile or bending load
σ_{1max}	maximum value of the first principal stress
σ_t	nominal tensile (axial) stress
σ_b	nominal bending stress
τ_{max}	maximum shear stress at the weld toe due to shearing load
τ_s	nominal shear stress

Appendix A

Approximating SCF formulas for the cruciform connections subjected to tensile, bending, and shearing loads

Independent variables: ρ, a, θ, t, T;

Normalized quantities: $X = \rho/(\rho + a)$, $Y = a/(a + t)$; $Z = T/a$;

Range of validity: $0 < \rho/a \leq 1.3$; $0 < a/t \leq 1.3$; $1 \leq T/a \leq 4$; $30° \leq \theta \leq 60°$.

In all formulas, θ is expressed in radians.

Exponent of the singular term for:

axial and bending load

$$n = \frac{-0.636620\theta - 0.093300\theta^2}{1 + 0.776350\theta + 0.040750\theta^{1.5} - 0.004990\theta^2 + 0.133650\theta^{2.5}}$$

shearing load

$$n_s = \frac{-\theta}{\theta + \pi}$$

Accuracy: maximum percentage error lower than 2.5% compared to the FEM results.

Tensile (axial) load

$$K_{tt} = X^n\left(A_0^t + A_1^t X + A_2^t X^2 + A_3^t X^3 + A_4^t X^4\right)\kappa_t \tag{A1}$$

where

$$A_0^t = A_{00}^t + A_{01}^t Y + A_{02}^t Y^2 + A_{03}^t Y^3 + A_{04}^t Y^4$$

$$A_{00}^t = 2.031 - 0.635\theta - 0.098\theta^4$$

$$A_{01}^t = 2.525 - 3.464\theta + 0.818\theta^4$$

$$A_{02}^t = -23.741 + 37.377\theta - 10.314\theta^4$$

$$A_{03}^t = 44.889 - 87.433\theta + 28.653\theta^4$$

$$A_{04}^t = -26.097 + 60.519\theta - 22.554\theta^4$$

$$A_1^t = A_{10}^t + A_{11}^t Y + A_{12}^t Y^2 + A_{13}^t Y^3 + A_{14}^t Y^4$$

$$A_{10}^t = -0.137 - 0.766\theta + 0.877\theta^4$$

$$A_{11}^t = -0.159 + 0.794\theta + 0.231\theta^4$$

$$A_{12}^{t} = 7.915 - 17.261\theta + 2.065\theta^{4}$$
$$A_{13}^{t} = -22.898 + 53.617\theta - 17.877\theta^{4}$$
$$A_{14}^{t} = 5.164 - 41.774\theta^{2} + 30.281\theta^{4}$$
$$A_{2}^{t} = A_{20}^{t} + A_{21}^{t}Y + A_{22}^{t}Y^{2} + A_{23}^{t}Y^{3} + A_{24}^{t}Y^{4}$$
$$A_{20}^{t} = -0.545 + 3.275\theta^{2} - 2.551\theta^{4}$$
$$A_{21}^{t} = -3.743 + 7.793\theta^{2} - 5.112\theta^{4}$$
$$A_{22}^{t} = 17.111 - 52.62\theta^{2} + 39.271\theta^{4}$$
$$A_{23}^{t} = -19.451 + 73.353\theta^{2} - 58.512\theta^{4}$$
$$A_{24}^{t} = 6.94 - 32.1\theta^{2} + 26.732\theta^{4}$$
$$A_{3}^{t} = A_{30}^{t} + A_{31}^{t}Y + A_{32}^{t}Y^{2} + A_{33}^{t}Y^{3} + A_{34}^{t}Y^{4}$$
$$A_{30}^{t} = 0.198 - 3.351\theta^{2} + 2.74\theta^{4}$$
$$A_{31}^{t} = 7.209 - 13.931\theta^{2} + 7.985\theta^{4}$$
$$A_{32}^{t} = -38.74 + 114.409\theta^{2} - 69.799\theta^{4}$$
$$A_{33}^{t} = 70.504 - 266.94\theta^{2} + 155.111\theta^{4}$$
$$A_{34}^{t} = -53.268 + 231.865\theta^{2} - 126.124\theta^{4}$$
$$A_{4}^{t} = A_{40}^{t} + A_{41}^{t}Y + A_{42}^{t}Y^{2} + A_{43}^{t}Y^{3} + A_{44}^{t}Y^{4}$$
$$A_{40}^{t} = 4.321 - 15.621\theta + 16.874\theta^{2} - 5.485\theta^{4}$$
$$A_{41}^{t} = -5.321 + 12.0\theta^{2} - 5.31\theta^{4}$$
$$A_{42}^{t} = 28.713 - 110.239\theta^{2} + 48.946\theta^{4}$$
$$A_{43}^{t} = -58.904 + 306.19\theta^{2} - 131.962\theta^{4}$$
$$A_{44}^{t} = 53.279 - 285.273\theta^{2} + 122.094\theta^{4}$$

$$\kappa_{t} = 1 + \left(\sqrt{Z} - 1\right)\left\{1 - \left[\left(-0.866 + 2.143\theta - 0.452\theta^{2}\right) + \left(2.49 + 15.12\theta - 9.25\theta^{2}\right)Y^{2}\right]X\right\}\cdot$$
$$\cdot Exp\left\{-\left[\left(9.675 - 11.567\theta + 4.966\theta^{2}\right)Y\right]^{2.7} - \left(3.61 - 4.05\theta + 1.66\theta^{2}\right)\right\}$$

Bending load

$$K_{tb} = X^{n}\left(A_{0}^{b} + A_{1}^{b}X + A_{2}^{b}X^{2} + A_{3}^{b}X^{3} + A_{4}^{b}X^{4}\right)\kappa_{b} \quad \text{(A2)}$$

where

$$A_{0}^{b} = A_{00}^{b} + A_{01}^{b}Y + A_{02}^{b}Y^{2} + A_{03}^{b}Y^{3} + A_{04}^{b}Y^{4}$$
$$A_{00}^{b} = 1.904 - 0.751\theta^{2} + 0.132\theta^{4}$$
$$A_{01}^{b} = -0.682 + 4.496\theta^{2} - 2.507\theta^{4}$$
$$A_{02}^{b} = -13.6 - 7.777\theta^{2} + 9.199\theta^{4}$$
$$A_{03}^{b} = 41.72 - 12.838\theta^{2} - 7.467\theta^{4}$$
$$A_{04}^{b} = -35.686 + 24.313\theta^{2} - 1.331\theta^{4}$$
$$A_{1}^{b} = A_{10}^{b} + A_{11}^{b}Y + A_{12}^{b}Y^{2} + A_{13}^{b}Y^{3} + A_{14}^{b}Y^{4}$$
$$A_{10}^{b} = -0.098 - 0.801\theta + 0.916\theta^{4}$$
$$A_{11}^{b} = 0.767 - 1.604\theta - 0.888\theta^{4}$$
$$A_{12}^{b} = -0.489 + 11.177\theta - 8.132\theta^{4}$$

$$A^b_{13} = -8.508 - 10.312\theta + 24.849\theta^4$$
$$A^b_{14} = 10.8 + 0.608\theta^2 - 19.056\theta^4$$
$$A^b_2 = A^b_{20} + A^b_{21}Y + A^b_{22}Y^2 + A^b_{23}Y^3 + A^b_{24}Y^4$$
$$A^b_{20} = -0.797 + 3.839\theta^2 - 2.973\theta^4$$
$$A^b_{21} = 2.411 - 14.024\theta^2 + 12.923\theta^4$$
$$A^b_{22} = -0.091 + 17.933\theta^2 - 23.494\theta^4$$
$$A^b_{23} = -13.593 + 25.528\theta^2 + 5.148\theta^4$$
$$A^b_{24} = 15.525 - 23.097\theta^2 - 6.363\theta^4$$
$$A^b_3 = A^b_{30} + A^b_{31}Y + A^b_{32}Y^2 + A^b_{33}Y^3 + A^b_{34}Y^4$$
$$A^b_{30} = 0.664 - 4.494\theta^2 + 3.458\theta^4$$
$$A^b_{31} = -4.748 + 24.121\theta^2 - 19.574\theta^4$$
$$A^b_{32} = 7.857 - 67.243\theta^2 + 55.928\theta^4$$
$$A^b_{33} = 2.829 + 135.344\theta^2 - 117.075\theta^4$$
$$A^b_{34} = 5.16 - 195.991\theta^2 + 157.608\theta^4$$
$$A^b_4 = A^b_{40} + A^b_{41}Y + A^b_{42}Y^2 + A^b_{43}Y^3 + A^b_{44}Y^4$$
$$A^b_{40} = -0.381 + 2.717\theta^2 - 2.003\theta^4$$
$$A^b_{41} = 1.613 - 14.683\theta^2 + 9.871\theta^4$$
$$A^b_{42} = 3.506 + 57.253\theta^2 - 35.678\theta^4$$
$$A^b_{43} = -3.197 - 194.43\theta^2 + 126.77\theta^4$$
$$A^b_{44} = -21.977 + 268.935\theta^2 - 178.897\theta^4$$
$$\kappa_b = 1 + \left(\sqrt{Z} - 1\right)\{1 - [(-0.73 + 1.85\theta - 0.2\theta^2) + (5.28 + 37.34\theta - 23.44\theta^2)Y^2]X\}\cdot$$
$$\cdot Exp\left\{-[(16.73 - 19.24\theta + 8.56\theta^2)Y]^{2.0} - (3.86 - 4.35\theta + 1.7\theta^2)\right\}$$

Shearing load

$$K_{ts} = X^{n_s}\left(A^s_0 + A^s_1X + A^s_2X^2 + A^s_3X^3 + A^s_4X^4\right)\kappa_s \tag{A3}$$

where

$$A^s_0 = A^s_{00} + A^s_{01}Y^2 + A^s_{02}Y^3 + A^s_{03}Y^4$$
$$A^s_{00} = 1.3265 + 0.3143\theta - 0.3\theta^2$$
$$A^s_{01} = -7.2 + 5.91\theta - 1.8355\theta^2$$
$$A^s_{02} = 15.53 - 28.5882\theta^2 + 17.0\theta^3$$
$$A^s_{03} = -11.6 + 18.3\theta^2 - 8.1207\theta^4$$
$$A^s_1 = -0.178 - 0.1095\theta^2 + \left(0.1422 + 0.2533\theta^2\right)Y^2$$
$$A^s_2 = A^s_{20} + A^s_{21}Y + A^s_{22}Y^2$$
$$A^s_{20} = 2.9022 - 8.3702\theta + 5.5285\theta^2$$
$$A^s_{21} = 1.0 - 1.3728\theta^2$$
$$A^s_{22} = -2.8318 + 5.1835\theta^2$$

$$A_3^s = A_{30}^s + A_{31}^s Y + A_{32}^s Y^3$$

$$A_{30}^s = -7.88 + 21.2254\theta - 13.2803\theta^2$$

$$A_{31}^s = -7.0 + 18.1182\theta - 12.0\theta^2$$

$$A_{32}^s = 11.7633 - 11.965\theta^2$$

$$A_4^s = A_{40}^s + A_{41}^s Y + A_{42}^s Y^2 + A_{43}^s Y^3$$

$$A_{40}^s = 8.1273 - 25.1758\theta + 22.4116\theta^2 - 5.4681\theta^3$$

$$A_{41}^s = 1.6654 - 17.8143\theta^2 + 17.908\theta^3$$

$$A_{42}^s = 8.7364 - 11.4037\theta^3$$

$$A_{43}^s = -18.3578 + 19.0842\theta^3$$

$$\kappa_s = 1 + \left(\sqrt{Z} - 1\right)\left\{1 - \left[\left(-0.40 + 0.67\theta + 0.70\theta^2\right) + \left(-4.17 + 18.54\theta - 6.94\theta^2\right)\left(\tfrac{2 \cdot Y}{1+Y}\right)^2\right] X^2\right\} \cdot$$

$$\cdot Exp\left\{-\left[\left(6.26 - 5.74\theta + 2.52\theta^2\right)\tfrac{2 \cdot Y}{1+Y}\right]^{2.0} - \left(3.84 - 3.31\theta + 1.23\theta^2\right)\right\}$$

Appendix B

Numerical SCF results for various weld angles θ for a cruciform joint subjected to tensile, bending, and shearing loads, while T/a = 1.

Table A1. K_{tt} values for a cruciform joint, while θ = 30° and T/a = 1. Tensile load.

θ = 30°	X															
Y	**0.010**	**0.018**	**0.025**	**0.032**	**0.050**	**0.079**	**0.100**	**0.150**	**0.200**	**0.250**	**0.300**	**0.350**	**0.400**	**0.450**	**0.500**	**0.562**
0.010	5.323	4.598	4.215	3.967	3.522	3.115	2.927	2.611	2.398	2.237	2.106	1.996	1.899	1.813	1.735	1.645
0.018	5.324	4.600	4.216	3.969	3.523	3.116	2.927	2.612	2.399	2.237	2.107	1.996	1.900	1.814	1.735	1.645
0.025	5.324	4.602	4.217	3.971	3.525	3.117	2.928	2.613	2.400	2.238	2.108	1.997	1.900	1.814	1.736	1.646
0.032	5.326	4.604	4.220	3.972	3.526	3.119	2.930	2.614	2.401	2.239	2.108	1.998	1.901	1.815	1.737	1.647
0.050	5.326	4.603	4.221	3.974	3.530	3.122	2.933	2.618	2.404	2.243	2.112	2.001	1.904	1.818	1.739	1.649
0.079	5.342	4.618	4.235	3.986	3.541	3.132	2.943	2.626	2.411	2.249	2.118	2.007	1.910	1.823	1.744	1.653
0.100	5.347	4.625	4.240	3.991	3.545	3.136	2.946	2.629	2.414	2.252	2.120	2.009	1.912	1.824	1.745	1.654
0.150	5.318	4.597	4.214	3.967	3.523	3.117	2.929	2.612	2.399	2.237	2.106	1.994	1.897	1.810	1.730	1.638
0.200	5.199	4.492	4.118	3.877	3.444	3.046	2.862	2.552	2.344	2.185	2.056	1.946	1.850	1.764	1.686	1.593
0.250	5.009	4.326	3.967	3.734	3.316	2.933	2.755	2.458	2.256	2.102	1.978	1.872	1.779	1.695	1.618	1.529
0.300	4.786	4.133	3.786	3.565	3.166	2.800	2.630	2.346	2.153	2.006	1.887	1.786	1.697	1.617	1.543	1.458
0.350	4.543	3.930	3.601	3.390	3.010	2.662	2.500	2.230	2.047	1.907	1.794	1.698	1.614	1.538	1.468	1.388
0.400	4.327	3.732	3.420	3.219	2.859	2.528	2.375	2.118	1.944	1.812	1.705	1.614	1.534	1.463	1.398	1.324
0.450	4.108	3.548	3.250	3.060	2.716	2.402	2.256	2.013	1.848	1.723	1.621	1.536	1.461	1.395	1.334	1.267
0.500	3.912	3.372	3.089	2.908	2.582	2.283	2.145	1.914	1.757	1.639	1.543	1.463	1.393	1.332	1.278	1.218
0.562	3.670	3.166	2.902	2.730	2.423	2.144	2.014	1.797	1.652	1.542	1.454	1.380	1.318	1.264	1.217	1.168

Table A2. K_{tt} values for a cruciform joint, while θ = 40° and T/a = 1. Tensile load.

θ = 40°	X															
Y	**0.010**	**0.018**	**0.025**	**0.032**	**0.050**	**0.079**	**0.100**	**0.150**	**0.200**	**0.250**	**0.300**	**0.350**	**0.400**	**0.450**	**0.500**	**0.562**
0.010	6.365	5.322	4.791	4.452	3.853	3.322	3.081	2.690	2.434	2.246	2.098	1.976	1.873	1.783	1.702	1.614
0.018	6.366	5.324	4.792	4.453	3.854	3.323	3.082	2.691	2.435	2.247	2.098	1.977	1.873	1.783	1.703	1.614
0.025	6.368	5.326	4.793	4.454	3.855	3.324	3.083	2.691	2.436	2.247	2.099	1.977	1.874	1.783	1.703	1.615
0.032	6.369	5.325	4.795	4.455	3.856	3.324	3.084	2.692	2.436	2.248	2.100	1.978	1.874	1.784	1.704	1.615
0.050	6.363	5.327	4.795	4.456	3.858	3.327	3.087	2.695	2.439	2.251	2.102	1.980	1.877	1.786	1.706	1.617
0.079	6.384	5.346	4.809	4.470	3.870	3.338	3.096	2.704	2.447	2.258	2.109	1.986	1.882	1.791	1.711	1.621
0.100	6.398	5.359	4.821	4.480	3.879	3.345	3.103	2.709	2.452	2.262	2.113	1.990	1.885	1.794	1.713	1.623
0.150	6.415	5.369	4.830	4.489	3.887	3.351	3.109	2.713	2.454	2.263	2.113	1.989	1.883	1.791	1.708	1.616
0.200	6.353	5.318	4.784	4.446	3.849	3.319	3.078	2.685	2.427	2.237	2.086	1.962	1.856	1.762	1.679	1.584
0.250	6.203	5.193	4.672	4.342	3.759	3.239	3.004	2.619	2.366	2.179	2.031	1.908	1.803	1.710	1.626	1.532
0.300	5.983	5.009	4.506	4.188	3.624	3.123	2.895	2.524	2.279	2.097	1.953	1.833	1.731	1.640	1.559	1.466
0.350	5.718	4.787	4.305	4.001	3.463	2.984	2.766	2.410	2.175	2.001	1.863	1.748	1.649	1.563	1.485	1.397
0.400	5.433	4.546	4.090	3.801	3.289	2.834	2.627	2.289	2.065	1.900	1.769	1.660	1.566	1.484	1.411	1.330
0.450	5.140	4.302	3.870	3.596	3.112	2.681	2.485	2.166	1.955	1.798	1.675	1.572	1.485	1.409	1.342	1.270
0.500	4.853	4.060	3.653	3.394	2.938	2.531	2.346	2.045	1.846	1.700	1.584	1.489	1.409	1.340	1.280	1.217
0.562	4.506	3.769	3.390	3.150	2.726	2.350	2.178	1.900	1.717	1.583	1.478	1.393	1.323	1.264	1.215	1.166

Table A3. K_{tt} values for a cruciform joint, while $\theta = 50°$ and $T/a = 1$. Tensile load.

$\theta = 50°$								X								
Y	**0.010**	**0.018**	**0.025**	**0.032**	**0.050**	**0.079**	**0.100**	**0.150**	**0.200**	**0.250**	**0.300**	**0.350**	**0.400**	**0.450**	**0.500**	**0.562**
0.010	7.054	5.749	5.097	4.688	3.982	3.373	3.103	2.676	2.405	2.211	2.061	1.941	1.840	1.754	1.678	1.596
0.018	7.055	5.750	5.097	4.689	3.983	3.373	3.104	2.676	2.405	2.211	2.061	1.941	1.840	1.754	1.679	1.596
0.025	7.056	5.751	5.099	4.690	3.984	3.374	3.104	2.677	2.406	2.211	2.062	1.941	1.841	1.754	1.679	1.597
0.032	7.059	5.753	5.100	4.691	3.985	3.375	3.105	2.678	2.406	2.212	2.062	1.942	1.841	1.755	1.680	1.597
0.050	7.042	5.749	5.098	4.691	3.986	3.377	3.108	2.680	2.409	2.214	2.065	1.944	1.843	1.757	1.681	1.599
0.079	7.065	5.766	5.113	4.704	3.997	3.387	3.116	2.687	2.415	2.220	2.070	1.949	1.848	1.761	1.686	1.603
0.100	7.084	5.780	5.126	4.716	4.007	3.395	3.124	2.693	2.421	2.225	2.074	1.953	1.851	1.764	1.688	1.605
0.150	7.121	5.811	5.151	4.740	4.027	3.411	3.138	2.705	2.429	2.232	2.079	1.956	1.853	1.764	1.686	1.600
0.200	7.105	5.800	5.142	4.730	4.018	3.402	3.128	2.694	2.417	2.218	2.064	1.939	1.834	1.743	1.663	1.574
0.250	7.012	5.724	5.074	4.668	3.963	3.353	3.082	2.651	2.376	2.177	2.023	1.897	1.792	1.700	1.618	1.526
0.300	6.837	5.579	4.943	4.548	3.860	3.264	2.999	2.577	2.306	2.110	1.958	1.834	1.729	1.638	1.556	1.465
0.350	6.588	5.376	4.763	4.381	3.718	3.143	2.887	2.478	2.215	2.025	1.877	1.756	1.654	1.565	1.486	1.398
0.400	6.296	5.136	4.551	4.186	3.551	3.000	2.755	2.363	2.111	1.929	1.787	1.671	1.573	1.488	1.413	1.331
0.450	5.975	4.873	4.318	3.971	3.368	2.845	2.612	2.240	2.001	1.827	1.693	1.583	1.491	1.412	1.344	1.270
0.500	5.640	4.599	4.075	3.747	3.178	2.685	2.465	2.114	1.888	1.725	1.599	1.497	1.413	1.342	1.281	1.217
0.562	5.214	4.251	3.766	3.464	2.938	2.483	2.280	1.956	1.749	1.601	1.487	1.397	1.323	1.263	1.214	1.165

Table A4. K_{tt} values for a cruciform joint, while $\theta = 60°$ and $T/a = 1$. Tensile load.

$\theta = 60°$								X								
Y	**0.010**	**0.018**	**0.025**	**0.032**	**0.050**	**0.079**	**0.100**	**0.150**	**0.200**	**0.250**	**0.300**	**0.350**	**0.400**	**0.450**	**0.500**	**0.562**
0.010	7.381	5.922	5.194	4.746	3.984	3.343	3.065	2.633	2.364	2.175	2.031	1.916	1.821	1.741	1.671	1.595
0.018	7.403	5.923	5.196	4.747	3.984	3.343	3.066	2.633	2.365	2.175	2.032	1.917	1.822	1.741	1.671	1.595
0.025	7.405	5.925	5.196	4.748	3.985	3.344	3.066	2.634	2.365	2.176	2.032	1.917	1.822	1.741	1.671	1.596
0.032	7.407	5.924	5.198	4.749	3.986	3.344	3.067	2.634	2.366	2.176	2.033	1.918	1.822	1.742	1.672	1.596
0.050	7.392	5.920	5.195	4.748	3.987	3.347	3.068	2.636	2.368	2.179	2.034	1.920	1.824	1.744	1.674	1.598
0.079	7.409	5.936	5.208	4.761	3.998	3.355	3.077	2.643	2.374	2.184	2.039	1.924	1.829	1.748	1.678	1.602
0.100	7.427	5.950	5.220	4.772	4.007	3.363	3.083	2.648	2.378	2.188	2.043	1.928	1.832	1.751	1.680	1.604
0.150	7.479	5.988	5.254	4.801	4.031	3.382	3.100	2.661	2.389	2.196	2.050	1.932	1.835	1.752	1.679	1.599
0.200	7.496	6.001	5.265	4.810	4.037	3.384	3.101	2.658	2.383	2.188	2.039	1.919	1.819	1.733	1.657	1.573
0.250	7.449	5.964	5.229	4.777	4.006	3.355	3.071	2.629	2.352	2.155	2.005	1.883	1.781	1.693	1.614	1.526
0.300	7.319	5.857	5.135	4.690	3.930	3.287	3.007	2.569	2.294	2.098	1.948	1.826	1.723	1.634	1.554	1.465
0.350	7.107	5.688	4.986	4.552	3.812	3.185	2.912	2.483	2.213	2.021	1.873	1.752	1.651	1.563	1.485	1.398
0.400	6.841	5.469	4.793	4.375	3.662	3.057	2.793	2.378	2.117	1.931	1.788	1.670	1.572	1.488	1.414	1.331
0.450	6.524	5.215	4.570	4.171	3.489	2.911	2.658	2.261	2.011	1.833	1.695	1.585	1.493	1.413	1.344	1.270
0.500	6.178	4.937	4.326	3.948	3.302	2.753	2.514	2.137	1.900	1.732	1.602	1.499	1.414	1.342	1.281	1.217
0.562	5.719	4.571	4.004	3.655	3.056	2.548	2.326	1.978	1.760	1.606	1.489	1.397	1.324	1.263	1.214	1.165

Table A5. K_{tb} values for a cruciform joint, while $\theta = 30°$ and $T/a = 1$. Bending load.

$\theta = 30°$								X								
Y	**0.010**	**0.018**	**0.025**	**0.032**	**0.050**	**0.079**	**0.100**	**0.150**	**0.200**	**0.250**	**0.300**	**0.350**	**0.400**	**0.450**	**0.500**	**0.562**
0.010	5.322	4.600	4.218	3.971	3.527	3.121	2.931	2.616	2.402	2.241	2.110	1.999	1.903	1.816	1.737	1.648
0.018	5.326	4.603	4.221	3.974	3.530	3.123	2.933	2.617	2.405	2.242	2.111	2.000	1.903	1.817	1.738	1.647
0.025	5.324	4.603	4.221	3.973	3.529	3.122	2.933	2.617	2.404	2.242	2.110	2.000	1.903	1.816	1.737	1.646
0.032	5.320	4.600	4.217	3.970	3.526	3.120	2.931	2.615	2.401	2.240	2.109	1.998	1.901	1.814	1.735	1.644
0.050	5.295	4.576	4.196	3.951	3.509	3.105	2.916	2.602	2.389	2.228	2.098	1.987	1.890	1.804	1.725	1.634
0.079	5.193	4.488	4.116	3.875	3.441	3.044	2.860	2.551	2.343	2.185	2.056	1.948	1.853	1.767	1.689	1.600
0.100	5.087	4.399	4.033	3.797	3.372	2.983	2.802	2.500	2.296	2.140	2.015	1.908	1.815	1.731	1.654	1.566
0.150	4.772	4.125	3.781	3.560	3.162	2.797	2.628	2.345	2.153	2.008	1.890	1.790	1.703	1.625	1.553	1.471
0.200	4.448	3.842	3.523	3.317	2.946	2.606	2.449	2.185	2.008	1.873	1.763	1.672	1.591	1.519	1.454	1.380
0.250	4.159	3.592	3.294	3.101	2.754	2.437	2.290	2.045	1.879	1.754	1.653	1.569	1.495	1.430	1.372	1.306
0.300	3.908	3.375	3.093	2.912	2.590	2.292	2.151	1.925	1.770	1.654	1.558	1.481	1.414	1.356	1.304	1.247
0.350	3.685	3.189	2.922	2.751	2.444	2.164	2.034	1.818	1.674	1.566	1.480	1.409	1.348	1.296	1.250	1.201
0.400	3.504	3.023	2.771	2.609	2.318	2.053	1.930	1.727	1.592	1.492	1.413	1.348	1.294	1.247	1.208	1.165
0.450	3.333	2.880	2.638	2.485	2.208	1.956	1.840	1.649	1.522	1.429	1.357	1.298	1.250	1.209	1.174	1.138
0.500	3.170	2.733	2.505	2.359	2.097	1.859	1.749	1.570	1.452	1.367	1.301	1.249	1.206	1.171	1.141	1.111
0.562	2.974	2.567	2.354	2.216	1.970	1.749	1.647	1.482	1.376	1.300	1.243	1.198	1.162	1.133	1.110	1.085

Table A6. K_{tb} values for a cruciform joint, while θ = 40° and T/a = 1. Bending load.

θ = 40°	X															
Y	**0.010**	**0.018**	**0.025**	**0.032**	**0.050**	**0.079**	**0.100**	**0.150**	**0.200**	**0.250**	**0.300**	**0.350**	**0.400**	**0.450**	**0.500**	**0.562**
0.010	6.370	5.331	4.799	4.460	3.862	3.330	3.089	2.697	2.440	2.252	2.103	1.981	1.876	1.786	1.706	1.616
0.018	6.380	5.341	4.807	4.468	3.868	3.336	3.094	2.701	2.444	2.255	2.106	1.983	1.879	1.787	1.706	1.617
0.025	6.387	5.347	4.812	4.472	3.872	3.339	3.097	2.704	2.446	2.256	2.107	1.984	1.879	1.788	1.707	1.617
0.032	6.391	5.350	4.815	4.475	3.874	3.341	3.099	2.705	2.447	2.257	2.107	1.983	1.879	1.787	1.706	1.616
0.050	6.388	5.347	4.812	4.472	3.872	3.339	3.096	2.702	2.444	2.253	2.103	1.979	1.874	1.782	1.700	1.609
0.079	6.323	5.294	4.763	4.427	3.832	3.304	3.063	2.673	2.416	2.227	2.078	1.954	1.849	1.757	1.675	1.583
0.100	6.239	5.224	4.700	4.367	3.781	3.259	3.022	2.636	2.382	2.195	2.047	1.925	1.821	1.729	1.648	1.557
0.150	5.930	4.963	4.465	4.149	3.592	3.095	2.870	2.503	2.262	2.083	1.942	1.825	1.726	1.638	1.560	1.473
0.200	5.543	4.640	4.174	3.879	3.358	2.894	2.684	2.341	2.115	1.949	1.818	1.709	1.617	1.536	1.465	1.385
0.250	5.156	4.317	3.884	3.609	3.125	2.694	2.499	2.181	1.972	1.819	1.698	1.599	1.516	1.443	1.380	1.310
0.300	4.799	4.016	3.614	3.364	2.913	2.512	2.331	2.033	1.844	1.701	1.591	1.504	1.429	1.365	1.310	1.248
0.350	4.487	3.754	3.378	3.139	2.719	2.347	2.179	1.906	1.729	1.601	1.502	1.422	1.356	1.300	1.252	1.201
0.400	4.211	3.522	3.169	2.946	2.552	2.204	2.048	1.795	1.633	1.516	1.426	1.355	1.297	1.248	1.208	1.165
0.450	3.966	3.317	2.984	2.774	2.405	2.080	1.934	1.700	1.550	1.444	1.364	1.301	1.251	1.209	1.174	1.138
0.500	3.724	3.115	2.803	2.606	2.261	1.958	1.822	1.606	1.471	1.376	1.305	1.250	1.206	1.170	1.141	1.111
0.562	3.451	2.886	2.598	2.416	2.099	1.821	1.698	1.505	1.385	1.303	1.243	1.198	1.162	1.133	1.109	1.085

Table A7. K_{tb} values for a cruciform joint, while θ = 50° and T/a = 1. Bending load.

θ = 50°	X															
Y	**0.010**	**0.018**	**0.025**	**0.032**	**0.050**	**0.079**	**0.100**	**0.150**	**0.200**	**0.250**	**0.300**	**0.350**	**0.400**	**0.450**	**0.500**	**0.562**
0.010	7.062	5.762	5.109	4.701	3.994	3.383	3.113	2.684	2.412	2.217	2.066	1.945	1.844	1.757	1.681	1.598
0.018	7.080	5.777	5.122	4.713	4.004	3.391	3.120	2.690	2.416	2.220	2.069	1.948	1.846	1.758	1.682	1.599
0.025	7.094	5.789	5.132	4.722	4.012	3.397	3.126	2.694	2.420	2.223	2.071	1.949	1.847	1.759	1.683	1.599
0.032	7.105	5.798	5.141	4.729	4.017	3.402	3.129	2.697	2.422	2.224	2.072	1.949	1.847	1.759	1.682	1.598
0.050	7.121	5.813	5.154	4.742	4.027	3.409	3.135	2.700	2.424	2.225	2.072	1.948	1.845	1.756	1.678	1.593
0.079	7.099	5.792	5.134	4.723	4.011	3.394	3.120	2.685	2.408	2.208	2.055	1.930	1.826	1.736	1.657	1.571
0.100	7.044	5.746	5.094	4.686	3.978	3.365	3.093	2.660	2.384	2.186	2.032	1.907	1.803	1.713	1.634	1.547
0.150	6.780	5.531	4.902	4.509	3.826	3.235	2.972	2.553	2.286	2.093	1.944	1.823	1.720	1.632	1.555	1.469
0.200	6.390	5.215	4.621	4.250	3.607	3.049	2.801	2.405	2.153	1.970	1.829	1.716	1.619	1.537	1.464	1.385
0.250	5.958	4.862	4.309	3.963	3.363	2.844	2.613	2.245	2.010	1.842	1.712	1.607	1.520	1.445	1.380	1.310
0.300	5.528	4.517	4.002	3.682	3.127	2.644	2.427	2.092	1.876	1.719	1.604	1.510	1.430	1.364	1.308	1.250
0.350	5.141	4.190	3.714	3.417	2.902	2.458	2.261	1.950	1.753	1.614	1.509	1.425	1.357	1.300	1.252	1.201
0.400	4.782	3.901	3.458	3.183	2.705	2.294	2.113	1.827	1.649	1.524	1.430	1.356	1.297	1.248	1.207	1.164
0.450	4.473	3.645	3.232	2.975	2.532	2.150	1.984	1.723	1.561	1.449	1.366	1.301	1.250	1.208	1.174	1.138
0.500	4.160	3.394	3.011	2.773	2.362	2.011	1.858	1.621	1.476	1.377	1.305	1.249	1.206	1.170	1.141	1.111
0.562	3.814	3.112	2.762	2.546	2.174	1.858	1.721	1.512	1.387	1.303	1.243	1.197	1.161	1.133	1.109	1.085

Table A8. K_{tb} values for a cruciform joint, while θ = 60° and T/a = 1. Bending load.

θ = 60°	X															
Y	**0.010**	**0.018**	**0.025**	**0.032**	**0.050**	**0.079**	**0.100**	**0.150**	**0.200**	**0.250**	**0.300**	**0.350**	**0.400**	**0.450**	**0.500**	**0.562**
0.010	7.417	5.938	5.211	4.762	3.999	3.355	3.076	2.642	2.372	2.181	2.036	1.921	1.825	1.744	1.673	1.597
0.018	7.442	5.959	5.229	4.778	4.011	3.365	3.084	2.648	2.376	2.185	2.040	1.923	1.827	1.745	1.675	1.598
0.025	7.464	5.976	5.242	4.791	4.021	3.372	3.091	2.652	2.380	2.188	2.042	1.925	1.828	1.746	1.675	1.598
0.032	7.481	5.989	5.254	4.801	4.029	3.378	3.096	2.656	2.382	2.190	2.043	1.926	1.829	1.746	1.675	1.597
0.050	7.521	6.021	5.281	4.825	4.047	3.392	3.107	2.663	2.387	2.192	2.044	1.925	1.827	1.744	1.671	1.592
0.079	7.532	6.029	5.287	4.829	4.048	3.389	3.103	2.656	2.377	2.180	2.030	1.910	1.810	1.725	1.651	1.570
0.100	7.504	6.006	5.265	4.809	4.030	3.372	3.085	2.638	2.360	2.162	2.011	1.891	1.790	1.704	1.629	1.546
0.150	7.307	5.844	5.122	4.675	3.915	3.271	2.990	2.551	2.277	2.082	1.933	1.814	1.714	1.628	1.552	1.469
0.200	6.953	5.560	4.872	4.447	3.721	3.106	2.838	2.418	2.156	1.970	1.827	1.712	1.617	1.535	1.464	1.385
0.250	6.517	5.212	4.566	4.168	3.487	2.910	2.659	2.265	2.019	1.846	1.713	1.607	1.519	1.445	1.380	1.310
0.300	6.058	4.847	4.247	3.877	3.244	2.705	2.472	2.112	1.886	1.727	1.606	1.508	1.432	1.367	1.308	1.250
0.350	5.611	4.487	3.932	3.591	3.006	2.514	2.300	1.967	1.761	1.618	1.510	1.425	1.357	1.300	1.251	1.201
0.400	5.201	4.160	3.647	3.331	2.793	2.339	2.143	1.840	1.654	1.526	1.429	1.356	1.297	1.248	1.207	1.164
0.450	4.838	3.867	3.392	3.099	2.602	2.185	2.006	1.730	1.563	1.449	1.365	1.301	1.250	1.208	1.174	1.138
0.500	4.475	3.581	3.143	2.874	2.418	2.037	1.874	1.626	1.477	1.377	1.304	1.249	1.205	1.170	1.141	1.111
0.562	4.067	3.259	2.864	2.621	2.212	1.874	1.730	1.514	1.387	1.303	1.242	1.197	1.161	1.133	1.109	1.085

Table A9. K_{ts} values for a cruciform joint, while θ = 30° and T/a = 1. Shearing load.

θ = 30°	X															
Y	0.010	0.018	0.025	0.032	0.050	0.079	0.100	0.150	0.200	0.250	0.300	0.350	0.400	0.450	0.500	0.562
0.010	2.705	2.493	2.372	2.292	2.142	1.996	1.925	1.802	1.715	1.647	1.589	1.539	1.494	1.453	1.414	1.367
0.018	2.704	2.491	2.371	2.291	2.140	1.995	1.924	1.801	1.714	1.646	1.588	1.538	1.493	1.452	1.413	1.367
0.025	2.702	2.489	2.369	2.290	2.139	1.993	1.923	1.800	1.713	1.645	1.587	1.537	1.492	1.451	1.412	1.366
0.032	2.698	2.486	2.367	2.287	2.137	1.991	1.921	1.798	1.711	1.643	1.586	1.536	1.491	1.449	1.410	1.364
0.050	2.683	2.514	2.360	2.277	2.129	1.984	1.914	1.791	1.705	1.637	1.580	1.530	1.485	1.444	1.405	1.359
0.079	2.657	2.451	2.334	2.257	2.108	1.964	1.895	1.774	1.689	1.621	1.565	1.515	1.471	1.430	1.391	1.346
0.100	2.640	2.432	2.315	2.236	2.089	1.947	1.878	1.758	1.673	1.607	1.551	1.502	1.458	1.417	1.379	1.334
0.150	2.568	2.365	2.251	2.175	2.032	1.894	1.827	1.710	1.628	1.563	1.508	1.461	1.418	1.379	1.342	1.298
0.200	2.487	2.291	2.180	2.106	1.968	1.834	1.769	1.656	1.576	1.513	1.461	1.415	1.374	1.336	1.301	1.259
0.250	2.407	2.216	2.110	2.038	1.904	1.775	1.712	1.603	1.525	1.465	1.414	1.370	1.331	1.295	1.261	1.222
0.300	2.331	2.146	2.043	1.973	1.844	1.718	1.658	1.552	1.478	1.419	1.370	1.328	1.291	1.257	1.225	1.188
0.350	2.259	2.081	1.981	1.913	1.788	1.666	1.607	1.505	1.433	1.377	1.330	1.290	1.254	1.222	1.193	1.159
0.400	2.193	2.020	1.922	1.857	1.735	1.617	1.560	1.462	1.392	1.338	1.293	1.255	1.221	1.191	1.164	1.134
0.450	2.130	1.962	1.867	1.804	1.686	1.572	1.516	1.421	1.354	1.302	1.259	1.223	1.192	1.164	1.139	1.112
0.500	2.071	1.907	1.815	1.754	1.639	1.528	1.474	1.382	1.317	1.268	1.227	1.194	1.165	1.139	1.117	1.094
0.562	1.998	1.840	1.751	1.692	1.581	1.475	1.423	1.335	1.274	1.228	1.190	1.160	1.134	1.112	1.094	1.074

Table A10. K_{ts} values for a cruciform joint, while θ = 40° and T/a = 1. Shearing load.

θ = 40°	X															
Y	0.010	0.018	0.025	0.032	0.050	0.079	0.100	0.150	0.200	0.250	0.300	0.350	0.400	0.450	0.500	0.562
0.010	3.221	2.908	2.723	2.619	2.391	2.186	2.088	1.921	1.805	1.715	1.641	1.577	1.521	1.471	1.425	1.372
0.018	3.219	2.899	2.722	2.606	2.390	2.185	2.087	1.920	1.804	1.714	1.640	1.577	1.521	1.470	1.424	1.371
0.025	3.217	2.897	2.720	2.604	2.388	2.184	2.086	1.919	1.803	1.713	1.639	1.576	1.520	1.469	1.423	1.370
0.032	3.210	2.895	2.718	2.601	2.386	2.182	2.084	1.917	1.801	1.712	1.638	1.574	1.519	1.468	1.422	1.369
0.050	3.212	2.886	2.711	2.594	2.379	2.175	2.078	1.911	1.796	1.706	1.633	1.570	1.514	1.464	1.417	1.364
0.079	3.162	2.859	2.688	2.573	2.361	2.157	2.063	1.896	1.781	1.692	1.619	1.557	1.501	1.451	1.406	1.353
0.100	3.154	2.841	2.667	2.553	2.341	2.141	2.045	1.881	1.767	1.679	1.607	1.544	1.490	1.440	1.395	1.342
0.150	3.075	2.769	2.600	2.489	2.282	2.086	1.993	1.833	1.722	1.637	1.566	1.506	1.452	1.404	1.360	1.309
0.200	2.977	2.681	2.518	2.410	2.210	2.020	1.930	1.776	1.668	1.585	1.517	1.459	1.407	1.361	1.319	1.270
0.250	2.874	2.588	2.430	2.325	2.133	1.950	1.863	1.714	1.611	1.531	1.466	1.410	1.361	1.317	1.277	1.231
0.300	2.771	2.495	2.342	2.242	2.056	1.880	1.796	1.653	1.554	1.477	1.415	1.362	1.316	1.275	1.237	1.196
0.350	2.671	2.405	2.258	2.161	1.982	1.812	1.732	1.594	1.499	1.426	1.367	1.317	1.274	1.236	1.202	1.164
0.400	2.576	2.319	2.177	2.084	1.912	1.748	1.671	1.539	1.448	1.379	1.323	1.276	1.236	1.201	1.170	1.137
0.450	2.485	2.237	2.101	2.011	1.845	1.687	1.613	1.486	1.400	1.334	1.282	1.239	1.202	1.170	1.143	1.114
0.500	2.397	2.159	2.027	1.940	1.780	1.629	1.557	1.436	1.354	1.293	1.244	1.204	1.171	1.143	1.120	1.094
0.562	2.292	2.064	1.938	1.855	1.702	1.558	1.491	1.377	1.301	1.244	1.201	1.166	1.138	1.114	1.095	1.074

Table A11. K_{ts} values for a cruciform joint, while θ = 50° and T/a = 1. Shearing load.

θ = 50°	X															
Y	0.010	0.018	0.025	0.032	0.050	0.079	0.100	0.150	0.200	0.250	0.300	0.350	0.400	0.450	0.500	0.562
0.010	3.716	3.275	3.038	2.882	2.600	2.337	2.213	2.005	1.863	1.755	1.669	1.596	1.533	1.478	1.428	1.373
0.018	3.713	3.273	3.036	2.881	2.598	2.335	2.212	2.004	1.862	1.755	1.668	1.595	1.532	1.477	1.427	1.372
0.025	3.709	3.271	3.035	2.880	2.597	2.334	2.211	2.003	1.861	1.754	1.667	1.594	1.532	1.476	1.427	1.371
0.032	3.707	3.268	3.032	2.878	2.595	2.333	2.209	2.001	1.860	1.753	1.666	1.593	1.531	1.475	1.426	1.370
0.050	3.701	3.262	3.023	2.870	2.587	2.326	2.203	1.996	1.854	1.748	1.661	1.589	1.526	1.471	1.421	1.366
0.079	3.667	3.265	3.001	2.849	2.569	2.309	2.186	1.981	1.841	1.734	1.649	1.577	1.515	1.460	1.410	1.355
0.100	3.647	3.215	2.981	2.829	2.552	2.293	2.172	1.967	1.828	1.723	1.637	1.566	1.504	1.449	1.400	1.345
0.150	3.561	3.139	2.911	2.762	2.491	2.239	2.120	1.920	1.785	1.682	1.598	1.528	1.468	1.414	1.366	1.312
0.200	3.450	3.041	2.820	2.676	2.413	2.169	2.054	1.861	1.729	1.630	1.549	1.482	1.423	1.372	1.326	1.274
0.250	3.325	2.931	2.719	2.580	2.326	2.091	1.980	1.794	1.668	1.572	1.496	1.431	1.376	1.327	1.283	1.235
0.300	3.197	2.818	2.613	2.480	2.237	2.010	1.904	1.726	1.605	1.514	1.441	1.380	1.328	1.283	1.242	1.198
0.350	3.070	2.706	2.509	2.381	2.148	1.931	1.829	1.659	1.544	1.458	1.389	1.332	1.284	1.242	1.205	1.166
0.400	2.946	2.597	2.408	2.286	2.062	1.854	1.757	1.594	1.485	1.404	1.340	1.287	1.243	1.205	1.173	1.138
0.450	2.827	2.492	2.311	2.193	1.979	1.780	1.687	1.533	1.430	1.354	1.295	1.246	1.207	1.173	1.145	1.115
0.500	2.711	2.390	2.216	2.104	1.898	1.709	1.621	1.474	1.378	1.307	1.253	1.210	1.174	1.145	1.120	1.095
0.562	2.570	2.266	2.102	1.995	1.802	1.623	1.541	1.405	1.317	1.254	1.206	1.169	1.139	1.115	1.095	1.074

Table A12. K_{ts} values for a cruciform joint, while $\theta = 60^\circ$ and $T/a = 1$. Shearing load.

$\theta = 60^\circ$								X							
Y	**0.010**	**0.018**	**0.025**	**0.032**	**0.050**	**0.079**	**0.100**	**0.150**	**0.200**	**0.250**	**0.300**	**0.350**	**0.400**	**0.450**	**0.500**	**0.562**
0.010	4.186	3.621	3.320	3.126	2.777	2.458	2.310	2.066	1.903	1.782	1.687	1.608	1.541	1.484	1.433	1.378
0.018	4.184	3.618	3.319	3.125	2.776	2.457	2.310	2.065	1.903	1.782	1.686	1.607	1.541	1.483	1.432	1.377
0.025	4.183	3.617	3.318	3.124	2.775	2.456	2.308	2.064	1.902	1.781	1.685	1.607	1.540	1.482	1.432	1.376
0.032	4.176	3.614	3.314	3.122	2.773	2.454	2.307	2.063	1.900	1.780	1.684	1.605	1.539	1.481	1.430	1.375
0.050	4.160	3.606	3.306	3.113	2.767	2.447	2.305	2.057	1.895	1.775	1.680	1.601	1.535	1.477	1.426	1.371
0.079	4.139	3.576	3.281	3.092	2.756	2.430	2.285	2.043	1.882	1.762	1.668	1.589	1.523	1.466	1.415	1.360
0.100	4.114	3.558	3.263	3.073	2.729	2.415	2.270	2.030	1.870	1.751	1.656	1.579	1.513	1.455	1.405	1.349
0.150	4.020	3.478	3.189	3.003	2.667	2.360	2.218	1.983	1.827	1.710	1.618	1.542	1.477	1.421	1.371	1.316
0.200	3.896	3.370	3.091	2.910	2.585	2.287	2.150	1.922	1.771	1.658	1.569	1.495	1.433	1.378	1.330	1.277
0.250	3.754	3.247	2.977	2.803	2.490	2.204	2.071	1.853	1.707	1.599	1.514	1.443	1.384	1.332	1.287	1.237
0.300	3.601	3.115	2.856	2.690	2.389	2.115	1.988	1.779	1.640	1.538	1.457	1.391	1.335	1.287	1.245	1.200
0.350	3.447	2.982	2.735	2.575	2.288	2.026	1.905	1.706	1.574	1.478	1.402	1.340	1.289	1.245	1.207	1.167
0.400	3.295	2.851	2.615	2.462	2.188	1.938	1.823	1.634	1.510	1.420	1.350	1.293	1.247	1.207	1.174	1.138
0.450	3.147	2.723	2.498	2.352	2.091	1.853	1.744	1.566	1.450	1.366	1.302	1.251	1.209	1.174	1.145	1.115
0.500	3.003	2.599	2.384	2.245	1.996	1.771	1.668	1.501	1.393	1.316	1.258	1.212	1.176	1.146	1.121	1.095
0.562	2.828	2.447	2.245	2.115	1.882	1.672	1.577	1.424	1.327	1.259	1.209	1.170	1.140	1.115	1.095	1.074

Appendix C

Graphical SCF results for various weld angles θ for a cruciform joint subjected to tensile, bending, and shearing loads.

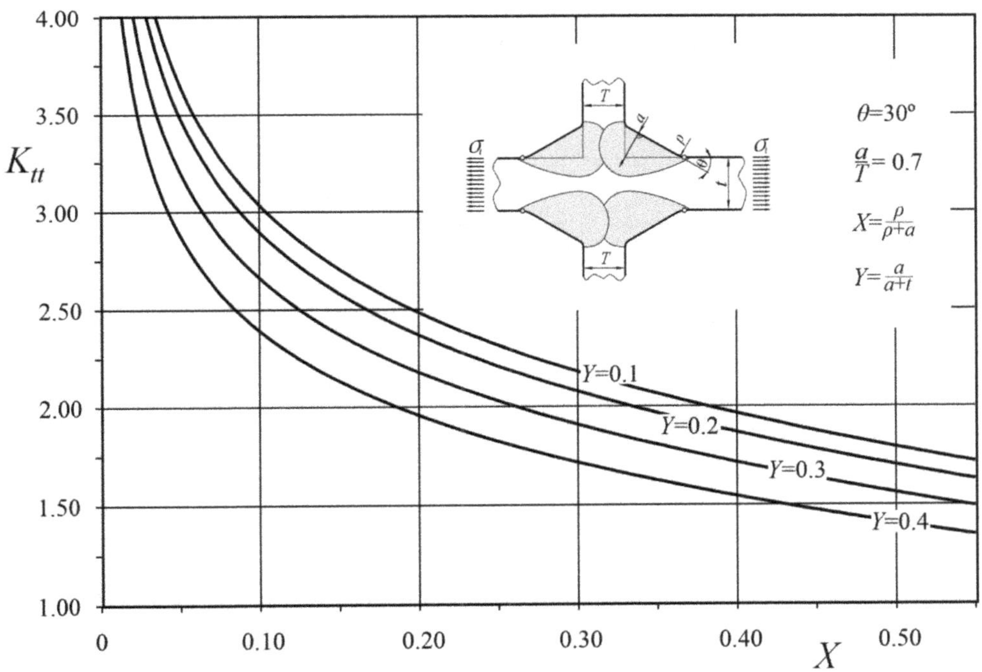

Figure A1. K_{tt} vs. X for $a/T = 0.7$ and $\theta = 30^\circ$. Tensile load.

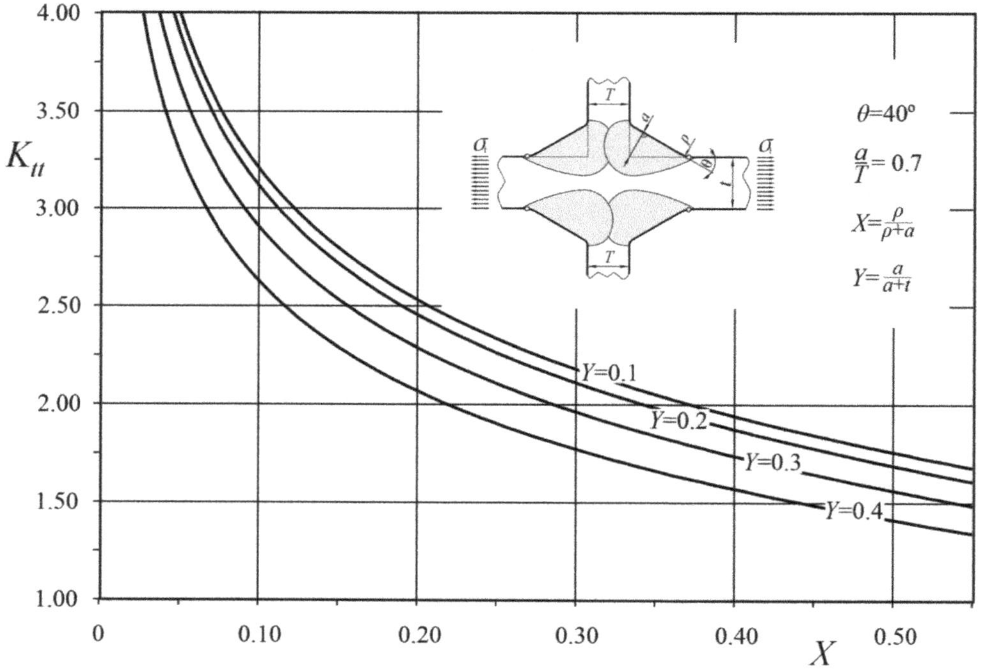

Figure A2. K_{tt} vs. X for $a/T = 0.7$ and $\theta = 40°$. Tensile load.

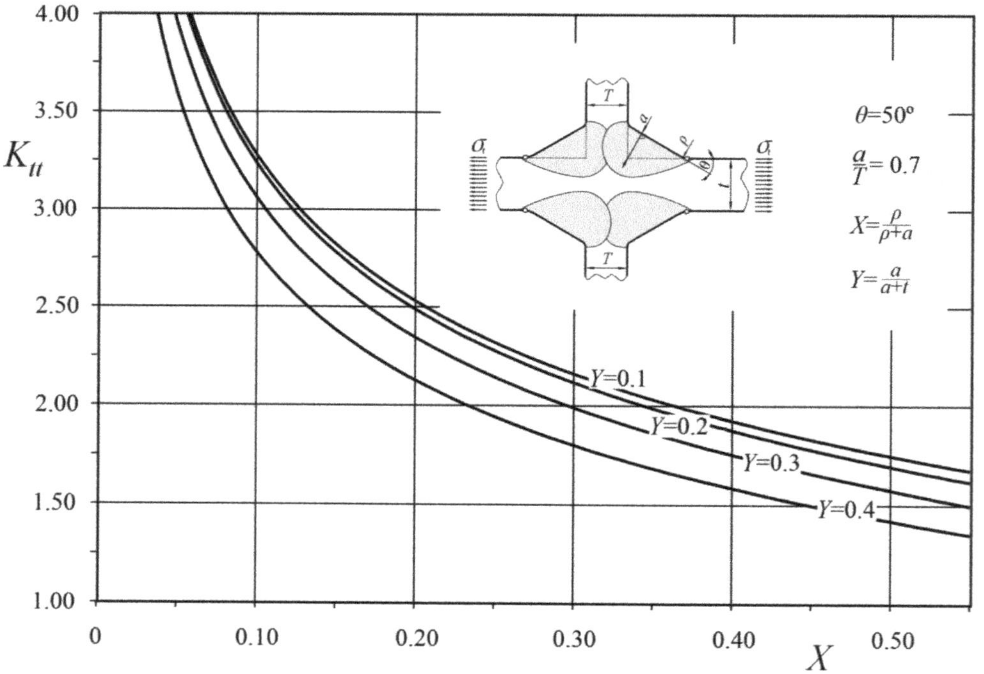

Figure A3. K_{tt} vs. X for $a/T = 0.7$ and $\theta = 50°$. Tensile load.

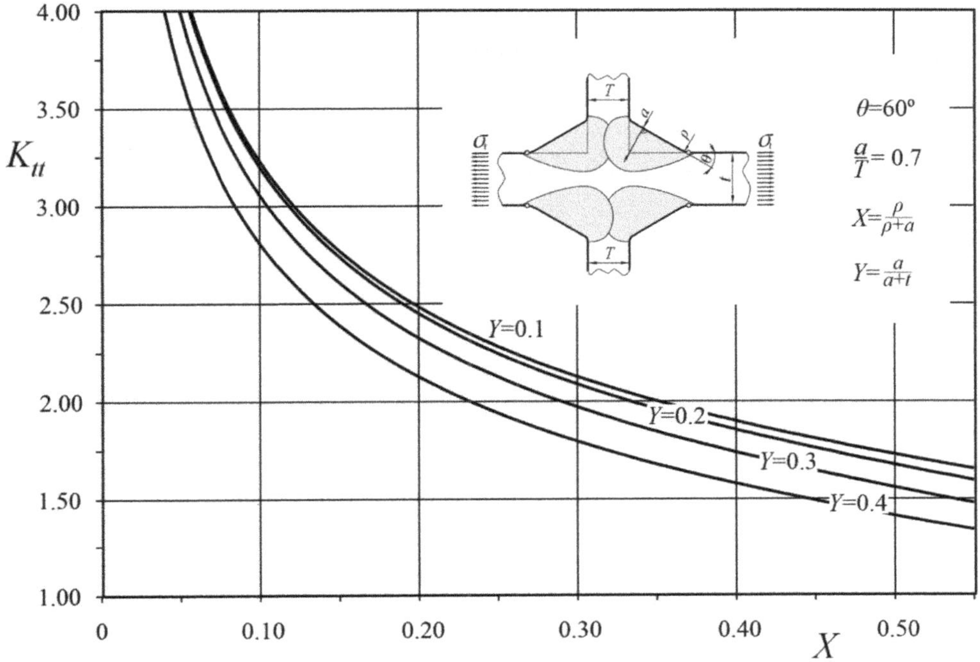

Figure A4. K_{tt} vs. X for $a/T = 0.7$ and $\theta = 60°$. Tensile load.

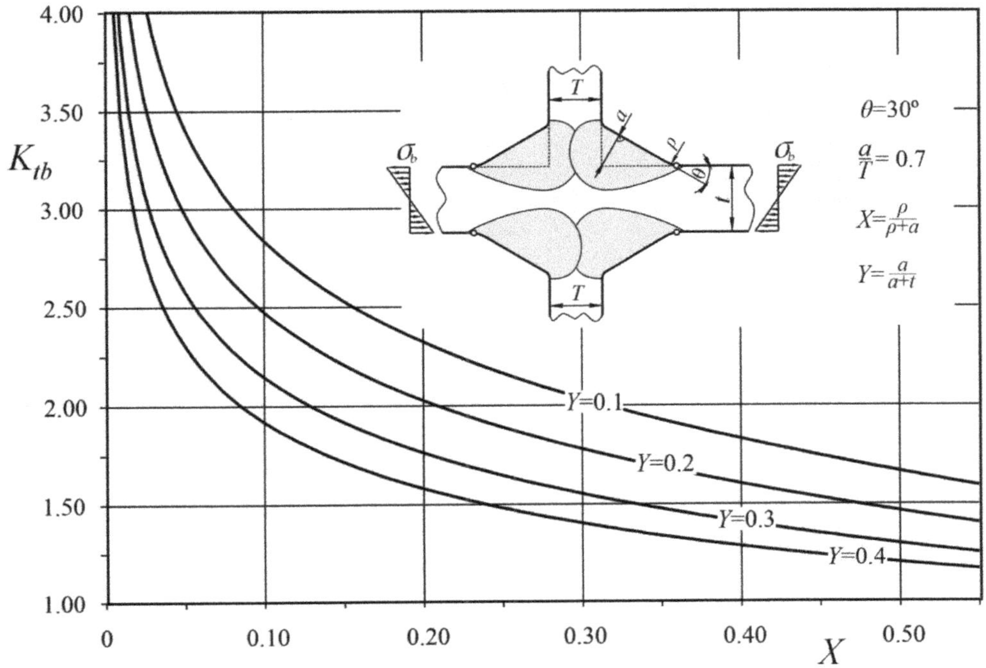

Figure A5. K_{tb} vs. X for $a/T = 0.7$ and $\theta = 30°$. Bending load.

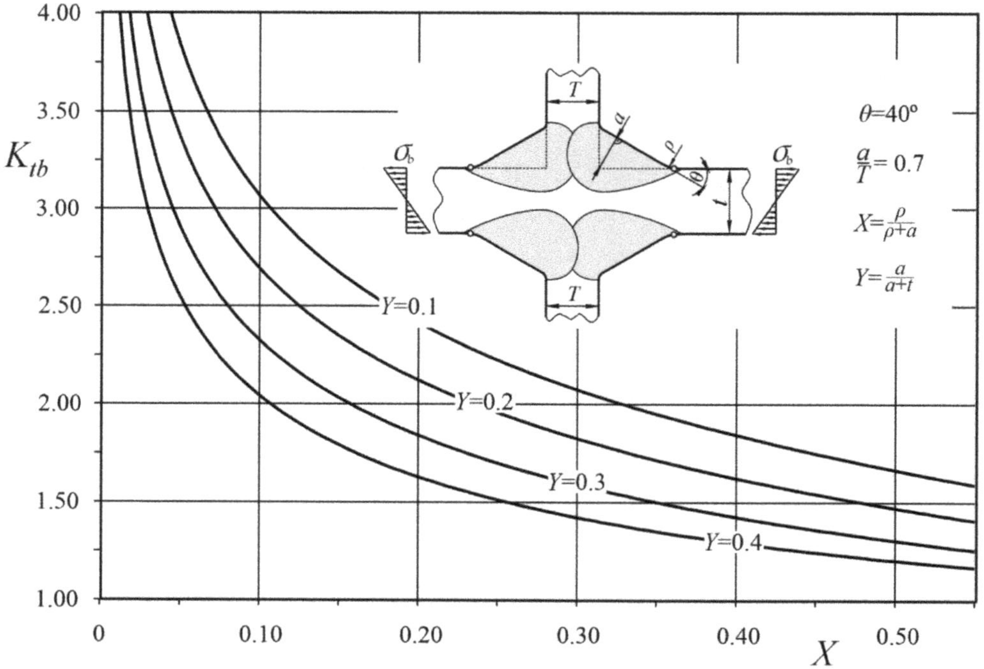

Figure A6. K_{tb} vs. X for a/T = 0.7 and θ = 40°. Bending load.

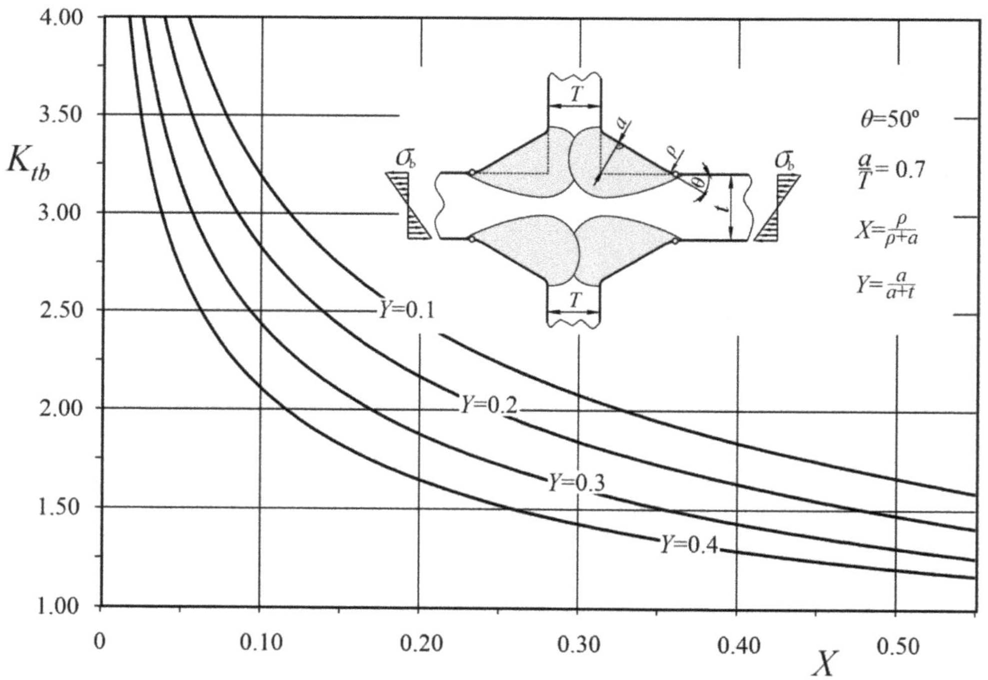

Figure A7. K_{tb} vs. X for a/T = 0.7 and θ = 50°. Bending load.

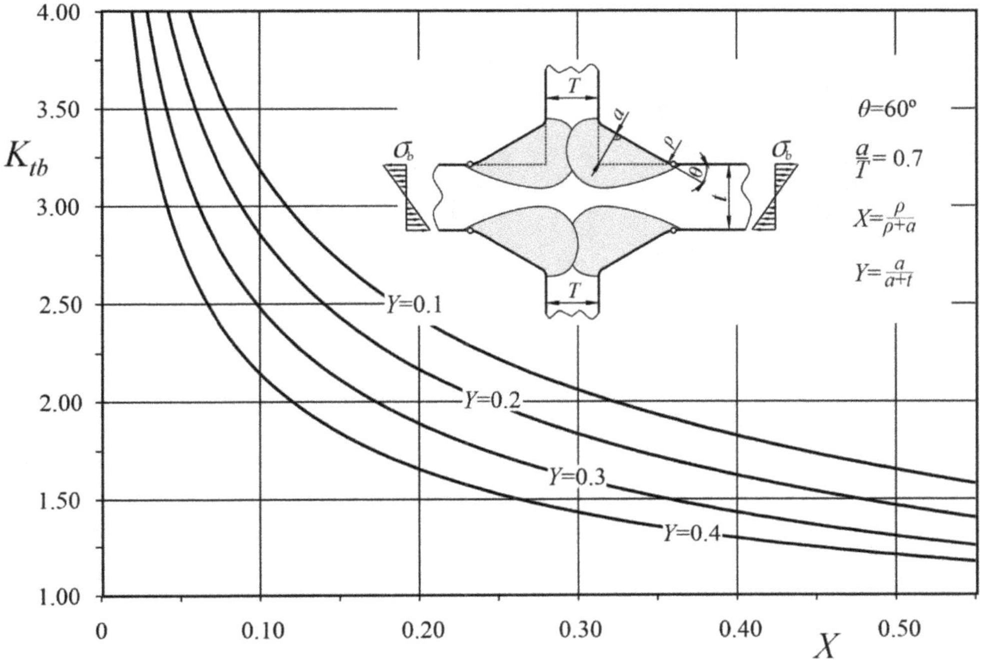

Figure A8. K_{tb} vs. X for a/T = 0.7 and θ = 60°. Bending load.

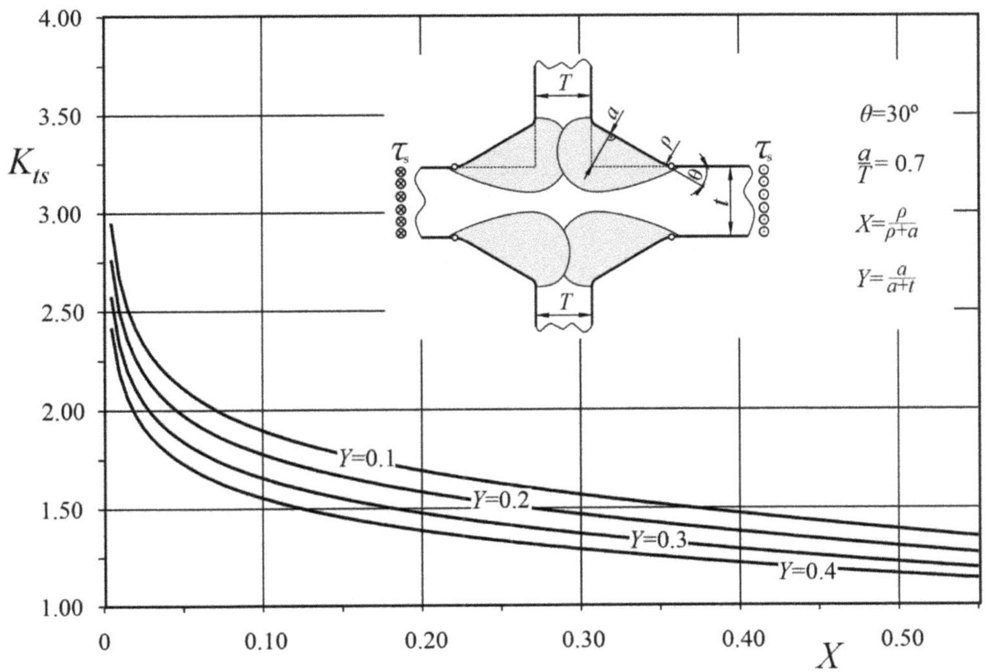

Figure A9. K_{ts} vs. X for a/T = 0.7 and θ = 30°. Shearing load.

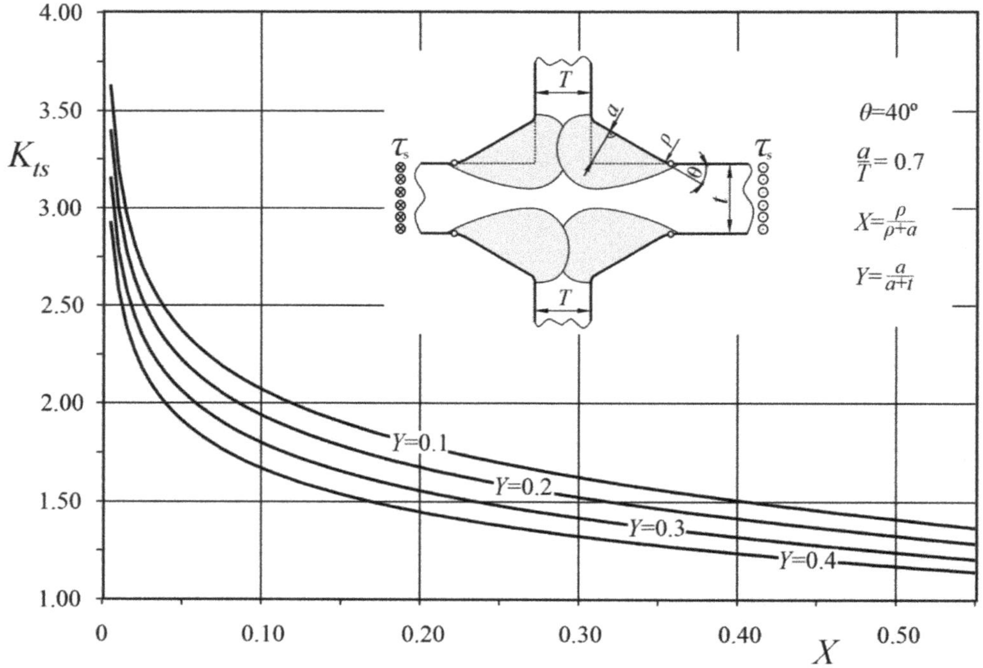

Figure A10. K_{ts} vs. X for $a/T = 0.7$ and $\theta = 40°$. Shearing load.

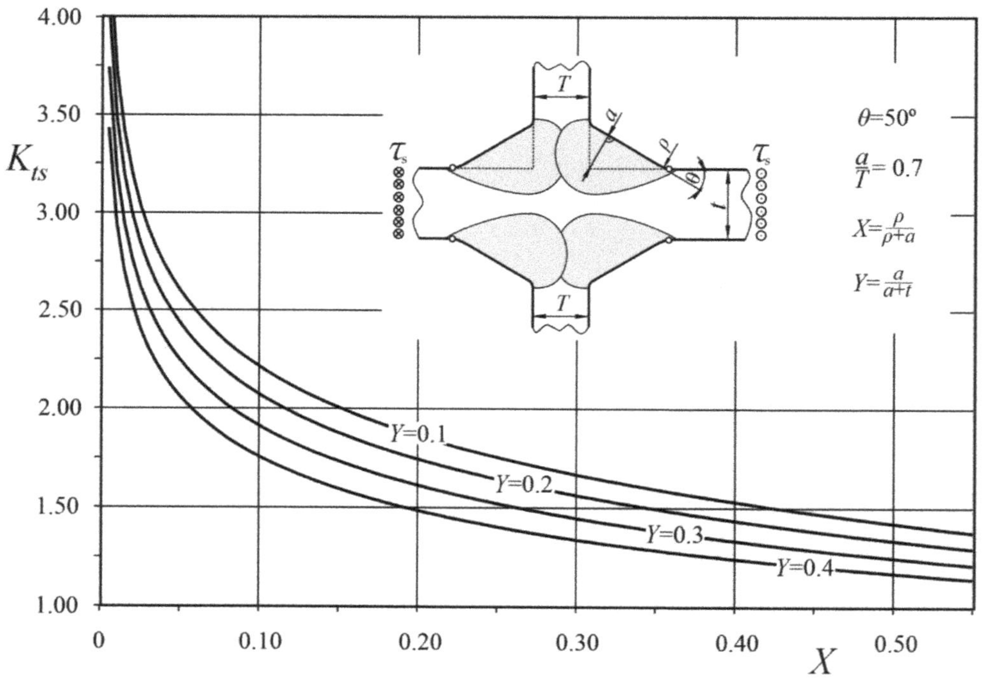

Figure A11. K_{ts} vs. X for $a/T = 0.7$ and $\theta = 50°$. Shearing load.

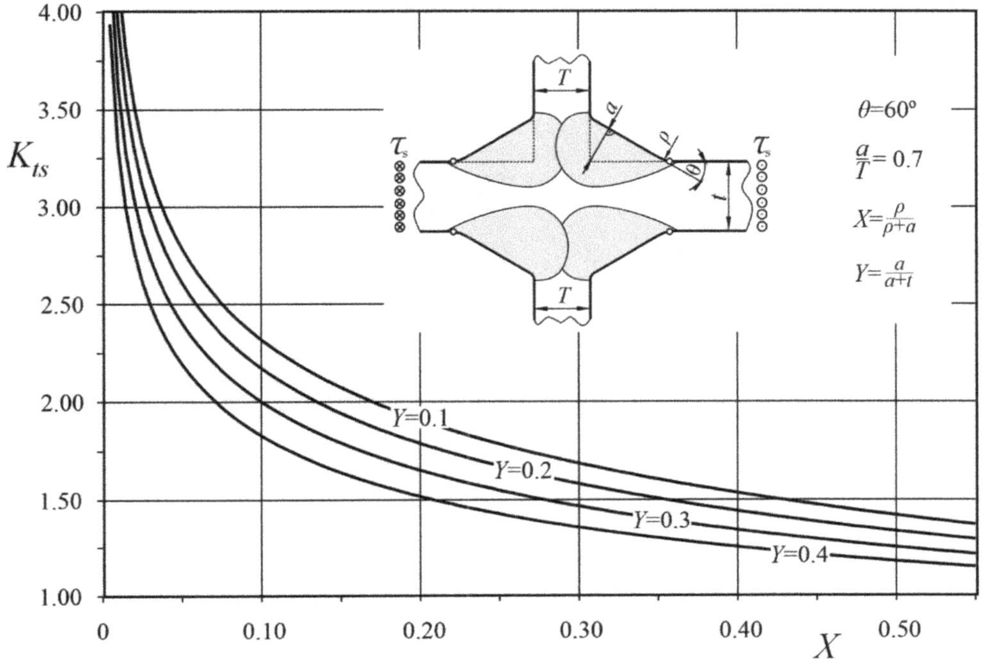

Figure A12. K_{ts} vs. X for a/T = 0.7 and θ = 60°. Shearing load.

References

1. Balasubramanian, V.; Guha, B. Influence of weld size on fatigue life prediction for flux cored arc welded cruciform joints containing lack of penetration defects. *Sci. Technol. Weld. Join.* **2000**, *5*, 99–104. [CrossRef]
2. Singh, P.J.; Achar, D.R.G.; Guha, B.; Nordberg, H. Influence of weld geometry and process on fatigue crack growth characteristics of AISI 304L cruciform joints containing lack of penetration defects. *Sci. Technol. Weld. Join.* **2002**, *7*, 306–312. [CrossRef]
3. Chung, H.Y.; Liu, S.H.; Lin, R.S.; Ju, S.H. Assessment of stress intensity factors for load-carrying fillet welded cruciform joints using a digital camera. *Int. J. Fatigue* **2008**, *30*, 1861–1872. [CrossRef]
4. Molski, K.L. Stress Concentration at Load-Carrying Fillet Welded Cruciform Joints Subjected to Tensile and Bending Loads. *Acta Mech. Autom.* **2019**, *13*, 245–250. [CrossRef]
5. Gurney, T.R. *Fatigue of Welded Structures*, 2nd ed.; Cambridge University Press: Cambridge, UK, 1979.
6. Fayard, J.L.; Bignonnet, A.; Dang Van, K. Fatigue design criteria for welded structures. *Fatigue Fract. Eng. Mater. Struct. (FFEMS)* **1996**, *19*, 723–729. [CrossRef]
7. Dong, P. A structural stress definition and numerical implementation for fatigue analysis of welded joints. *Int. J. Fatigue* **2001**, *23*, 865–876. [CrossRef]
8. Singh, P.J.; Guha, B.; Achar, D.R.G. Fatigue life prediction using two stage model for AISI 304L cruciform joints, with different fillet geometry, failing at toe. *Fatigue Fract. Eng. Mater. Struct. (FFEMS)* **2003**, *8*, 69–75. [CrossRef]
9. Radaj, D.; Sonsino, C.M.; Fricke, W. Recent developments in local concepts of fatigue assessment of welded joints. *Int. J. Fatigue* **2009**, *31*, 2–11. [CrossRef]
10. Lotsberg, I.; Sigurdsson, G. Hot Spot Stress S-N Curve for Fatigue Analysis of Plated Structures. *J. Offshore Mech. Arct. Eng.* **2006**, *128*, 330–336. [CrossRef]
11. Park, W.; Miki, C. Fatigue assessment of large-size welded joints based on the effective notch stress approach. *Int. J. Fatigue* **2008**, *30*, 1556–1568. [CrossRef]
12. Schijve, J. Fatigue predictions of welded joints and the effective notch stress concept. *Int. J. Fatigue* **2012**, *45*, 31–38. [CrossRef]
13. Tchoffo Ngoula, D.; Beier, H.T.H.; Vormwald, M. Fatigue crack growth in cruciform welded joints: Influence of residual stresses and of the weld toe geometry. *Int. J. Fatigue* **2017**, *101*, 253–262. [CrossRef]
14. Zerbst, U.; Madia, M.; Schork, B. Fracture mechanics based determination of the fatigue strength of weldments. *Procedia Struct. Integr.* **2016**, *1*, 10–17. [CrossRef]
15. Monahan, C.C. *Early Fatigue Cracks Growth at Welds*; Computational Mechanics Publications: Southampton, NY, USA, 1995.
16. Remes, H.; Varsta, P. Statistics of Weld Geometry for Laser-Hybrid Welded Joints and its Application within Notch Stress Approach. *Weld. World* **2010**, *54*, R189–R207. [CrossRef]

17. Wang, Y.; Luo, Y.; Tsutsumi, S. Parametric Formula for Stress Concentration Factor of Fillet Weld Joints with Spline Bead Profile. *Materials* **2020**, *13*, 4639. [CrossRef]
18. Randić, M.; Pavletić, D.; Turkalj, G. Multiparametric Investigation of Welding Techniques on Toe Radius of High Strength Steel at Low-Temperature Levels Using 3D-Scanning Techniques. *Metals* **2019**, *9*, 1355. [CrossRef]
19. Weinert, J.; Gkatzogiannis, S.; Engelhardt, I.; Knoedel, P.; Ummenhofer, T. Investigation of corrosive influence on the fatigue behaviour of HFMI-treated and as-welded transverse non-load-carrying attachments made of mild steel S355. *Int. J. Fatigue* **2021**, *151*, 106225. [CrossRef]
20. Tarasiuk, P.; Molski, K.L.; Szymaniuk, A. Fatigue designing of welded agricultural wheels. *Eksploat. Niezawodn.* **2013**, *15*, 123–128.
21. *EN 1993-1-9:2005*; Eurocode 3: Design of Steel Structures—Part. 1–9: Fatigue. European Committee for Standardization (CES): Brussels, Belgium, 2005.
22. Hobbacher, A.F. The new IIW recommendations for fatigue assessment of welded joints and components—A comprehensive code recently updated. *Int. J. Fatigue* **2009**, *31*, 50–58. [CrossRef]
23. Fricke, W. *IIW Recommendations for the Fatigue Assessment of Welded Structures by Notch Stress Analysis*; Woodhead Publishing: Cambridge, UK, 2012; pp. 2–41.
24. Fricke, W. IIW guideline for the assessment of weld root fatigue. *Weld. World* **2013**, *57*, 753–791. [CrossRef]
25. *EN ISO 9692-1:2013*; Welding and Allied Processes—Types of Joint Preparation—Part. 1: Manual Metal Arc Welding, Gas-Shielded Metal Arc Welding, Gas Welding, TIG Welding and Beam Welding of Steels. International Organization for Standardization: Geneva, Switzerland, 2013.
26. Haagensen, P.J.; Maddox, S.J. *Recommendations on Methods for Improving the Fatigue Strength of Welded Joints*; Woodhead Publishing: Cambridge, UK, 2013; ISBN 9781782420644.
27. *EN ISO 5817:2014*; Welding—Fusion-welded joints in steel, nickel, titanium and their alloys (beam welding excluded)— Quality levels for imperfections. International Organization for Standardization: Geneva, Switzerland, 2014.
28. Peterson, R.E. *Stress Concentration Design Factors*, 2nd ed.; Wiley: New York, NY, USA, 1974.
29. Ushirokawa, O.; Nakayama, E. Stress concentration factor at welded joints. *Ishikawajima Harima Eng. Rev.* **1983**, *23*, 351–355.
30. Tsuji, I. Estimation of stress concentration factor at weld toe of non-load carrying fillet welded joints. *Trans. West. Jpn. Soc. Nav. Archit.* **1990**, *80*, 241–251.
31. Iida, K.; Uemura, T. Stress concentration factor formulas widely used in Japan. *Fatigue Fract. Eng. Mater. Struct.* **1996**, *19*, 779–786. [CrossRef]
32. Molski, K.L.; Tarasiuk, P.; Glinka, G. Stress concentration at cruciform welded joints under axial and bending loading modes. *Weld. World* **2020**, *64*, 1867–1876. [CrossRef]
33. Hasebe, N.; Sugimoto, T.; Nakamura, T. Stress concentration of longitudinal shear problems. *J. Eng. Mech.* **1987**, *113*, 1358–1367. [CrossRef]
34. Molski, K.L.; Tarasiuk, P. Stress Concentration Factors for Butt-Welded Plates Subjected to Tensile, Bending and Shearing Loads. *Materials* **2020**, *13*, 1798. [CrossRef]
35. Molski, K.L.; Tarasiuk, P. Stress Concentration Factors for Welded Plate T-Joints Subjected to Tensile, Bending and Shearing Loads. *Materials* **2021**, *14*, 546. [CrossRef]

materials

MDPI

Article

Effect of Residual Stress and Microstructure on the Fatigue Crack Growth Behavior of Aluminum Friction Stir Welded Joints

Yanning Guo [1,*] and Peiyao Li [2]

1 School of Civil Engineering, Xi'an University of Architecture and Technology, Xi'an 710055, China
2 School of Aeronautics, Northwestern Polytechnical University, Xi'an 710072, China
* Correspondence: guoyanning226@163.com

Abstract: Friction stir welding (FSW) has been adopted in the aerospace industry for fabricating structural alloys due to the low melting point and high thermal conductivity of aviation aluminum alloys. However, welding residual stresses can lead to secondary deformation in friction stir welded (FSWed) structures. Additionally, microstructural characteristics impact the crack growth rates and directions in these structures. Therefore, it is necessary to investigate the effects of residual stress and microstructure on the fatigue responses of FSWed joints. In this paper, we studied the fatigue crack growth behavior of two homogeneous and dissimilar FSWed joints with varying welding parameters, namely 2024-T3 and 7075-T6. The residual stresses were measured with the X-ray diffraction method. The dislocations and precipitates in different zones of the FSWed joints were analyzed via transmission electron microscopy (TEM). The results demonstrated that the residual stress significantly affected the fatigue crack growth rate and direction; the tensile residual stress promoted fatigue crack growth and offset the decrease in the fatigue crack growth rate that occurred due to grain refinement. The results of the microstructural analysis indicated that dislocation density and sliding resistance increased with the decrease in rotational speed and led to a decreased rate of fatigue crack propagation.

Keywords: aluminum alloys; friction stir welding; fatigue crack growth; residual stress; microstructure

1. Introduction

AA 2xxx and AA 7xxx aluminum alloys are widely used in aircraft structures due to their low density, high strength, and high stiffness [1]. Specifically, the 2024-T3 alloy is primarily employed in the fuselage for the rivets, skin, wings, skeleton, etc. The 7075-T6 alloy serves as the essential material in wing spars, baffle frames, and wing panels. However, traditional welding methods cause several defects in the 2024-T3 and 7075-T6 alloys [2]. On the other hand, friction stir welding (FSW) stands out as a typical solid-state joining technology, which is capable of fabricating high-quality welded joints with minimal welding deformation [3,4]. Therefore, FSW is being increasingly adopted in industries such as aerospace, shipping, railway, electronics, and others [5,6].

During the welding process, changes in local temperature and deformation induce plastic deformation, which occurs in the welded zone (WZ) and thereby leads to the formation of residual stress, which is a self-balancing system that is not dependent on external loads [3,7]. The residual stress and the external loads can result in secondary deformations, as well as the redistribution of residual stress, which significantly influences the integrity and safety of welded structures [8–10]. Previous studies have shown that the maximum residual stress in the longitudinal direction (LD) is generated in the heat-affected zone (HAZ), while the minimum compressive residual stress originates on the advancing side just beyond the WZ [11,12]. Furthermore, another study indicated that in the 2024/7075 dissimilar welded joints, the longitudinal residual stress on the 7075 side was higher than that on the 2024 side [13]. This is related to the position within the welded joint, as the heat input on the advancing side is relatively higher than that on the retreating side. Additionally, compared to the 2024 alloy, 7075 aluminum exhibits higher mechanical properties, including yield strength and hardness.

Due to the strong local thermal coupling effect in the FSW process, the microstructure of the welded joint becomes diversified and non-uniform [14,15]. The WZ, which is composed of fine equiaxed grains experiences the highest thermal cycle and the most intense plastic deformation during the welding process, causing dynamic recrystallization in this region [16]. In the thermo-mechanically affected zone (TMAZ), plastic deformation induces the formation of several dislocations within the crystal structure. Additionally, the increase in temperature leads to dynamic recovery, thereby resulting in the formation of a large number of sub-grains [17]. A previous study demonstrated that the HAZ exclusively underwent the thermal cycle that results in a partial coarsening of the grains, which typically leads to a morphology that closely resembles that of the base metal [18].

Previous studies have also demonstrated that the fatigue crack growth rate in FSWed joints is influenced by the interaction between residual stress distribution and the microstructure [19,20]. The high level of compressive residual stress near the welded zone can induce crack closure effects, while tensile stress increases the effective stress intensity factor range [21,22]. Fratini et al. [23,24] noted that stress relief did not alter the hardness and microstructural properties of FSWed structures, but the closure phenomena resulting from residual stress did affect the growth rates. Ilman et al. [25] stated that the effect of the microstructure on the fatigue crack growth rate was not significant when compared to residual stress. However, in the study by Anderson-Wedge et al. [26], the changes in crack growth were attributed to the depletion of strengthening precipitates in thermally affected zones rather than to the residual stresses.

Considering the varying conclusions from previous studies, this paper aims to investigate the coupling effect between residual stress and microstructure on the fatigue crack growth behavior of 2024 and 7075 aluminum alloys. The residual stresses were measured with the X-ray diffraction method, and the microstructure was observed with TEM. The finite element model was built using ANSYS and FRANC3D to calculate the fatigue crack growth rate, and the results were compared with those of the subsequent experiments.

2. Experimental Section

Rectangular plates of 2024-T3 and 7075-T6 alloys that were 87.5 mm × 400 mm × 8 mm in size were joined using the FSW-RL31-010 machine (Beijing FSW Technology Co., Ltd., Beijing, China) at Northwestern Polytechnical University in Xi'an, China. The shoulder diameter was 20 mm, and the probe diameters were 5 mm and 7.5 mm. Table 1 displays the chemical compositions of the 2024-T3 and 7075-T6 alloys, which were produced by Shanghai Miandi Metal Group Co., Ltd. (Shanghai, China) The welding parameters are shown in Table 2.

Table 1. Chemical compositions of the 2024-T3 and 7075-T6 alloys.

Material	Chemical Composition (%)								
2024-T3	Cu 3.8–4.9	Si 0.5	Fe 0.5	Mn 0.3–0.9	Mg 1.2–1.8	Zn 0.25	Cr 0.1	Ti 0.15	Al Base
7075-T6	Si 0.4	Fe 0.5	Mn 0.3	Cu 1.2–2.0	Mg 2.1–2.9	Cr 0.18–0.28	Zn 5.1–6.1	Ti 0.2	Al Base

Table 2. Welding parameters.

Specimen	Material	Rotational Speed (rpm)	Welding Speed (mm/min)
1	2024 homogeneous	400	150
2	2024 homogeneous	200	150
3	7075 homogeneous	400	150
4	7075 homogeneous	400	100
5	2024/7075 dissimilar	400	150

3. Residual Stress Measurement

The residual stress distribution of the 2024 specimen (400 rpm-150 mm/min) was analyzed using the X-350A X-ray diffraction setup at Xi'an Jiaotong University in Xi'an, China (Figure 1a). The stress distributions in both the longitudinal direction (LD) and the transverse direction (TD) were measured separately. The scan line is illustrated in Figure 1b.

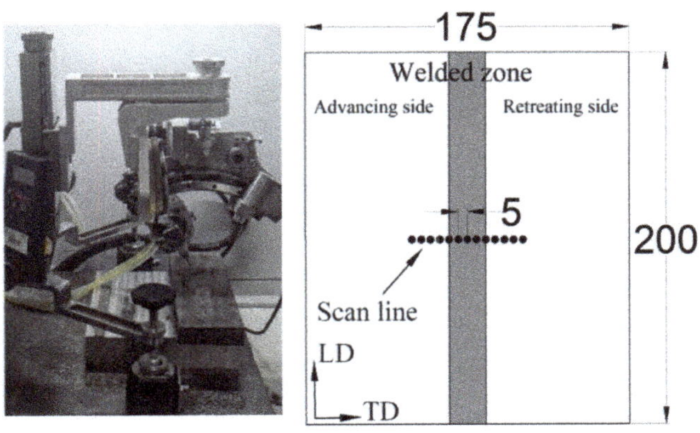

Figure 1. Residual stress measurement: (**a**) X-ray diffraction method; (**b**) scan line (mm).

The residual stress profile of the 2024 FSWed specimen (400 rpm-150 mm/min) is shown in Figure 2. The LD stress profile exhibits a "double peak" near the welded zone, with a peak stress of 103.2 MPa on the advancing side and a maximum stress of 93.5 MPa on the retreating side. This difference arises due to the comparatively higher heat input on the advancing side. In the TD, the residual stress values are relatively uniform, with a peak stress of 27.65 MPa on the advancing side.

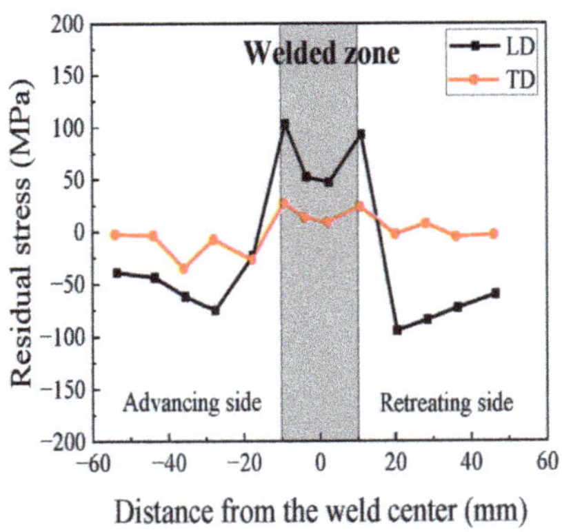

Figure 2. Results of residual stress measurement.

4. Fatigue Crack Growth Experiment

The fatigue crack growth experiments in this study were conducted using an INSTRON 810 testing machine at Northwestern Polytechnical University, Xi'an, China, with a maximum loading force of 26 kN (Figure 3a). This setup ensured symmetrical load distribution with a load error of less than ±1% and minimal variation in the indicating value. The testing machine was equipped with an accurate counting device to meet the testing requirements. The compact tension (C(T)) specimens were machined according to the ASTM-647 standard [27]; the size of the specimen is shown in Figure 3b. It can be seen that the cracks in the C(T) specimens were parallel to the welded joints. To facilitate fixture fixation and crack growth observation, the surfaces of the specimen were polished before the experiment (Figure 3c). Fatigue tests were performed at R = 0.1 for all the C(T) specimens, with a fatigue load frequency of 20 Hz. Load ranges of 6.21 kN for the 2024 and 7075 homogeneous specimens, 8.04 kN for the 2024/7075 dissimilar specimens, and 3.26 kN for the base metal were applied.

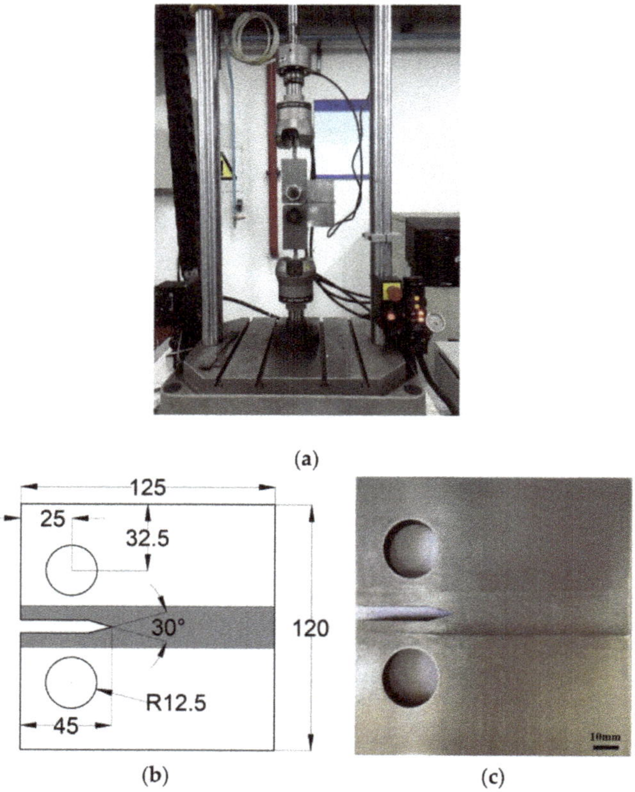

Figure 3. (**a**) Fatigue crack growth experiment; (**b**) size of the C(T) specimens (mm); (**c**) C(T) specimen.

The results of the fatigue crack growth rate experiment of the 2024 homogeneous FSWed specimens with different rotational speeds are presented in Figure 4. It can be seen that the crack growth rate of the 2024 base metal is higher than that of the welded specimen. Under the same welding speed, the fatigue crack growth rate of the 400 rpm specimen is higher than that the of 200 rpm specimen. As seen in Figure 4b, the crack path of 200 rpm was deflected towards the loading hole. With further crack propagation, the crack crossed through the WZ into the thermo-mechanically affected zone (TMAZ). Figure 4c shows that the crack growth path of the 400 rpm specimen remained within the WZ. According to previous studies on the dynamic tensile properties of the WZ, the increase in rational speed reduces the size of the particles in the WZ and makes them more uniform [28]. This enhances the mechanical properties of the WZ and reduces the fatigue crack growth rate in this region. Therefore, it can be concluded that tensile longitudinal residual stress promotes the crack growth and that it offsets the decrease in fatigue crack growth rate by grain refinement.

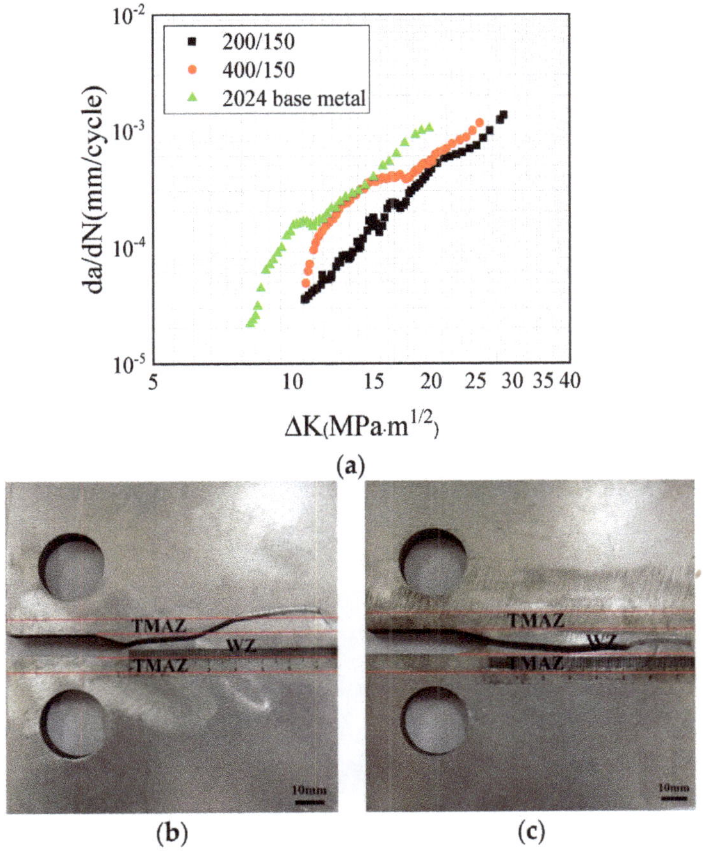

Figure 4. Fatigue crack growth results of the 2024 specimen: (**a**) crack growth rates; (**b**) crack growth path at 200 rpm; (**c**) crack growth path at 400 rpm.

Figure 5a displays the fatigue crack growth results of the 7075 homogeneous FSWed specimens with different welding speeds. It is evident that welding speed exerts a significant influence on the crack growth rate. At the initial stage, the crack growth rate of the 7075 base metal was much higher than that of the FSWed specimen. Notably, the fatigue crack growth rate of the specimen with a welding speed of 100 mm/min was higher than that of the specimen with a speed of 150 mm/min. In a previous study, it was demonstrated that the mechanical properties of the 100 mm/min and 150 mm/min specimens were similar [29]; however, the longitudinal residual stress of the 100 mm/min specimen was higher than that of the 150 mm/min specimen. Consequently, the reduction in longitudinal residual stress contributed to the lower fatigue crack growth rate.

Figure 6a shows the fatigue crack growth rate curve of the 2024/7075 dissimilar FSWed specimen with a welding speed of 150 mm/min and rational speed of 400 rpm. It was compared with the 2024 and 7075 homogeneous FSWed specimens under the same welding parameters. The results indicated that, during the initial stage of crack growth, the fatigue crack growth rate of the 2024/7075 dissimilar specimens was between that of the 2024 and 7075 homogenous specimens. With further crack propagation, the dissimilar crack growth rate gradually approached that of the 2024 homogeneous specimen. This could be attributed to the lower hardness of the 2024 material, which facilitates crack propagation and ultimately allows the crack to extend into the 2024 side.

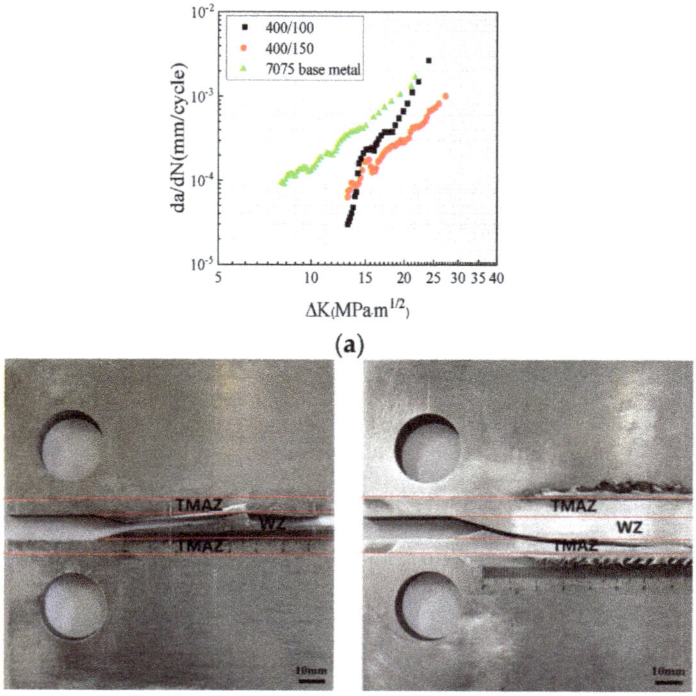

(b) (c)

Figure 5. Fatigue crack growth results of the 7075 specimen: (**a**) crack growth rates; (**b**) crack growth path at 100 mm/min; (**c**) crack growth path at 150 mm/min.

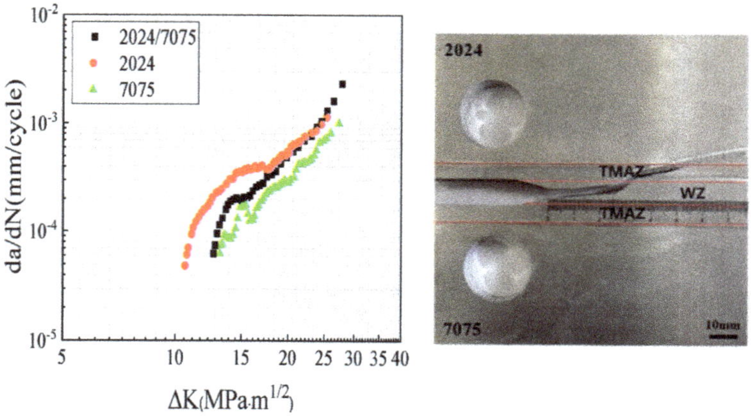

(a) (b)

Figure 6. Fatigue crack growth results of the 2024/7075 dissimilar specimen: (**a**) crack growth rates; (**b**) crack growth path.

5. Numerical Analysis of Fatigue Crack Growth

The residual stress distribution of the FSWed specimens was calculated using ANSYS software, and the stress results were subsequently imported into FRANC3D for crack growth and post-processing analysis. The calculation process is illustrated in Figure 7.

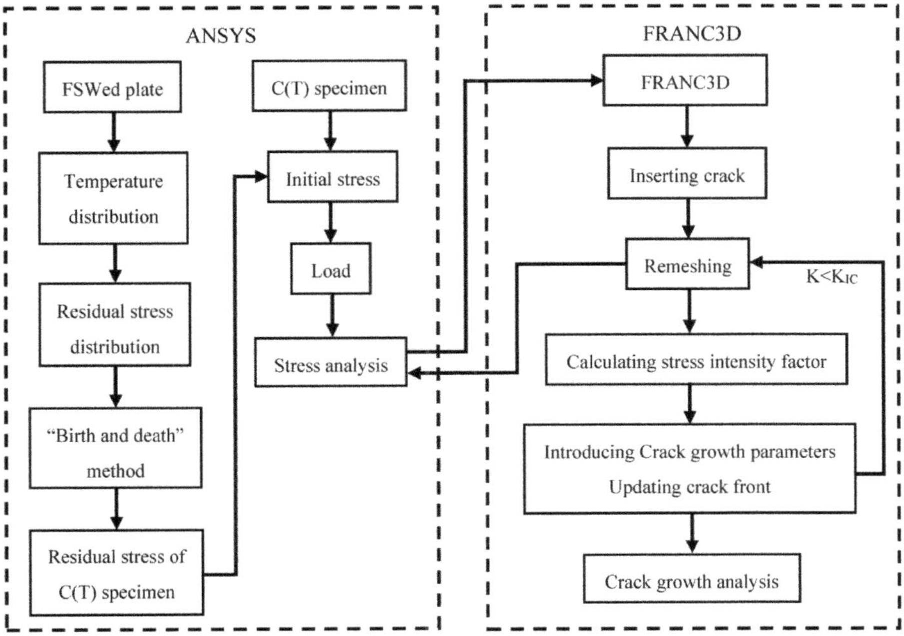

Figure 7. Flow chart of the process for calculating the fatigue crack growth behavior.

5.1. Residual Stress Distribution of the Fatigue Specimen

The coupled thermo-mechanical model was employed to calculate the residual stress distribution of the welded plates [28]. Considering the influence of the mechanical properties on the fatigue crack growth rate, the model was divided into seven zones, which represent the base metal (BM), the advancing HAZ, the advancing TMAZ, the WZ, the retreating TMAZ, and the retreating HAZ, respectively. The thermo-mechanical properties of the alloys are shown in Tables 3–5 [13]. For the 2024/7075 dissimilar welded zone, the parameters were derived as the mean of those for the 2024 and 7075 specimens. Figure 8a shows the size and position of the C(T) specimen. The "birth and death element" method was used to calculate the residual stress of the C(T) specimen (Figure 8b). The FSWed plate model includes 105,992 elements, while the C(T) specimen contains 40,768 elements (Figure 8c).

Table 3. Thermo-mechanical properties of the 2024 and 7075 aluminum alloys.

	Temperature (°C)	**25**	**100**	**200**	**300**	**400**	**500**
2024	Specific heat (J/(kg·K))	900	910	950	1020	1100	1100
	Density (kg/m^3)	2770	2760	2730	2710	2690	2670
	Thermal conductivity (W/(m·K))	126	129	153	178	187	194
	Poisson's ratio	0.33	0.33	0.33	0.33	0.33	0.33
	Young's modulus (MPa)	7.2×10^4	7.1×10^4	6.3×10^4	5.3×10^4	4.5×10^4	4.3×10^4
	Coefficient of thermal expansion ($\times 10^{-5}$/K)	2.11	2.29	2.38	2.47	2.65	2.87
7075	Specific heat (J/(kg·K))	835.4	897	974	1012.5	1128	1205
	Density (kg/m^3)	2800	2800	2800	2800	2800	2800
	Thermal conductivity (W/(m·K))	114.8	128.4	142.2	152.7	160.8	166.7
	Poisson's ratio	0.33	0.33	0.33	0.33	0.33	0.33
	Young's modulus (MPa)	7.3×10^4	7.2×10^4	6.4×10^4	5.6×10^4	4.8×10^4	4.8×10^4
	Coefficient of thermal expansion ($\times 10^{-5}$/K)	2.3	2.44	2.66	2.87	3.09	3.2

Table 4. Mechanical parameters of the 2024 FSWed specimen.

	Temperature (°C)	25	100	200	300	400	500
2024 base metal	Yield stress (MPa)	306	261	152	57	13	5
	Plastic modulus (MPa)	530	400	330	233	95	65
2024 advancing HAZ	Yield stress (MPa)	326	278	162	60	13	5
	Plastic modulus (MPa)	424	289	103	63	35	15
2024 advancing TMAZ	Yield stress (MPa)	205	174	101	38	9	3
	Plastic modulus (MPa)	820	750	600	370	210	90
WZ	Yield stress (MPa)	229	198	113	42	9	3
	Plastic modulus (MPa)	960	870	720	270	105	70
2024 retreating TMAZ	Yield stress (MPa)	211	180	105	39	9	3
	Yield stress (MPa)	820	750	600	370	210	90
7075 retreating HAZ	Yield stress (MPa)	313	267	155	58	13	5
	Plastic modulus (MPa)	424	289	103	63	35	15

Table 5. Mechanical parameters of the 7075 FSWed specimen.

	Temperature (°C)	25	100	200	300	400	500
7075 base metal	Yield stress (MPa)	455	389	278	47	33	20
	Plastic modulus (MPa)	250	210	150	50	15	10
7075 advancing HAZ	Yield stress (MPa)	381	326	233	39	27	16
	Plastic modulus (MPa)	300	260	190	90	45	25
7075 advancing TMAZ	Yield stress (MPa)	252	215	154	26	18	11
	Plastic modulus (MPa)	990	870	775	300	145	90
WZ	Yield stress (MPa)	259	221	158	26	18	11
	Plastic modulus (MPa)	940	830	680	230	75	50
7075 retreating TMAZ	Yield stress (MPa)	264	226	161	79	18	11
	Plastic modulus (MPa)	990	870	775	300	145	90
7075 retreating HAZ	Yield stress (MPa)	371	317	226	38	26	16
	Plastic modulus (MPa)	300	260	190	90	45	25

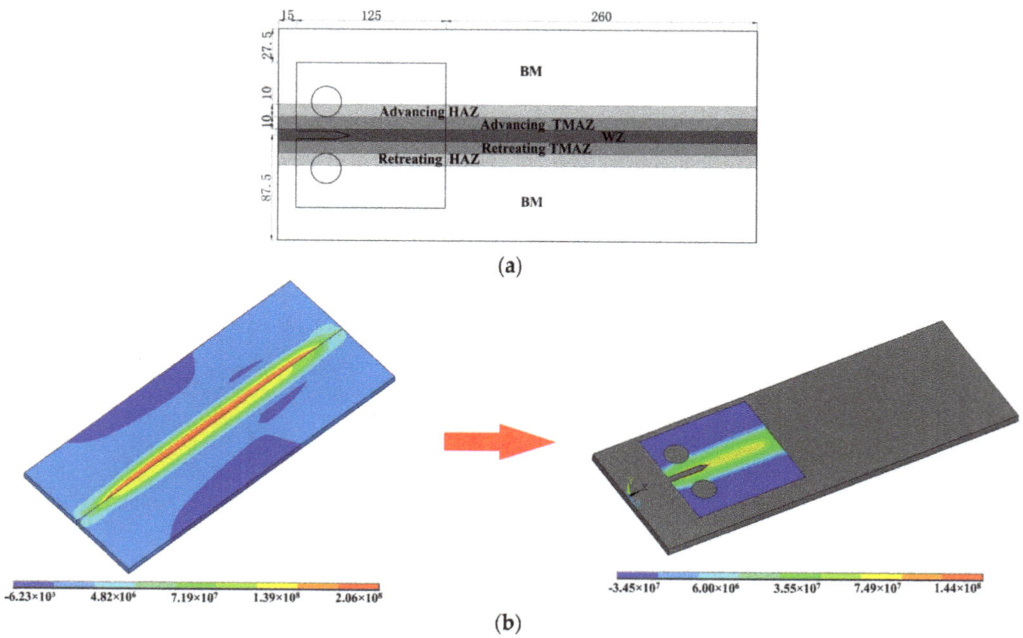

Figure 8. *Cont.*

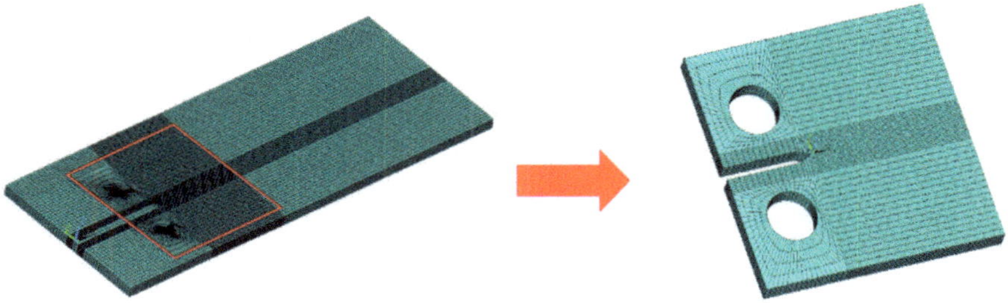

(c)

Figure 8. Residual stress model of the C(T) specimen: (**a**) size of the specimen (mm); (**b**) "birth and death element" method (Pa); (**c**) finite element meshing.

The rate of heat generation can be obtained using the following equation [30]:

$$\dot{q} = \int_{r_0}^{R_0} 2\pi\omega \cdot r^2 \mu(T) p(T) dr = \frac{2}{3}\pi\omega\mu(T)p(T)(R_0^3 - r_0^3) \quad (1)$$

The boundary conditions are illustrated in Figure 9. The surface heat input from the shoulder was applied to the top surface of the specimen with a radius of 10 mm. The volume heat from the pin was generated as a body heat source with dimensions of 5 mm in diameter and 7.5 mm in length. As the temperature varied over time during welding, the coefficient of friction (μ) was assumed to be 0.6, and the pressure of the welding tool (p) was 10 kN. The ambient temperature was set at 25 °C. The heat transfer coefficient of the top and side surfaces was assumed to be 20 W/(m^2 °C), while that of the bottom surface was 300 W/(m^2 °C).

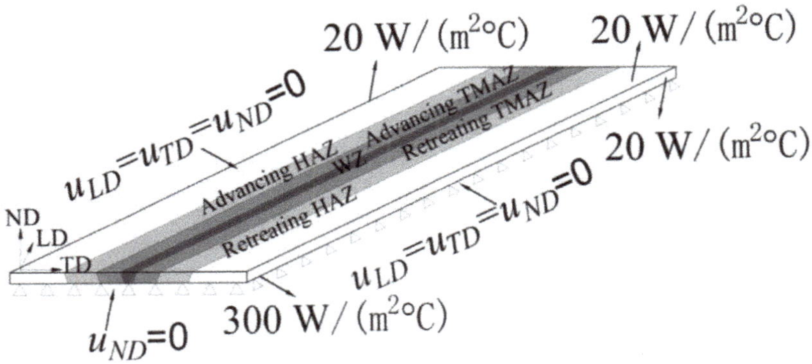

Figure 9. Boundary conditions.

A comparative analysis of the numerical and measured residual stress profiles is shown in Figure 10. The numerical simulation of residual stress also exhibits a "double-peak" pattern. In the advancing TMAZ, the maximum LD stress reaches 126 MPa, which is 22.5 MPa higher than that of the measurements. Moreover, for the TD, the numerical and measured residual stress values are close, with a difference of 5.3 MPa in the retreating TMAZ. It can be seen that the numerical simulation results are in agreement with the X-ray diffraction measurements. Thus, the residual stress model presented in this paper is deemed reliable.

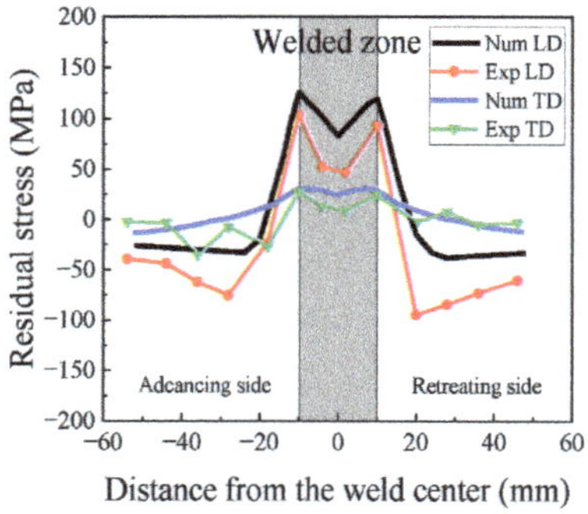

Figure 10. Comparison of the residual stress profiles of the numerical and X-ray diffraction results.

The longitudinal residual stress distribution of the 2024 C(T) specimen is shown in Figure 11a. Path 1 represents the normal direction in the middle of the specimen, while Path 2 highlights its crossing path. Upon comparison with the welded plate, it is evident that the residual stress in the C(T) specimen was redistributed after being removed from the welded plate. Figure 11b illustrates that the residual stress along Path 1 decreased from top to bottom. Figure 11c shows the longitudinal residual stress of the 2024 C(T) specimen with different rotational speeds in Path 2. It can be seen that for the same welding speed (150 mm/min), the maximum longitudinal residual stress in Path 2 increased with the increasing rotational speed. The impact of the welding speed on the residual stress of the 7075 was investigated, and the result is displayed in Figure 11d. The maximum longitudinal residual stress in Path 2 decreased with increasing welding speed.

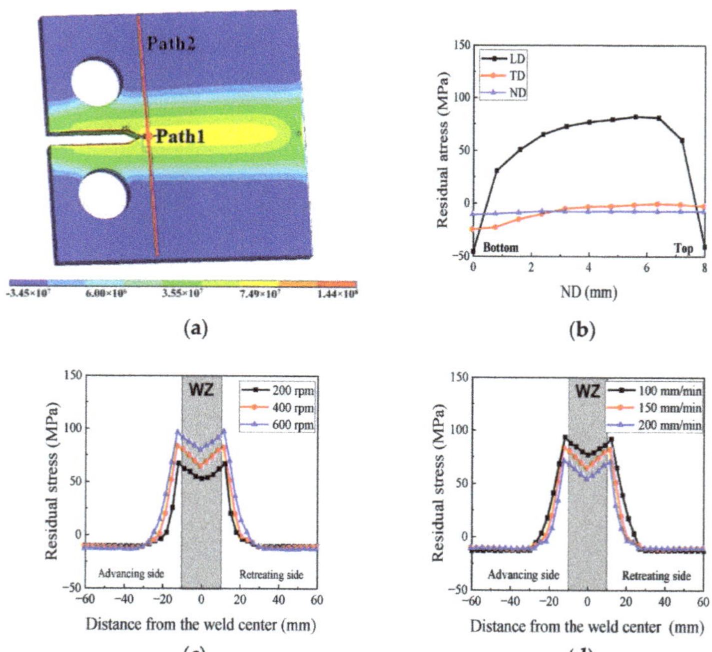

Figure 11. Residual stress of the C(T) specimen: (**a**) longitudinal residual stress distribution of 2024 (400 rpm-150 mm/min) (Pa); (**b**) residual stress in Path1; (**c**) effect of rotational speed on LD residual stress in Path 2; (**d**) effect of welding speed on LD residual stress in Path 2.

5.2. Calculation Results of Fatigue Crack Growth

In this paper, fatigue crack propagation in the FSW joints was designed using the Walker model, which incorporates the effects of residual stress and stress relaxation. The Walker model has a simple form, requires few parameters, and is capable of predicting crack growth during stable propagation. The results of the study in Ref. [24] demonstrated that the Walker model exhibited good consistency with the fatigue crack growth tests.

$$\frac{da}{dN} = C\left[(1-R)^m \Delta K\right]^n \tag{2}$$

where a is the crack length; N is the number of propagation cycles; $\frac{da}{dN}$ is the crack growth rate; C is the coefficient of the Walker model; m and n are the exponents; and ΔK is the stress intensity factor range.

According to the crack growth rates of the 2024 and 7075 base metal specimens, the material constants C, m, and n were determined via data analysis and linear fitting (Table 6). For the dissimilar welded specimens, the average values of the material constants from the 2024 and 7075 base metals were utilized. Figure 12 presents the comparison of the fatigue specimens and the results of the numerical calculation, respectively. It is shown that the crack growth predictions were in agreement with the experimental results.

Table 6. Parameters of the Walker model.

Material	C	m	n
2024	2.71×10^{-10}	4.360	3.375
7075	6.79×10^{-10}	3.957	3.108
2024/7075	4.75×10^{-10}	4.159	3.242

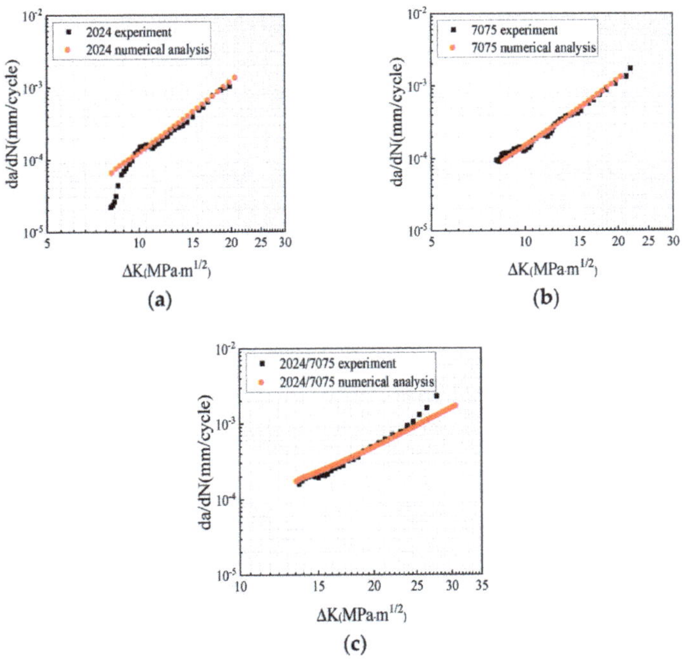

Figure 12. Comparison of the experimental results and numerical analysis: (**a**) 2024 specimen; (**b**) 7075 specimen; (**c**) 2024/7075 dissimilar welded specimen.

The residual stress with a rotational speed of 400 rpm and a welding speed of 150 mm/min was applied to the C(T) specimen. The fatigue crack growth model was designed with the same load conditions as those in the experiments. To assess the effects of the mechanical properties and residual stress on fatigue crack growth behavior, the fatigue crack growth rates under the following five conditions were compared. These conditions, respectively, represent 2024 "base metal", "Mechanical properties" (only considering the mechanical properties of the FSWed specimen), "Longitudinal direction" (only considering the residual stress of the longitudinal direction), "Transverse direction" (only considering the residual stress of the transverse direction), and "Normal direction" (only considering the residual stress of the normal direction). The results shown in Figure 13a demonstrate that the 2024 base metal exhibited the highest fatigue crack growth rate. However, the crack growth rate decreased when the residual stresses of the transverse and normal directions were considered in the calculations. This decrease occurred because of the presence of compressive stress in these two directions, although the stress level was quite low. On the other hand, the tensile stress in the longitudinal direction led to an increase in the crack growth rate.

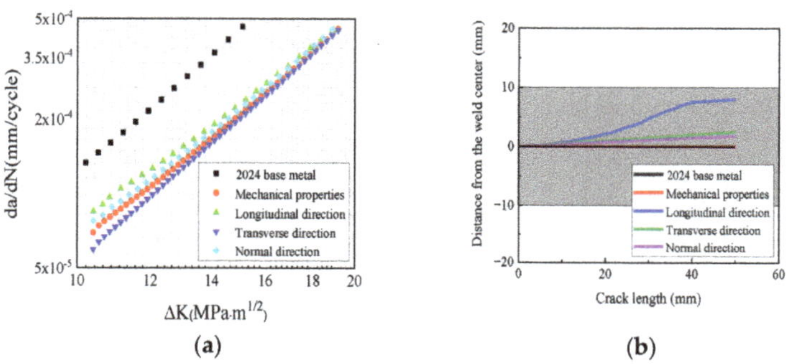

Figure 13. Comparison of the fatigue crack growth rate in the 2024 specimen: (**a**) curve of the fatigue crack growth rate; (**b**) path of the fatigue crack growth.

The comparison of the fatigue crack growth paths is presented in Figure 13b. The crack growth path of the "Mechanical properties" condition aligns with that of the base metal, indicating that the mechanical properties of the FSWed specimen did not influence the fatigue crack deflection. When residual stress was taken into account, all the cracks in the specimen deflected upward. However, the crack deflection angle that resulted from the longitudinal residual stress was larger than that observed in the other directions. This difference can be attributed to the higher residual stress value in the longitudinal direction and the greater stress gradient along the thickness of the specimen as compared to the transverse and normal directions.

The comparison of the fatigue crack growth rates for the 2024 specimen under different rotational speeds is shown in Figure 14a. The results demonstrate that in the 2024 FSWed specimen, the fatigue crack growth rate increased with the increasing rotational speed. This occurred because the residual stress in the longitudinal direction in the 600 rpm specimen was higher than that in the other directions. Thus, the changes in mechanical properties that occurred due to grain refinement in the welded joint were counteracted, which thereby led to an increased rate of crack growth. In Figure 14b, the fatigue crack growth rate of the 7075 specimens decreased with the increasing welding speed. Hence, for the 7075 fatigue specimens, the fatigue crack growth rate was also predominantly influenced by the residual stress.

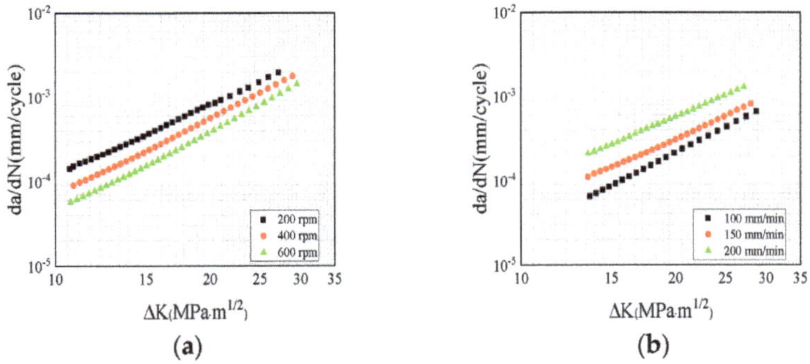

Figure 14. Effect of the welding parameters on fatigue crack growth: (**a**) effect of rotational speed (2024 specimen); (**b**) effect of welding speed (7075 specimen).

6. Microstructure Analysis

The fracture morphologies of the 2024 (400 rpm-150 mm/min) specimens in the fatigue propagation zone are presented in Figure 15. These specimens exhibited fatigue striations. Notably, the crack at the front exhibited an inclination rather than expanding vertically. This phenomenon occurred as the residual stress in the specimen gradually decreased from the top to the bottom, and because the grain size at the bottom of the WZ was smaller than that at the top [13]. As a result, the rate of fatigue crack propagation at the top was faster than that at the bottom.

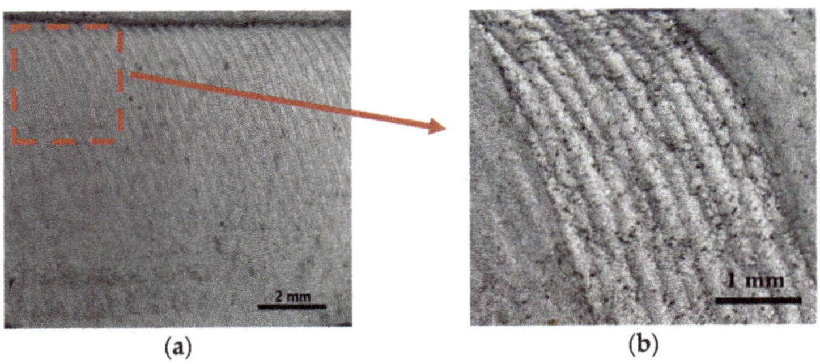

Figure 15. (**a**) Fracture morphology of the 2024 400 rpm-150 mm/min specimen; (**b**) local enlarged image.

In Figure 16, the TEM diagrams of the 2024 welded joints with different rotational speeds are shown using the Talos F200X at Xi'an University of Architecture and Technology in Xi'an, China. The TEM images of the WZ confirm the presence of needle-shaped S-phase (Al_2CuMg) precipitates [31]. In the WZ (Figure 16a,d), most of the dislocations appeared in the form of distortion and entanglement, which resulted from the significant distortion caused by the rotation of the stirring probe. In the TMAZ, the S-phase was also present, with more dislocation entanglement observed around the grain boundaries (Figure 16b,e). This corresponds to the residual stress measurements shown in Figure 2. In the HAZ of the 400 rpm-150 mm/min specimen (Figure 16c), the precipitated S-phase increased due to the influence of the welding thermal cycling, while the extent of dislocation entanglement decreased relatively. In Figure 16f, fewer dislocation cell substructures and $AlCu_3$ precipitates (remnants from the aging treatment process) were observed in the base material.

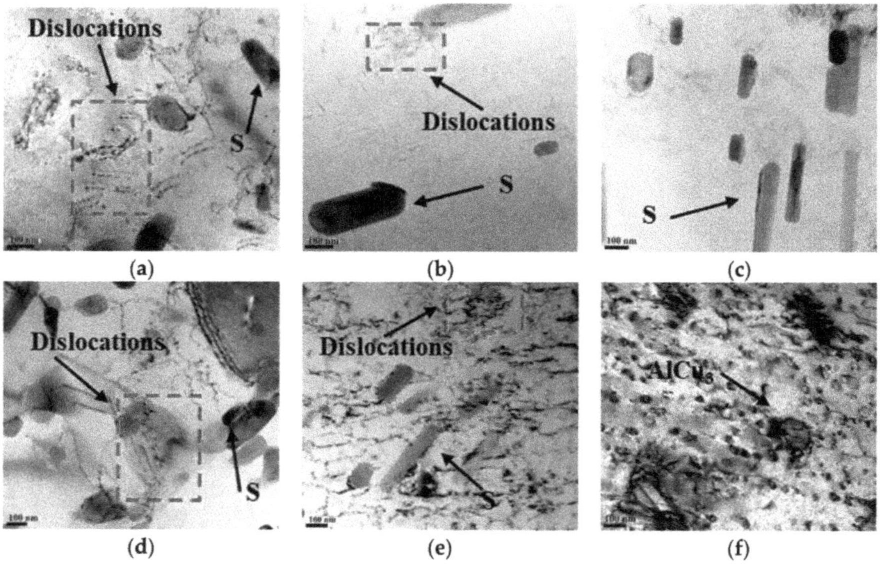

Figure 16. TEM diagrams of the welded joints of the 2024 specimen: (**a**) 400 rpm-150 mm/min WZ; (**b**) 400 rpm-150 mm/min advancing TMAZ; (**c**) 400 rpm-150 mm/min HAZ; (**d**) 200 rpm-150 mm/min WZ; (**e**) 200 rpm-150 mm/min advancing TMAZ; (**f**) base metal.

A comparison between the 200 rpm and 400 rpm specimens revealed that the 200 rpm specimen exhibited a higher dislocation density and a greater presence of precipitated phases in the WZ and TMAZ. During the process of fatigue crack propagation, these precipitated phases not only pinned dislocations but also facilitated the generation of additional dislocations, thereby leading to dislocation entanglement. As a result, the dislocation density and sliding resistance increased within the welding joint, which thus resulted in a decreased rate of fatigue crack propagation. This explains why the fatigue crack growth rate of the 200 rpm specimen is lower than that of the 400 rpm from a microstructural perspective, as shown in Figure 5a. Additionally, when compared to the WZ, the S-phase precipitates in the TMAZ and HAZ were coarser; this is a phenomenon that is well-documented in the literature [32,33]. Therefore, the crack predominantly propagated toward the TMAZ, as shown in Figures 4 and 5.

7. Conclusions

In this study, the effect of residual stress and microstructure on the fatigue crack growth behavior of 2024-T3 and 7075-T6 friction stir welded joints with different welding parameters was analyzed. The key findings are as follows:

(1) The welding parameters significantly influenced the fatigue crack growth rate. For the 2024 specimens, the fatigue crack growth rate increased with the higher rotational speed, while for the 7075 specimens, it decreased with the higher welding speed.
(2) The longitudinal residual stress had the most significant impact on crack deflection. Additionally, the tensile residual stress in the welded joints accelerated crack growth, partially counteracting the reduction in the fatigue crack growth rate due to grain refinement. Conversely, the compressive residual stress reduced the stress intensity factor, thereby resulting in a diminished crack driving force.
(3) The microstructural characteristic of the welded joints aligns with the distribution of residual stress. It was revealed that the precipitated phase both pinned dislocations and induced the generation of additional dislocations, thereby resulting in dislocation plugging. Consequently, as the rotational speed decreased, the dislocation density and

sliding resistance increased, which thus led to a decreased rate of fatigue crack propagation. Furthermore, the TMAZ and HAZ contained coarser precipitates compared to the WZ, causing the crack to deflect towards the TMAZ.

Author Contributions: Conceptualization, Y.G.; Methodology, Y.G.; Software, Y.G.; Resources, P.L.; Data curation, P.L.; Writing—original draft, Y.G.; Funding acquisition, Y.G. All authors have read and agreed to the published version of the manuscript.

Funding: This research received external funding from the Post-doctoral Research Project in Shaanxi Province (No. 2023BSHEDZZ266).

Data Availability Statement: Data are contained within the article.

Conflicts of Interest: The authors declare no conflict of interest.

References

1. Rajendran, C.; Kumar, M.V.; Sonar, T.; Mallieswaran, K. Investigating the Effect of PWHT on microstructural features and fatigue crack growth behavior of friction stir welded AA2024-T6 aluminum alloy joints. *Forces Mech.* **2022**, *8*, 100107. [CrossRef]
2. Thomas, W.M.; Nicholas, E.D.; Needham, J.C.; Murch, M.G.; Temple-Smith, P.; Dawes, C.J. Friction Stir Butt Welding. International Patent Application No. PCT/GB92102203; Great Britain Patent Application No. 9125978.8, 6 December 1991.
3. Oliveira, J.P.; Duarte, J.F.; Inácio, P.; Schell, N.; Miranda, R.M.; Santos, T.G. Production of Al/NiTi composites by friction stir welding assisted by electrical current. *Mater. Des.* **2017**, *113*, 311–318. [CrossRef]
4. Ke, W.C.; Oliveira, J.P.; Ao, S.S.; Teshome, F.B.; Chen, L.; Peng, B.; Zeng, Z. Thermal process and material flow during dissimilar double-sided friction stir spot welding of AZ31/ZK60 magnesium alloys. *J. Mater. Res. Technol.* **2022**, *17*, 1942–1954. [CrossRef]
5. Zhang, X.; Ma, Y.E.; Wang, Z.; Guo, Y. The effect of residual stress on mixed-mode crack propagation behaviour in friction stir welded 7075-T6 aluminium alloy panel under biaxial loading. *Fatigue Fract. Eng. Mater. Struct.* **2021**, *44*, 805–821. [CrossRef]
6. Das, J.; Banik, S.R.; Reddy, S.R.S.K.; Sankar, M.R.; Robi, P.S. Review on process parameters effect on fatigue crack growth rate in friction stir welding. *Mater. Today Proc.* **2019**, *18*, 3061–3070. [CrossRef]
7. Luan, G.H.; North, T.H.; Guo, D.L. Characterizations of frictions stir welding on aluminum alloy. *Trans. China Weld. Inst.* **2002**, *23*, 62–66.
8. Milan, M.T.; Bose Filho, W.W.; Tarpani, J.R.; Malafaia, A.M.D.S.; Silva, C.P.O.; Pellizer, B.C.; Pereira, L.E. Residual stress evaluation of AA2024-T3 friction stir welded joints. *J. Mater. Eng. Perform.* **2007**, *16*, 86–92. [CrossRef]
9. Dattoma, V.; Giorgi, M.D.; Nobile, R. An experimental study on residual stress relaxation in low-cycle fatigue of Inconel 718 Plus. *Fatigue Fract. Eng. Mater. Struct.* **2022**, *46*, 310–324. [CrossRef]
10. Liu, R.F.; Young, M.C. Influence on structure and fracture mechanics evaluation of a BWR feedwater nozzle following weld overlay repair. *J. Mech.* **2023**, *39*, 113–134. [CrossRef]
11. He, X.C.; Gu, F.S.; Ball, A. A review of numerical analysis of friction stir welding. *Prog. Mater. Sci.* **2014**, *65*, 1–66. [CrossRef]
12. Mishra, R.S.; Ma, Z.Y. Friction stir welding and processing. *Mater. Sci. Eng. R Rep.* **2005**, *50*, 1–78. [CrossRef]
13. Guo, Y.N.; Ma, Y.E.; Zhang, X.S.; Li, J. Study on residual stress distribution of 2024-T3 and 7075-T6 aluminum dissimilar friction stir welded joints. *Eng. Fail. Anal.* **2020**, *118*, 104911. [CrossRef]
14. Cai, B.; Zheng, Z.Q.; He, D.Q.; Li, S.C.; Li, H.P. Friction stir weld of 2060 Al-Cu-Li alloy: Microstructure and mechanical properties. *J. Alloys Compd.* **2015**, *649*, 19–27. [CrossRef]
15. Liu, M.; Tao, X.; Di, Z.; Qin, M.; Liu, Z.; Bai, S. Effect of Pre-Rolling on Microstructure and Fatigue Crack Propagation Resistance of a Third-Generation Al-Li Alloy. *Materials* **2023**, *16*, 7540. [CrossRef]
16. Wang, F.F.; Li, W.Y.; Shen, J.; Hu, S.Y.; dos Santos, J.F. Effect of tool rotational speed on the microstructure and mechanical properties of bobbin tool friction stir welding of Al-Li alloy. *Mater. Des.* **2015**, *86*, 933–940. [CrossRef]
17. Prangnell, P.B.; Heason, C.P. Grain structure formation during friction stir welding observed by the stop action technique. *Acta Mater.* **2005**, *53*, 3179–3192. [CrossRef]
18. Threadgill, P.L.; Leonard, A.J.; Shercliff, H.R.; Withers, P.J. Friction stir welding of aluminium alloys. *Int. Mater. Rev.* **2009**, *54*, 49–93. [CrossRef]
19. Vuherer, T.; Milčić, M.; Glodež, S.; Milčić, D.; Radović, L.; Kramberger, J. Fatigue and fracture behavior of Friction Stir Welded AA-2024-T351 joints. *Theor. Appl. Fract. Mech.* **2021**, *114*, 103027. [CrossRef]
20. Lan, Z.; Zhong, H.L.; Li, S.C.; Zhao, H.; Chen, J.; Qi, L. Microstructure, mechanical properties and fatigue crack growth behavior of friction stir welded joint of 6061-T6 aluminum alloy. *Int. J. Fatigue* **2020**, *135*, 105556.
21. Pasta, S.; Reynolds, A.P. Residual stress effects on fatigue crack growth in a Ti-6Al-4V friction stir weld. *Fatigue Fract. Eng. Mater. Struct.* **2008**, *31*, 569–580. [CrossRef]
22. Bussu, G.; Irving, P.E. The role of residual stress and heat affected zone properties on fatigue crack propagation in friction stir welded 2024-T351 aluminium joints. *Int. J. Fatigue* **2003**, *25*, 77–88. [CrossRef]
23. Fratini, L.; Zuccarello, B. An analysis of through-thickness residual stresses in aluminium FSW butt joints. *Int. J. Mach. Tools Manuf.* **2006**, *46*, 611–619. [CrossRef]

24. Fratini, L.; Pasta, S.; Reynolds, A.P. Fatigue crack growth in 2024-T351 friction stir welded joints: Longitudinal residual stress and microstructural effects. *Int. J. Fatigue* **2009**, *31*, 495–500. [CrossRef]
25. Ilman, M.N.; Kusmono, P.; Iswanto, T. Fatigue crack growth rate behaviour of friction-stir aluminium alloy AA2024-T3 welds under transient thermal tensioning. *Mater. Des.* **2013**, *50*, 235–243. [CrossRef]
26. Anderson-Wedge, K.; Stubblefield, G.; Zhu, N.; Long, B.; Daniewicz, S.R.; Allison, P.; Sowards, J.; Rodriguez, O.; Amaro, R. Characterization of the evolution of 2219-T87 aluminum as a function of the friction stir welding process. *Int. J. Fatigue* **2021**, *142*, 105954. [CrossRef]
27. *ASTM-647-2011*; Standard Test Method for Measurement of Fatigue Crack Growth Rates 1 [S]. ASTM International: West Conshohocken, PA, USA, 2008.
28. Buglioni, L.; Tufaro, L.N.; Svoboda, H.G. Thermal cycles and residual stresses in FSW of aluminum alloys: Experimental measurements and numerical models. *Procedia Mater. Sci.* **2015**, *9*, 87–96. [CrossRef]
29. Guo, Y.N.; Ma, Y.E.; Wang, F. Dynamic fracture properties of 2024-T3 and 7075-T6 aluminum friction stir welded joints with different welding parameters. *Theor. Appl. Fract. Mech.* **2019**, *104*, 102372. [CrossRef]
30. Chen, C.M.; Kovacevic, R. Finite element modeling of friction stir welding-thermal and thermo mechanical analysis. *Int. J. Mach. Tools Manuf.* **2003**, *43*, 1319–1326. [CrossRef]
31. Niu, P.; Li, W.; Yang, C.; Chen, Y.; Chen, D. Low cycle fatigue properties of friction stir welded dissimilar 2024-to-7075 aluminum alloy joints. *Mater. Sci. Eng. A* **2022**, *832*, 142423. [CrossRef]
32. Genevois, C.; Deschamps, A.; Denquin, A.; Doisneau-Cottignies, B. Quantitative investigation of precipitation and mechanical behaviour for AA2024 friction stir welds. *Acta Mater.* **2005**, *53*, 2447–2458. [CrossRef]
33. Zhang, Z.; Xiao, B.L.; Ma, Z.Y. Hardness recovery mechanism in the heat-affected zone during long-term natural aging and its influence on the mechanical properties and fracture behavior of friction stir welded 2024Al-T351 joints. *Acta Mater.* **2014**, *73*, 227–239. [CrossRef]

materials

MDPI

Article

Characteristics and Quality of Flame-Retarded Ramie Fabrics for the Development of Functional Textiles

Asri Peni Wulandari [1,2,*], Erlin Karlina [3], Eric Tanudjaja [3], Abdul Rohmat [4], Joko Kusmoro [1,2], Muhammad Fadhlillah [5], Karlina Somantri [6], Roni Sahroni [6] and Widya Fatriasari [7]

[1] Department of Biology, Faculty of Mathematics and Natural Science, Padjadjaran University, Sumedang 45363, Indonesia; joko.kusmoro@unpad.ac.id

[2] Center for Study of Bioprospection of Natural Fiber and Bioresources, Faculty of Mathematics and Natural Science, Padjadjaran University, Bandung 40132, Indonesia

[3] PT Dymatic Chemicals Indonesia, Bandung 40552, Indonesia; erlinkarlina512@gmail.com (E.K.); eric.tanudjaja@dymatic.co.id (E.T.)

[4] PT Haramai Jaya Nusantara, Bandung 40132, Indonesia; abdul.rohmat@rajantara.com

[5] Department of Chemistry, Faculty of Mathematics and Natural Science, Padjadjaran University, Sumedang 45363, Indonesia; fadhlillah@unpad.ac.id

[6] Bandung Polytechnic of Textile Technology, Bandung 40272, Indonesia; karlina.somantri@gmail.com (K.S.); roni.sahroni@stttekstil.ac.id (R.S.)

[7] Research Center for Biomass and Bioproduct, National Research and Innovation Agency, Bogor 16911, Indonesia; widy003@brin.go.id

* Correspondence: asri.peni@unpad.ac.id

Abstract: Cellulose fabric testing for flame-retardant studies is frequently necessary in various textile applications. Natural cellulose material from ramie (*Boehmeria nivea*) is being promoted as an alternative raw material for the development of fire-resistant fabrics. This research aims to optimize the coating process of ramie fabric using a phosphorus-based flame retardant (FR) to enhance its flame-retardant characteristics. The FR treatment involves bleaching the fabric with H_2O_2; followed by fabric finishing using a formula comprising 3% (v/v) hydroxymethyl resin; phosphoric acid (2%); and two formulations of the flammable agent Flamatic DM-3072N: 40% (v/v) and 50% (v/v), applied using the pad-dry-cure method. The flame-retardant properties of the treated fabric are evaluated through flammability testing based on the ASTM D6413-08 standard, limiting oxygen index (LOI) analysis, and micrograph surface structure analysis with SEM. The results indicate that ramie fabric treated with the FR-50% material exhibits superior fire resistance, preventing fire spread on the fabric with a char length of 15–30 mm and a LOI value of 29. These findings highlight the potential of FR-treated ramie fabrics for various industries, including the automotive and protective clothing industries.

Keywords: natural fiber; ramie fabric; flammability; flame retardant; coating; functional textile

1. Introduction

Wear performance represents one of the most rapidly expanding segments within the textile industry. Market expansion is fueled by the introduction of novel advancements in fabric technology. The evolution of the fabric sector has spurred the creation of high-tech protective textiles characterized by enhanced tensile strength, cut resistance, abrasion resistance, and durability.

Natural fibers such as cotton and linen are frequently employed as raw materials in textile manufacturing [1,2], owing to their extensive presence in the market due to their remarkable attributes of softness, comfort, and hydrophobicity [3,4].

The utilization of natural fibers in various applications is still hindered by their mechanical quality, particularly their low strength compared to synthetic fibers [5]. This inherent weakness in natural fibers stems from their polar nature and abundant hydrogen bonds, rendering them susceptible to water absorption, particularly in high-humidity (hydrophilic) environments. The absorption of water disrupts molecular chain interactions, leading to heightened water diffusivity and potential uptake of water vapor from the atmosphere [6]. The ingress of water into fibers can lead to weathering, diminished compatibility of fiber composites, an increase in weight, and susceptibility to degradation [7].

Despite these disadvantages, natural fibers or cellulose fibers also possess several advantages. For example, cotton fiber and pineapple fiber are non-flammable materials with an LOI value of approximately 18% [8–10]. Several methods have been explored to enhance the resistance of cotton-based textiles, including the sol-gel technique [11,12], layer-by-layer (LbL) assembly [13,14], plasma processing [15,16], flame-retardant finishing technology [17,18], coating technology [19], and surface-grafting treatment [20].

Renewable resources, such as DNA [21], chitosan [22,23], protein [24], starch [15,25], and phytic acid (PA) [16,22,26,27] have been employed as flame retardants over the past decade. The results demonstrate a significant impact of the mentioned renewable resources in enhancing the fire resistance of natural fiber-based fabrics. PA, easily extractable from various plant parts except leaves [28], contains six phosphate groups and twelve hydroxyl groups, enabling interaction with almost all types of reactive groups [22].

Dittenber (2012) [29] reported that ramie fiber, a cellulose-based natural fiber, exhibits the highest mechanical properties among other natural fibers. In our previous research, the modification of ramie fibers through cross-linking methods by adding CA (citric acid) and SHP (Sodium Hypophosphite) was found to enhance the mechanical strength of ramie fibers by up to 20 times [30]. This highlights the potential of enhancing ramie fibers through refinement methods for various applications. In this study, a highly strengthened ramie fabric will be developed for fire-resistant functional applications.

Several methods reported for enhancing the fire resistance properties of modified ramie fabric have utilized the Schiff base reaction between aldehyde groups and amino groups to pretreat the fabric surface. In this regard, the utilization of Polyethyleneimine (PEI) acts as the positive charge, while ammonium polyphosphate (APP) serves as the negative charge [31]. Another method involves using a polyethylenimine (PEI) solution and phosphoric acid zirconium [32]. Additionally, a method using an adhesive layer of rosin benzoxazine containing a catalyst mixture in the form of methyl tosylate and iron acetylacetonate with the fire inhibitors being polyethyleneimine (PEI) and poly bis phenolic phenyl phosphate ester [33]. Although these methods significantly improve fire resistance, the extensive use of finishing additives and organic solvents still limits their application prospects. These methods are conducted with multilayer module units assembled layer by layer, requiring the assembly of 15–30 bilayers to achieve satisfactory fire-resistant modifications, thus constraining their widespread use in the industry.

In this research, further investigation was conducted on ramie fabric as a cellulose material, dissolving it using a simpler and more practical method without layering lamination processes. Additionally, testing was conducted using a non-ionic and non-toxic substance, Flamatic DM-3072N, as a nitrogen derivative product containing organophosphorus compounds. The flame-retardant capability will be assessed through characterization of ramie fabric and quality flammability tests.

2. Materials and Methods

2.1. *Materials*

The fine-fiber woven ramie fabric (KR-T01) is a specially woven fabric made from ramie fiber, produced through the spinning and fabric provision processes conducted by PT Haramai Jaya Nusantara, Bandung, Indonesia. The characteristics of the plain fabric are shown in Table 1. The chemicals and flame retardant (FR) DM-3072N (a nitrogen derivate containing an organophosphorus compound) were provided by PT Dymatic Chemicals Indonesia, Bandung, Indonesia. FR DM-3072N was chosen due to its suitability for durable flame-retardant finishing of cotton fabric, providing the fabric with excellent flame-retardant effect, non-toxicity, and good compatibility [34].

Table 1. Characterization of woven ramie fabric (KR-T01).

No.	Characterization	Woven Fabric	Standards Test
1.	Weave Type	Plain	SNI 0458:2013; SNI ISO 7211-1:2010
2.	Thread Number (tex)	25.26	SNI ISO 7211-5:2010
3.	Strand Yarn (strands/inch)	Warp: 52 Weft: 57	SNI ISO 7211-2:2010
4.	Fabric Thickness (mm)	0.28	SNI ISO 5084:2010
5.	Fabric Weight (g/m)	186.97	SNI ISO 3801:2010
6.	Fabric Width (cm)	159.64	SNI ISO 22198:2010
7.	Change in Size After Washing (%)	−3.21	SNI ISO 6330-2015
8.	Fabric Tensile Strength/2.5 cm (N)	Warp: 249.76 Weft: 207.38	SNI 0276-2009
9.	Fabric Tear Strength (n)	Warp: 55.52 Weft: 37.08	SNI 0521-2008

2.2. Characterization of Ramie Fabric

Ramie fabric samples, before being treated with FR, were characterized based on the SNI 0458:2013 standard: SNI ISO 7211-1:2010 for fabric construction/woven types; the thread number test with SNI ISO 7211-5:2010, the thread total test (SNI ISO 7211-2:2010); the fabric thickness test (SNI ISO 5084:2010), the fabric weight test (SNI ISO 3801:2010), fabric width (SNI ISO 22198:2010), the size change test after washing (SNI ISO 6330:2015), fabric tensile strength/2.5 cm (SNI 0276:2009), as well as fabric tear strength testing (SNI 0521:2008).

2.2.1. Peroxide Treatment

The bleaching process was conducted according to Repon et al. (2021) [35] with some modifications. This process aims to eliminate lignin and other impurities from the ramie fabric, resulting in the loss of certain cellulose fibers along with lignin content [36]. In brief, the treatment with H_2O_2 [Merck] was carried out using the following recipe: hydrogen peroxide at a concentration of 4% (w/v), Sodium hydroxide at 2% (w/v) [Sigma-Aldrich, Merck KGaA, Darmstadt, Germany], and a stabilizing agent at 0.1% (w/v) [DIMASTAB CEB], with a material to liquor ratio of 1:40. The process was conducted at 100 °C for 30 min. Subsequently, after bleaching, the fabric underwent a cold wash with aquades at room temperature for 10 min followed by drying at 120 °C for 3 min.

2.2.2. Preparation of Flame-Retardant Fabric

The process of finishing the ramie fabric with a flame retardant refers to the Raymond Li (2015) method [37] with modifications. The ramie fabric samples were coated with basal recipes based on 3% hydroxyl methyl resin [DYMAFIN DM-3511A] and 2% phosphoric acid with variation of Flamatic DM-3072N at concentrations of 40% (v/v) and 50% (v/v). These concentrations were selected based on their superior flame-retardant properties compared to other variations tested previously. The treatment process utilized the pad-dry-cure method at temperatures ranging from 150 to 170 °C for 1–5 min. The FR process was evaluated using the standard test method for the flame resistances of textiles (the vertical test) [Method: ASTM D6432].

2.3. Flammability Testing and Evaluation

The combustion resistance of the treated specimens was evaluated using the vertical test method, specifically ASTM D6413. The specimens used for this test were sized 10 × 3 cm. Vertical testing was carried out to evaluate the flame spread characteristics on the textile specimens. A flame source was ignited beneath the specimen for a duration of 10 s and then removed, allowing the material to self-extinguish. The evaluation of combustion was performed according to predefined criteria outlined in Table 2.

Table 2. The evaluation of combustion was conducted based on predefined criteria.

Criteria Conditions	Classification		
	B1	B2	B3
Char length (cm)	≤15	≤20	No special treatment
After flame time (s)	≤5	≤15	
After glow time (s)	≤15	≤30	

Notes: Char length: when measuring the fire resistance of textiles, the distance from the edge of the fabric, which is directly exposed to fire to the furthest point of visible fabric damage after the specified tearing force has been applied. After flame time: the length of time that a material continues to burn after cessation (natural or induced) of the material's combustion. Afterglow time: the incandescent time continues after removal of the ignition source and/or cessation of ignition.

The flammability assessment aimed to evaluate the fire resistance of the fabric samples treated with FR formulations intended for potential development as a functional textile. During this stage, the burning time and char length were examined. Cheng et al. (2016) [38] categorized fabrics into three groups based on the vertical burning test according to GB 8624-2012.

2.4. Limiting Oxygen Index (LOI)

The limiting oxygen index (LOI) represents the minimum oxygen fraction required in a mixed environment of oxygen (O_2) and nitrogen (N_2) to sustain vertical combustion. The test was conducted using a Dynisco Plastics LOI analyzer, following ASTM Standard Method D2863 [39]. The sample dimensions were 52 mm × 140 mm. The sample was ignited from the top end using a butane flame (within 20 s) and allowed to burn downwards. The test was repeated by adjusting the volumetric flow rate of O_2 and N_2 until a specific burning time of 3 min or a burning length of 50 mm was achieved. LOI is calculated as follows:

$$\text{LOI}\ (\%) = \frac{[O_2]}{[O_2]+[N_2]} \times 100$$

where $[O_2]$ is the volumetric discharge of oxygen and $[N_2]$ is the volumetric discharge of nitrogen.

2.5. Fourier-Transform Infrared (FTIR) Analysis

Fourier-transform infrared (FTIR) analysis was conducted on the ramie fabric samples with FR coating treatment and without FR coating treatment. The analysis was performed using a Perkin Elmer FTIR (Model: System 2000, PerkinElmer Corporation, Waltham, MA, USA) employing the ATR (attenuated total reflectance) method with spectrum two specs, resolution 4 cm^{-1}, and 16 scans. This FTIR analysis generated data in the form of absorbance or transmittance graphs. The FTIR analysis was conducted quantitatively by observing the spectrum's shape at specific peaks and the intensity of the FTIR graph. The tests were carried out in the wave number range 400–4000 cm^{-1}, with a resolution set at 2 cm^{-1} to examine the absorption peak of the OH group (hydroxyl) at the wave number 3000–3650 cm^{-1} and the absorption of -COOR′ (ester) at the wave number 1710–1780 cm^{-1}.

2.6. NMR Analysis

Nuclear magnetic resonance (NMR) analysis was used to obtain information about the structure and conformation of chemical compounds. NMR spectroscopy serves as a reliable method for elucidating the structure of organic compounds [40]. The NMR analysis was conducted utilizing a Bruker AV III 600 spectrometer (Billerica, MA, USA). Signals obtained from NMR spectroscopy provide information about interactions between nuclei and electrons as well as nuclear interactions, aiding in the determination of chemical compound structures [41]. The resulting NMR spectrum comprises one or more resonance peaks at specific frequencies.

2.7. Micrograph Surface Structure Analysis

Micrograph surface structure analysis was conducted to observe the morphology of ramie fabric samples with and without FR coating, carried out with an acceleration voltage of 10 kV. The surface structure of the ramie fabric was observed using a scanning electron microscope (SEM) equipped with a FE-SEM Thermo Scientific Quattro S completed with an EDS detector, WetSTEM, heating stage, and tensile stage. The samples were coated with gold using a Cressington 208HR High-Resolution sputter coater (Cressington Scientific Instruments, Watford, UK). Magnification ranged from 100 to 10,000 times until surface objects were visible.

3. Results and Discussions

3.1. The Structural Properties of the Fabrics

The ramie fabric utilized in this study is a woven material identified by the code KR-T01. The property characterization findings of the ramie fabric, determined through testing according to the Indonesian National Standard (SNI), are presented in Table 1.

3.2. Flame Retardancy and Flammability

To assess the fire-resistant properties of the treated ramie fabric, we conducted limiting oxygen index (LOI) and vertical flammability test on ramie fabrics treated with various concentrations of the Flamatic retardant. The results are presented in Figure 1 and Table 3. As shown in Figure 1, the untreated ramie fabric exhibited no fire resistance during the vertical flammability test, evidenced by a flame time of 20 s and a glow time of 78 s.

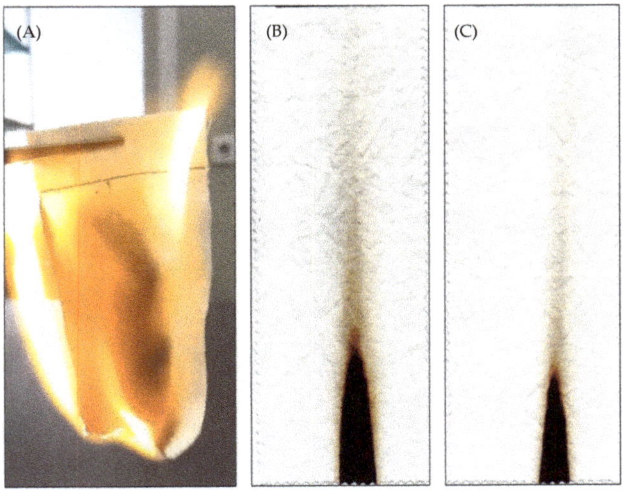

Figure 1. Vertical flammability test of: (**A**) the untreated ramie fabric (**B**) 40%—FR (**C**) 50%—FR.

Additionally, when the concentration of the Flamatic retardant is set to 40% and 50%, the flame time and glow time for all treated ramie fabrics are reduced to 0 s, while the char length ranges from 15 to 30 mm. This indicates that the treated ramie fabrics successfully passed the vertical flammability test (Table 3). Various natural materials, including cotton and wool, treated with different FR methods and materials, have exhibited signs of improved fire resistance. The findings of this study suggest that ramie fabric treated with Flamatic-DM-3072N demonstrates superior quality as a fire-resistant functional textile.

The phosphorus compounds contained in the flame retardant act as an acid donor or dehydrogenation catalyst. Meanwhile, the nitrogen compounds act as blowing agents, releasing the necessary gas (N_2) for foam formation [42]. The Phosphorus–Nitrogen synergistic flame retardant was more widely employed due to its excellent efficiency to achieve optimal performance of FR [43,44].

Moreover, these compounds have broader applications and are currently gaining increased attention. The flame-retardant mechanism of the traditional phosphorus flame retardant is to form phosphoric acid to promote cationic crosslinking after combustion. The addition of nitrogen leads to the emission of substantial amounts of non-toxic, non-combustible gases when heated. This synergistic effect of nitrogen and phosphorus contributes significantly to achieving the desired flame-retardant properties [45,46].

Table 3. Vertical flammability test of ramie fabric and compared to other natural fiber material.

No.	Material	After Flame Times (s)	After Glow Times (s)	Dripping	Ignition	Char Length (mm)	Reference
1.	Ramie fabric (untreated)	20	78	No	No	-	this work
2.	Ramie fabric 40%—FR	0	0	No	No	30	this work
3.	Ramie fabric 50%—FR	0	0	No	No	15	this work
4.	Cotton fabric (ASMPEA treated)	0	0	No	No	48–64	[47]
5.	Cotton fabric (untreated)	6.9	44.6	-	-	-	[48]
6.	Cotton fabric (350 g/L-SPPTMS)	0	0	-	-	81	[48]
7.	Wool fabric (PA treated)	5–15	15–30	-	-	150–200	[49]

According to the classification of FR fabric based on the vertical burning test from GB 8624-2012 [38], the ramie fabric which treated with the Flamatic retardant at 40–50% could be classified with a B1 classification: char length $\leq$ 15 cm, after flame time $\leq$ 5 s, after glow time $\leq$ 15 s.

3.3. Tensile Characterization

The tensile strength test was only conducted on KR-T01—50% FR due to its lower char length and having better optimization results compared to KR-T01—40% FR. In addition, the KR-T01—40% FR sample was only used to assess its flame-retardant characteristics. The characterization of ramie fabric based on tensile strength revealed changes in the fabric's tensile strength before and after treatment (Table 4). The results showed that the tensile strength of KR-T01—50% FR was reduced by more than 10 MPa (40.60 $\pm$ 3.32) compared to the KR-T01—Non FR (untreated) (56.69 $\pm$ 5.84). This indicates that the mechanical properties of ramie fabric treated with 50% FR were weakened due to the flame-retardant finishing.

Table 4. Characterization of woven ramie fabric based on tensile strength.

No.	Material Fabric	Tensile Strength (MPa)
1.	KR-T01—non-FR (untreated)	56.69 $\pm$ 5.84
3.	KR-T01—50% FR	40.60 $\pm$ 3.32

3.4. Limiting Oxygen Index (LOI)

The method widely used to assess the combustion ability or flammability of polymers is the LOI. Generally, materials with an oxygen index < 22 are classified as flammable, those with an oxygen index between 22 and 27 are considered combustible, and those with an oxygen index > 27 are considered refractory [50]. Below is a comparison of LOI data between fabrics crafted from ramie fiber and fabrics made from other fibers or commercial fabrics (Table 5).

Table 5. Limiting oxygen index (LOI) value for fabrics of different types and with different treatments.

No.	Material (Trademark)	Treatment	LOI (%)	Reference
1.	Ramie fabric	Untreated	22	this work
2.	Ramie fabric	Finished flamatic 50%	29	this work
3.	Cotton fabric	Spptms	28.5	[48]
4.	Cotton fabric	Coated T/C blend	27.1	[51]
5.	Cotton fabric	Finished	27.4	[52]
6.	Wool fabric	Bio-based phytic acid	23.6–27	[49]

The LOI value of untreated ramie fabric is 22%, an indication of its combustible properties. Conversely, the results of the LOI analysis of fabrics treated with FR (KR-T01—50% FR) demonstrate that fire resistance can be enhanced with a flame-retarding agent, yielding an LOI of 29%. This places the fabric in the category of refractory properties, comparable to cotton material treated with SPPTMS [48]. These results showed the advantages of using ramie material for the production of functional textiles with fire-resistant and heat-resistant characteristics.

Woven ramie fabric treated with a flame-retardant coating with the formula Flamatic DM-3072N showed potential as a raw material for producing functional textiles with flame retardant properties. Possible products derived from this flame-retardant ramie fabric include flame-retardant clothing, curtains, wall coverings, interior fabric-based products, and more.

3.5. Structural Reaction Analysis

3.5.1. FTIR Spectra

In Figure 2, the FTIR spectra of ramie fabric 50%—FR are presented. The near-infrared absorption signal at 3335 cm^{-1} corresponds to the oscillation of the NH_4^+ group [53], while at 2898 cm^{-1} shows CH and CH_2 stretching [54]. Peak absorption caused by the presence of P–O, C–N, and N–H groups can be seen at 1287.6 cm^{-1}, 1406.8 cm^{-1}, and 1650.0 cm^{-1} [55,56]. The signal intensity of the treated sample is slightly weaker in terms of C–O bonds, and this may be due to the replacement of -OH to form P-O-C bonds by covalent interactions in cellulose [57]. The absorption peak is seen at 1662 cm^{-1}, indicating the presence of N–H vibrations [58,59]. The side shift at the absorption peak is 1662 cm^{-1} and 1272 cm^{-1} corresponds to changes in the structure of the P-O and P-O-C groups [45]. The small peaks observed at 800 and 850 cm^{-1} may be the result of various salts [60]. Based on the results of this FTIR analysis, it can be confirmed that the FR treatment with the DM-3072N coating material was successfully impregnated into the ramie fabric.

The cellulose component of natural fibers generally does not exhibit flame-retardant properties and fire protection mechanisms [61]. The results of FR treatment show that a fabric made from natural ramie fiber can withstand fire due to the binding process and the presence of mineral salts in the FR formula.

3.5.2. NMR

The ^{13}C NMR (101 MHz, $CDCl_3$) test results confirm the presence of an additional peak in the ramie fabric after treatment. Based on the results, NMR analysis of the FR fabric after treatment shows the presence of chemical shift/additional peaks at δ 51.92 ppm (Figure 3b). The peak shift indicates the possible presence of CH_2-N or CH-N which may have contributions in the fire-retardant effect of the ramie fabric (Figure 4). Further research using a ^{31}P NMR test is needed to characterize the presence of the main flame-retardant element, phosphorus, in the treated fabric.

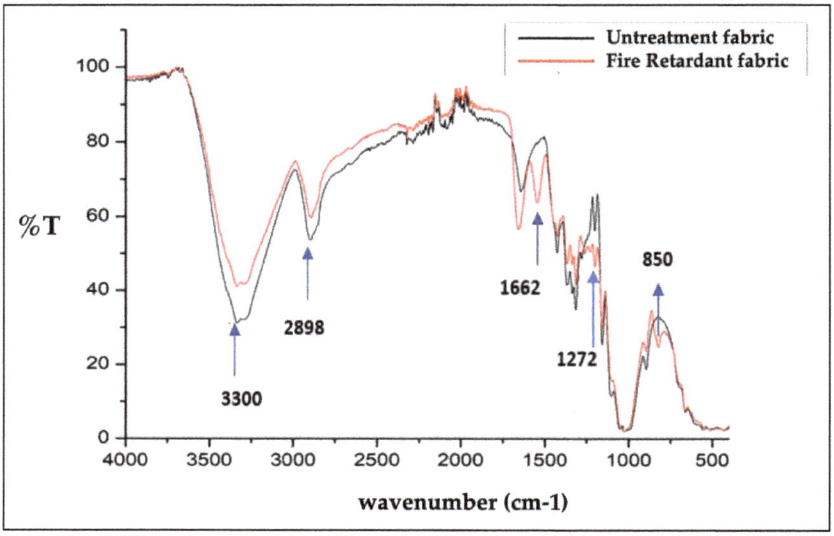

Figure 2. FTIR spectra of ramie fabric untreated and treated with FR formula.

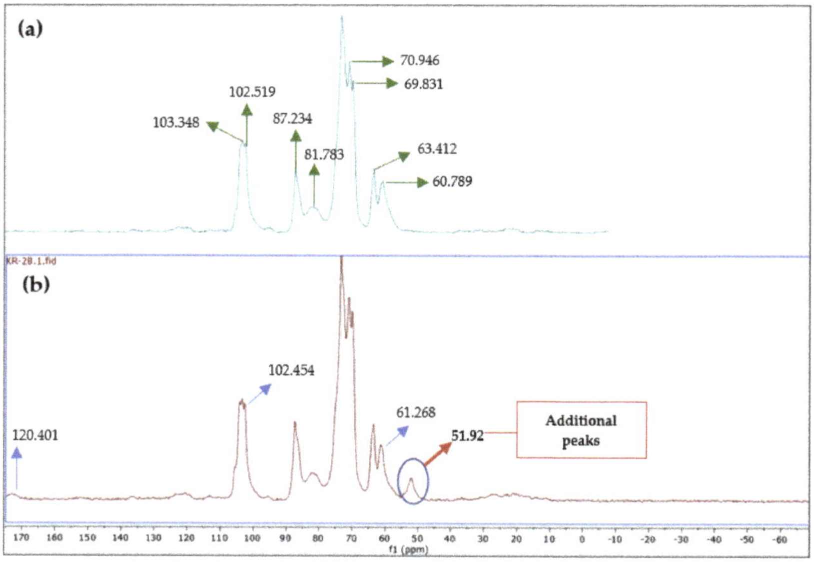

Figure 3. NMR analysis of ramie fabric (**a**) before treatment, and (**b**) with FR treatment.

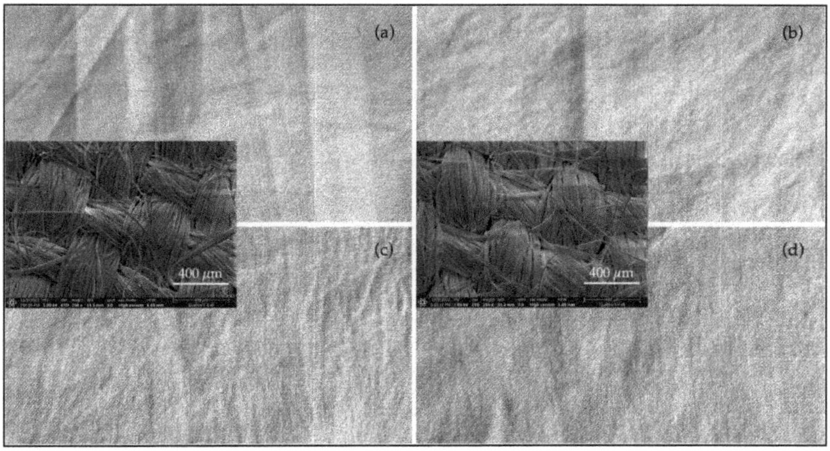

Figure 4. Optical microscope image: (**a**,**c**) KR-T01 ramie fabric without FR treatment, (**b**,**d**) ramie fabric with FR treatment (top image of fabric condition before washing and bottom image of fabric condition after washing).

3.6. Morphological Characterization

The optical microscope images shown in Figure 4 revealed that ramie fabrics retained their fabric texture after treatment. Furthermore, FR-coated samples exhibited smoother surfaces compared to uncoated fabrics, as the active FR filled the gaps in the woven fabric structure. The weight of the ramie fabric before and after 50% FR treatment was 4.38 g and 4.88 g, respectively. With increasing concentrations of FR, the treated ramie fabric acquires weight gain and shows improved flame-retardancy behavior, which is indicated by the increased LOI and decreased char length [62].

The change in fabric size after washing also did not have a significant effect on the woven fabric. After washing, the woven fabric exhibited a shrinkage (reduction in size) of 3.21%. The ramie fabric that has undergone processing is in accordance with the standard testing value for the change in fabric size after washing, which is ±3% (ISO 6330). The results of this research showed that woven ramie fabric, formed into threads woven perpendicularly, maintains stability and resists deformation during the treatment process.

Testing with a scanning electron microscope (SEM) showed that the control ramie fabric exhibited a tightly woven structure with a dense arrangement of fibers. The fabric demonstrates a cross-weaving technique characterized by regular grooves and a dense configuration of warp and weft threads. The warp count comprises two strands of intersecting threads, while the weft count consists of a single strand. The untreated ramie fabric displays a relatively smooth surface structure without significant alterations.

4. Conclusions

This discovery reveals a lamination method utilizing non-toxic organophosphorus material, offering a practical, streamlined process that eliminates the need for layering lamination on ramie fabric. The inclusion of an FR agent content ranging from 40 to 50% enhances the flame-retardant properties of ramie fabric. The application of this resin demonstrates remarkable flammability characteristics, as evidenced by the vertical flame test, yielding a char length of 15–30 mm. These findings signify a heightened level of safety, particularly in containing fire spread on ramie fabric and extinguishing fires in potentially hazardous conditions. The simplicity of this process makes it a practical method for potentially transforming the characteristics of ramie fabric, which could be applied to various types of cellulose-based materials. The research findings reveal the promising development of flame-retardant ramie fabric, showcasing optimal flame-retardant properties with versatile application across industrial sectors. These include security industries necessitating protection against fire risks, such as flame-retardant attire for military personnel, firefighters, and workers in the oil and gas industry. Moreover, flame-retardant ramie fabric can also find utility in fire blankets and furniture materials like those for chairs, carpets, lampshades, curtain, and wall panels. These applications play a pivotal role in impeding fire spread and extending response time during emergencies.

Author Contributions: A.P.W.: conceptualization, writing—original draft, and writing—review and editing; E.K. and E.T.: methodology and validation; A.R., J.K. and M.F.: resources and project administration; K.S. and R.S.: formal analysis and investigation; W.F.: resources and visualization. All authors have read and agreed to the published version of the manuscript.

Funding: This research was funded by the Matching Fund Program from the Ministry of Education, Culture, Research, and Technology, Republic of Indonesia, in collaboration with PT Berkah Rami Neforindo and Universitas Padjadjaran, grant numbers 241/E1/KS.06.02/2022 and 0391/E.E1/KS.03.00/2023.

Institutional Review Board Statement: Not applicable.

Informed Consent Statement: Not applicable.

Data Availability Statement: Data are contained within the article.

Conflicts of Interest: Author Erlin Karlina and Eric Tanudjaja was employed by the company PT Dymatic Chemicals Indonesia; author Abdul Rohmat was employed by the company PT Haramai Jaya Nusantara. The remaining authors declare that the research was conducted in the absence of any commercial or financial relationships that could be construed as a potential conflicts of interest.

Abbreviations

FR—flame retardant; CA—citric acid; SHP—Sodium Hypophosphite; H_2O_2—Hydrogen peroxide; NaOH—Sodium hydroxide; Resin—flammable organic substrate, produced to provide and preserve structure, stability, and viscosity.

References

1. Bilir, M.Z.; Gürcüm, B.H. Ballistic wearable electronic vest design. *J. Ind. Text.* **2018**, *47*, 1769–1790. [CrossRef]
2. Sezer Hicyilmaz, A.; Altin, Y.; Bedeloglu, A. Polyimide-coated fabrics with multifunctional properties: Flame retardant, UV protective, and water proof. *J. Appl. Polym. Sci.* **2019**, *136*, 47616. [CrossRef]
3. Abou-Okeil, A.; El-Sawy, S.M.; Abdel-Mohdy, F.A. Flame retardant cotton fabrics treated with organophosphorus polymer. *Carbohydr. Polym.* **2013**, *92*, 2293–2298. [CrossRef]
4. Fiss, B.G.; Hatherly, L.; Stein, R.S.; Friščić, T.; Moores, A. Mechanochemical Phosphorylation of Polymers and Synthesis of Flame-Retardant Cellulose Nanocrystals. *ACS Sustain. Chem. Eng.* **2019**, *7*, 7951–7959. [CrossRef]
5. Ahmad, J.; Zhou, Z. Mechanical properties of natural as well as synthetic fiber reinforced concreate: A review. *Constr. Build. Mater.* **2022**, *333*, 127353. [CrossRef]
6. Gurunathan, T.; Mohanty, S.; Nayak, S.K. A review of the recent developments in biocomposites based on natural fibres and their application perspectives. *Compos. Part A Appl. Sci. Manuf.* **2015**, *77*, 1–25. [CrossRef]
7. Hardjono, H.; Permatasari, D.A.; Sari, V.A. The influence of citric acid addition on the characteristics of biodegradable plastic film form kepok banana peel starch (Musa Acuminata Balbisiana Colla). *J. Bahan Alam Terbarukan* **2016**, *5*, 22–28. [CrossRef]
8. Han, Y.; Zhang, X.; Wu, X.; Lu, C. Flame Retardant, Heat Insulating Cellulose Aerogels from Waste Cotton Fabrics by in Situ Formation of Magnesium Hydroxide Nanoparticles in Cellulose Gel Nanostructures. *ACS Sustain. Chem. Eng.* **2015**, *3*, 1853–1859. [CrossRef]
9. Suhas; Gupta, V.; Carrott, P.; Singh, R.; Chaudhary, M.; Kushwaha, S. Cellulose: A review as natural, modified and activated carbon adsorbent. *Bioresour. Technol.* **2016**, *216*, 1066–1076. [CrossRef]
10. Abdelrahman, M.S.; Khattab, T.A. Development of One-Step Water-Repellent and Flame-Retardant Finishes for Cotton. *ChemistrySelect* **2019**, *4*, 3811–3816. [CrossRef]
11. Hamawand, I.; Sandell, G.; Pittaway, P.; Chakrabarty, S.; Yusaf, T.; Chen, G.; Seneweera, S.; Al-Lwayzy, S.; Bennett, J.; Hopf, J. Bioenergy from Cotton Industry Wastes: A review and potential. *Renew. Sustain. Energy Rev.* **2016**, *66*, 435–448. [CrossRef]
12. Bentis, A.; Boukhriss, A.; Grancaric, A.M.; El Bouchti, M.; El Achaby, M.; Gmouh, S. Flammability and combustion behavior of cotton fabrics treated by the sol gel method using ionic liquids combined with different anions. *Cellulose* **2019**, *26*, 2139–2153. [CrossRef]

13. Jiang, Z.; Wang, C.; Fang, S.; Ji, P.; Wang, H.; Ji, C. Durable flame-retardant and antidroplet finishing of polyester fabrics with flexible polysiloxane and phytic acid through layer-by-layer assembly and sol–gel process. *J. Appl. Polym. Sci.* **2018**, *135*, 46414. [CrossRef]
14. Lin, D.; Zeng, X.; Li, H.; Lai, X.; Wu, T. One-pot fabrication of superhydrophobic and flame-retardant coatings on cotton fabrics via sol-gel reaction. *J. Colloid. Interface Sci.* **2019**, *533*, 198–206. [CrossRef]
15. Palaskar, S.S.; Desai, A.N.; Shukla, S.R. Development of multifunctional cotton fabric using atmospheric pressure plasma and nano-finishing. *J. Text. Inst.* **2016**, *107*, 405–412. [CrossRef]
16. Pan, Y.; Liu, L.; Zhang, Y.; Song, L.; Hu, Y.; Jiang, S.; Zhao, H. Effect of genipin crosslinked layer-by-layer self-assembled coating on the thermal stability, flammability and wash durability of cotton fabric. *Carbohydr. Polym.* **2019**, *206*, 396–402. [CrossRef] [PubMed]
17. Caschera, D.; Toro, R.G.; Federici, F.; Riccucci, C.; Ingo, G.M.; Gigli, G.; Cortese, B. Flame retardant properties of plasma pre-treated/diamond-like carbon (DLC) coated cotton fabrics. *Cellulose* **2015**, *22*, 2797–2809. [CrossRef]
18. Poon, C.K.; Kan, C.W. Effects of TiO_2 and curing temperatures on flame retardant finishing of cotton. *Carbohydr. Polym.* **2015**, *121*, 457–467. [CrossRef] [PubMed]
19. Xie, K.; Gao, A.; Zhang, Y. Flame retardant finishing of cotton fabric based on synergistic compounds containing boron and nitrogen. *Carbohydr. Polym.* **2013**, *98*, 706–710. [CrossRef]
20. Lin, D.; Zeng, X.; Li, H.; Lai, X. Facile fabrication of superhydrophobic and flame-retardant coatings on cotton fabrics via layer-by-layer assembly. *Cellulose* **2018**, *25*, 3135–3149. [CrossRef]
21. Qiu, S.; Li, Y.; Li, G.; Zhang, Z.; Li, Y.; Wu, T. Robust Superhydrophobic Sepiolite-Coated Polyurethane Sponge for Highly Efficient and Recyclable Oil Absorption. *ACS Sustain. Chem. Eng.* **2019**, *7*, 5560–5567. [CrossRef]
22. Zhao, H.B.; Chen, M.; Chen, H.B. Thermally Insulating and Flame-Retardant Polyaniline/Pectin Aerogels. *ACS Sustain. Chem. Eng.* **2017**, *5*, 7012–7019. [CrossRef]
23. Ye, D.-D.; Wang, T.; Liao, W.; Wang, H.; Zhao, H.-B.; Xu, S.; Wang, Y.-Z. Ultrahigh-Temperature Insulating and Flame retardant Aerogels from Cationic Amylopectin and Clay via a Facile Route. *ACS Sustain. Chem. Eng.* **2019**, *7*, 11582–11592. [CrossRef]
24. Ortelli, S.; Malucelli, G.; Blosi, M.; Zanoni, I.; Costa, A.L. NanoTiO_2 @DNA complex: A novel eco, durable, fire retardant design strategy for cotton textiles. *J. Colloid. Interface Sci.* **2019**, *546*, 174–183. [CrossRef] [PubMed]
25. Liu, Y.; Wang, Q.Q.; Jiang, Z.M.; Zhang, C.J.; Li, Z.F.; Chen, H.Q.; Zhu, P. Effect of chitosan on the fire retardancy and thermal degradation properties of coated cotton fabrics with sodium phytate and APTES by LBL assembly. *J. Anal. Appl. Pyrolysis* **2018**, *135*, 289–298. [CrossRef]
26. Laufer, G.; Kirkland, C.; Morgan, A.B.; Grunlan, J.C. Intumescent multilayer nanocoating, made with renewable polyelectrolytes, for flame-retardant cotton. *Biomacromolecules* **2012**, *13*, 2843–2848. [CrossRef]
27. Wang, X.; Lu, C.; Chen, C. Effect of chicken-feather protein-based flame retardant on flame retarding performance of cotton fabric. *J. Appl. Polym. Sci.* **2014**, *131*, 1–8. [CrossRef]
28. Choi, K.; Seo, S.; Kwon, H.; Kim, D.; Park, Y.T. Fire protection behavior of layer-by-layer assembled starch–clay multilayers on cotton fabric. *J. Mater. Sci.* **2018**, *53*, 11433–11443. [CrossRef]
29. Dittenber, D.B.; GangaRao, H.V. Critical review of recent publications on use of natural composites in infrastructure. *Compos. Part A Appl. Sci. Manuf.* **2012**, *43*, 1419–1429. [CrossRef]
30. Wulandari, A.P.; Awis, V.P.D.; Budiono, R.; Kusmoro, J.; Hidayat, S.S.; Masruchin, N.; Lubis, M.A.R.; Fatriasari, W.; Rachmawati, U. Tensile Strength Improvements of Ramie Fiber Threads through Combination of Citric Acid and Sodium Hypophosphite Cross-Linking. *Materials* **2023**, *16*, 4758. [CrossRef]
31. Yan, H.; Cheng, J.; Fang, Z. Method for Improving Flame Retardant Property of Modified Ramie Fabric. Chinese Patent CN111764154A, 13 October 2020.
32. Yan, H.; Cheng, J.; Fang, Z.; Zhao, L. A Kind of Three Flame-Retardant Modified Ramie Fabrics of Component LBL Self-Assembly and Preparation Method Thereof. Chinese Patent CN105755836B, 28 November 2017.
33. Yan, H.; Li, N.; Cheng, J.; Fang, Z. Flame-Retardant Modified Ramie Fabric/Benzoxazine Colophony Laminate and Preparation Method Thereof. Chinese Patent CN106079723B, 31 October 2018.
34. Sun, Z.; Hou, Y.; Hu, Y.; Hu, W. Effect of Additive Phosphorus-Nitrogen Containing Flame Retardant on Char Formation and Flame Retardancy of Epoxy Resin. *Mater. Chem. Phys.* **2018**, *214*, 154–164. [CrossRef]
35. Repon, M.R.; Siddiquee, N.A.; Jalil, M.A.; Mikucioniene, D.; Karim, M.R.; Islam, T. Flame retardancy enhancement of jute fabric using chemical treatment. *Tekstilec* **2021**, *64*, 70–80. [CrossRef]
36. Klunkin, W.; Hinmo, S.; Thipchai, P.; Rachtanapun, P. Effect of Bleaching Processes on Physicochemical and Functional Properties of Cellulose and Carboxymethyl Cellulose from Young and Mature Coconut Coir. *Polymers* **2023**, *15*, 3376. [CrossRef] [PubMed]
37. *ASTM D6413*; Standard Test Method for Flame Resistance of Textiles (Vertical Test). ASTM International: West Conshohocken, PA, USA, 2015.
38. Cheng, X.-W.; Guan, J.-P.; Chen, G.; Yang, X.-H.; Tang, R.-C. Adsorption and Flame Retardant Properties of Bio-Based Phytic Acid on Wool Fabric. *Polymers* **2016**, *8*, 122. [CrossRef] [PubMed]
39. Dayrit, F.M.; de Dios, A.C. 1H and 13C NMR for the Profiling of Natural Product Extracts: Theory and Applications. In *Spectroscopic Analyses-Developments and Applications*; BoD–Books on Demand: Norderstedt, Germany, 2017. [CrossRef]

40. Ismail, A.F.; Rana, D.; Matsuura, T.; Foley, H.C. Membrane Characterization. In *Carbon-Based Membranes for Separation Processes*; Springer: New York, NY, USA, 2011. [CrossRef]
41. Castellano, A.; Colleoni, C.; Iacono, G.; Mezzi, A.; Plutino, M.R.; Malucelli, G.; Rosace, G. Synthesis and characterization of a phosphorous/nitrogen based sol-gel coating as a novel halogen- and formaldehyde-free flame retardant finishing for cotton fabric. *Polym. Degrad. Stab.* **2019**, *162*, 148–159. [CrossRef]
42. Passauer, L.; Fischer, S.; Bender, H.; Tech, S.; Wagenführ, A. Verwendung Wasserlöslicher Stickstoff- und Phosphorhaltiger Polysaccharidderivate als Flammschutzmittel, Holzverbundwerkstoff mit Verbesserten Flammschutzeigenschaften sowie Verfahren zu Dessen Herstellung. Deutsches Patent-und Markenamt, DE102012204238B4, 16 March 2012.
43. Luo, Y.; Geng, Z.; Zhang, W.; He, J.; Yang, R. Strategy for Constructing Phosphorus-Based Flame-Retarded Polyurethane Elastomers for Advanced Performance in Long-Term. *Polymers* **2023**, *15*, 3711. [CrossRef] [PubMed]
44. Salmeia, K.A.; Jovic, M.; Ragaisiene, A.; Rukuiziene, Z.; Milasius, R.; Mikucioniene, D.; Gaan, S. Flammability of Cellulose-Based Fibers and the Effect of Structure of Phosphorus Compounds on Their Flame Retardancy. *Polymers* **2016**, *8*, 293. [CrossRef] [PubMed]
45. Chen, M.J.; Shao, Z.B.; Wang, X.L.; Chen, L.; Wang, Y.Z. Halogen-free flame-retardant flexible polyurethane foam with a novel nitrogen-phosphorus flame retardant. *Ind. Eng. Chem. Res.* **2012**, *51*, 9769–9776. [CrossRef]
46. Wu, D.H.; Zhao, P.H.; Zhang, M.; Liu, Y.Q.; Li, T.K.; Liao, D.X. Synthesis and Flame Retardant Properties of a Novel Monomolecular P-N Intumescent Flame Retardant. *Chin. J. Synth. Chem.* **2013**, *21*, 530–533.
47. Chang, S.C.; Sachinvala, D.; Sawhney, P.; Parikh, D.V.; Jarrett, W.; Grimm, C. Epoxy phosphonate crosslinkers for providing flame resistance to cotton textiles. *Polym. Adv. Technol.* **2007**, *18*, 611–619. [CrossRef]
48. Kong, D.; Liu, J.; Zhang, Z.; Wang, S.; Lu, Z. Preparation of synergistic silicon, phosphorus and nitrogen flame retardant based on cyclosiloxane and its application to cotton fabric. *Cellulose* **2021**, *28*, 8115–8128. [CrossRef]
49. Rusmirović, J.D.; Lukić, V.; Kovačević, T.M.; Bogosavljević, M.; Brzić, S.; Marinković, A.D.; Stevanović, T. Fireproof phosphorylated kraft lignin/polyesterbased composites: Green material for rocket propellant thermal protection systems. *Sci. Tech. Rev.* **2019**, *69*, 16–22. [CrossRef]
50. Chang, S.; Slopek, R.P.; Condon, B.; Grunlan, J.C. Surface coating for flame-retardant behavior of cotton fabric using a continuous layer-by-layer process. *Ind. Eng. Chem. Res.* **2014**, *53*, 3805–3812. [CrossRef]
51. He, W.-L.; Huang, Y.-T.; Gu, L.; Shen, J.-C.; Cheng, X.-W.; Guan, J.-P. Fabrication of P/N/B-Based Intumescent Flame-Retardant Coating for Polyester/Cotton Blend Fabric. *Materials* **2022**, *15*, 6420. [CrossRef] [PubMed]
52. Liu, Z.; Xu, M.; Wang, Q.; Li, B. A novel durable flame retardant cotton fabric produced by surface chemical grafting of phosphorus- and nitrogen-containing compounds. *Cellulose* **2017**, *24*, 4069–4081. [CrossRef]
53. Cilurzo, F.; Selmin, F.; Minghetti, P.; Montanari, L. The effects of bivalent inorganic salts on the mucoadhesive performance of a polymethylmethacrylate sodium salt. *Int. J. Pharm.* **2005**, *301*, 62–70. [CrossRef]
54. Nguyen, M.M.; Al-Abdul-Wahid, M.S.; Fontenot, K.R.; Graves, E.E.; Chang, S.C.; Condon, B.D.; Grimm, C.C.; Lorigan, G.A. Understanding the mechanism of action of triazine-phosphonate derivatives as flame retardants for cotton fabric. *Molecules* **2015**, *20*, 11236–11256. [CrossRef]
55. Wan, C.; Liu, M.; He, P.; Zhang, G.; Zhang, F. A novel reactive flame retardant for cotton fabric based on a thiourea-phosphoric acid polymer. *Ind. Crops Prod.* **2020**, *154*, 112625. [CrossRef]
56. Jia, Y.; Lu, Y.; Zhang, G.; Liang, Y.; Zhang, F. Facile synthesis of an eco-friendly nitrogen-phosphorus ammonium salt to enhance the durability and flame retardancy of cotton. *J. Mater. Chem. A* **2017**, *5*, 9970–9981. [CrossRef]
57. Pethsangave, D.A.; Khose, R.V.; Wadekar, P.H.; Some, S. Novel Approach toward the Synthesis of a Phosphorus-Functionalized Polymer-Based Graphene Composite as an Efficient Flame Retardant. *ACS Sustain. Chem. Eng.* **2019**, *7*, 11745–11753. [CrossRef]
58. Ling, C.; Guo, L. Preparation of a flame-retardant coating based on solvent-free synthesis with high efficiency and durability on cotton fabric. *Carbohydr. Polym.* **2020**, *230*, 115648. [CrossRef] [PubMed]
59. Miller, F.A.; Wilkins, C.H. Infrared Spectra and Characteristic Frequencies of Inorganic Ions. *Anal. Chem.* **1952**, *24*, 1253–1294. [CrossRef]
60. Basak, S.; Samanta, K.K.; Chattopadhyay, S.K. Fire retardant property of cotton fabric treated with herbal extract. *J. Text. Inst.* **2015**, *106*, 1338–1347. [CrossRef]
61. Liao, Y.; Chen, Y.; Wan, C.; Zhang, G.; Zhang, F. An eco-friendly NP flame retardant for durable flame-retardant treatment of cotton fabric. *Int. J. Biol. Macromol.* **2021**, *187*, 251–261. [CrossRef]
62. Wang, H.; Guo, S.; Zhang, C.; Qi, Z.; Li, L.; Zhu, P. Flame Retardancy and Thermal Behavior of Wool Fabric Treated with a Phosphorus-Containing Polycarboxylic Acid. *Polymers* **2021**, *13*, 4111. [CrossRef]

materials

MDPI

Article

Metallurgical and Mechanical Properties of Stellite 6 Deposition Developed Through Friction Surfacing Technique

Mohammad Faseeulla Khan [1], Ramachandran Damodaram [2,*], Hussain Altammar [1] and Gangaraju Manogna Karthik [3,*]

[1] Department of Mechanical Engineering, College of Engineering, King Faisal University, Al-Ahsa 31982, Saudi Arabia; fmohammad@kfu.edu.sa (M.F.K.); haltammar@kfu.edu.sa (H.A.)

[2] Department of Mechanical Engineering, Sri Sivasubramaniya Nadar College of Engineering, Chennai 603110, India

[3] Department of Mechanical Engineering, Indian Institute of Technology (BHU), Varanasi 221005, India

* Correspondence: damodaramr@ssn.edu.in (R.D.); gmkarthik.mec@iitbhu.ac.in (G.M.K.)

Abstract: Friction surfacing (FS) is a solid-state process for depositing metallurgically bonded coatings for corrosion and wear protection. It is particularly attractive for depositing coatings in materials that are difficult to fusion deposit. Stellite 6 is one such material, which is widely used as a protective coating on steel structures to combat wear and corrosion. In the current study, Stellite 6 was successfully friction-surfaced on low carbon steel plates without using any preheating. The microstructures and wear behavior of Stellite 6 coatings produced using FS were comparatively investigated with those produced using the plasma transferred arc (PTA) process. The PTA coatings showed a cast microstructure consisting of γ-dendrites and an inter-dendritic carbide network. On the other hand, the FS coatings showed a wrought microstructure with dynamically recrystallized grains and fine, uniformly distributed carbide particles. The FS coatings also showed uniform composition across the coating thickness and were undiluted, while the PTA coatings showed significant dilution as well as strong local variations in chemistry. The FS coatings exhibited a 22% increase in hardness (550 HV) compared to the PTA coatings (450 HV). Pin-on-disc dry sliding wear tests showed that the FS coatings (1.205 mm^3) were more wear resistant compared to the PTA coatings (6.005 mm^3), highlighting their superior mechanical performance. This study uniquely demonstrates the feasibility of depositing Stellite 6 coatings using FS without the need for preheating or post-deposition heat treatments, while achieving superior microstructural refinement, hardness, and wear resistance compared to conventional PTA coatings.

Keywords: Stellite 6; friction surfacing; plasma transferred arc; coatings; dilution; wear

1. Introduction

Low carbon steel (LCS) is a mixture of a low ratio of carbon to iron; it contains less than 0.3% of carbon, which results in no improvement in strength, hardness, and wear properties with heat treatment. These steels are relatively weak and soft but are ductile [1]. Due to its softness and ductility, LCS is mainly used in machinery and welding parts at a low cost; for example, it is used in automobile body parts, food cans, bridge components, and structural components like I-beams [1]. However, due to their low hardness and poor wear resistance, LCS requires efficient surface engineering procedures to increase its wear resistance [2].

Stellite 6 is a cobalt-based alloy that consists of complex carbides. It performs exceptionally well in wear, erosion, and corrosion resistance [3–7]. Stellite 6 is used in high-temperature and high-pressure scenarios. Due to its good corrosion resistance, it is primarily used in light water nuclear reactors and valve applications [5,8–10]. Along with preventing erosion and corrosion, it also exhibits excellent wear resistance due to its CoCr alloy matrix. The dendritic carbides present in the CoCr alloy matrix improve the toughness and wear resistance of the Stellite 6 coating [3–7,11–13].

Recent studies have investigated various methods to enhance the wear resistance of Stellite 6 coatings. Cinca and Guilemany (2013) demonstrated that cold gas spraying, a solid-state process, effectively deposits Stellite 6 coatings with notable wear resistance under abrasive and sliding conditions [14]. Wei et al. (2024) found that adding titanium and nickel to laser-cladded Stellite 6 coatings led to the formation of TiC particles, improving wear resistance by approximately 13% and significantly enhancing thermal fatigue performance [15]. Felix-Martinez et al. (2024) investigated Stellite 6/WC-6Co composite coatings produced via laser cladding and observed that continuous wave deposition resulted in higher hardness and superior wear resistance at both room and elevated temperatures, attributed to increased diffusion of tungsten and carbon into the matrix [16]. Additionally, research by Manukonda et al. (2024) indicated that incorporating TiC into Stellite 6 coatings on stainless steel substrates via laser cladding significantly improved wear resistance due to the hardening effect of TiC particles [17]. Collectively, these studies highlight the effectiveness of both fusion and solid-state processes, as well as compositional modifications, in enhancing the wear resistance of Stellite 6 coatings.

In short, Stellite 6 coatings are widely used to improve mechanical properties and wear resistance when deposited on various alloys. These coatings can be applied using different techniques, such as gas tungsten arc coating (GTAC), plasma transferred arc (PTA) coating, laser melting, and cladding, etc. [2,9,11,12,18–25]. A comparative study in [9] analyzed various techniques for applying Stellite 6 coatings, including high velocity oxygen fuel (HVOF) spraying (a thermal spraying process) and GTAW. HVOF spraying is commonly used in hydraulic machinery due to its superior performance in water-based corrosion resistance [5,12,22,26–30]. Studies have shown that depositing Stellite 6 coatings using these techniques significantly enhances the wear properties of the base metal while reducing the coefficient of friction. However, these coating techniques involve high temperatures and pressures, making them more suitable for hard and tough alloys.

Friction surfacing (FS) is a solid-state coating technique that applies thick coatings for wear prevention. Unlike fusion-based and thermal spraying processes, FS does not involve melting, thereby eliminating issues related to dilution between the base and weld metal. Fusion-based welding techniques suffer from hot cracking over prolonged periods, leading to material degradation [2]. Therefore, FS provides a more reliable alternative by effectively mitigating these issues, offering significant improvements in wear resistance and durability. This makes FS an ideal method for depositing Stellite 6 coatings onto LCS to enhance wear resistance. Furthermore, FS has gained global acceptance due to its effectiveness [2,24–26]. However, limited research exists on FS coatings of Stellite 6 and their wear performance [2].

In the present study, the authors have attempted, for the first time, to understand the significance of FS coatings of Stellite 6 in wear characteristics. Stellite 6 coatings were deposited on LCS using the FS technique, a solid-state coating method, and compared with those produced using the PTA process. To evaluate wear resistance, pin-on-disc dry wear tests were conducted on both FS and PTA coatings, analyzing the coefficient of friction (μ) and wear surface morphology. This study provides new insights into the wear behavior of FS coatings and their microstructural characteristics, which have not been extensively explored in the existing literature. Unlike conventional fusion-based methods, FS offers superior coating integrity with minimal dilution, improving long-term durability. This research advances surface engineering strategies for enhancing the wear resistance of LCS, making FS a viable alternative to traditional techniques.

2. Materials and Methods

2.1. Coating Process

Friction surfacing (FS) is a solid-state surface technique where a consumable rod to be deposited as the coating material will be fixed into the rotating chuck/vertical head. The rotating consumable rod will be brought in contact with a stationary substrate under a constant predefined axial load. The consumable rod material is softened because of frictional heating at the rubbing interface between the rod and the substrate. After this, the substrate will be traversed and a constant feed and the softened consumable rod at one end will deposit over the substrate as surface coating. The schematic of the FS is given in Figure 1. In the present work, Stellite 6 was used as the consumable rod material. The consumable rod is 10 mm in diameter and 150 mm in height, obtained through casting. Hot rolled LCS plate of 12 mm thickness is used as substrate material. Before the FS process, the substrate surface was mechanically cleaned to remove oxides, contaminants, and surface irregularities. Then, the substrate was firmly fixed to avoid any movement during the coating process, ensuring uniform deposition. The chemical composition of Stellite 6 (consumable rod) and LCS (substrate) is given in Table 1.

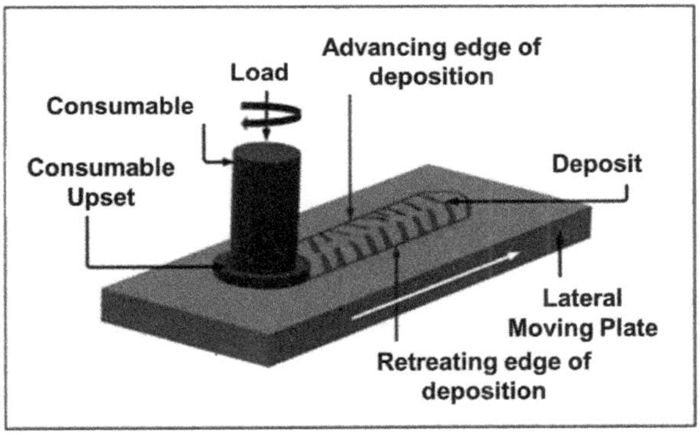

Figure 1. Schematic of the friction surfacing.

Table 1. Chemical composition of the Stellite 6 consumable rod and LCS substrate.

Materials	Composition (Wt%)								
	Cr	W	C	Fe	Co	Si	Mn	P	S
Stellite 6 consumable rod	27.50	5.00	3.30	2.90	bal.	-	-	-	-
LCS substrate	-	-	0.12	bal.	-	0.36	0.42	0.02	0.01

To study and compare the effect of solid-state FS Stellite 6 coating on wear characteristics, a fusion-based surface coating technique, PTA, was used to deposit Stellite 6 coatings. PTA is very similar to the plasma transfer arc welding principle. However, here the objective of the process is to apply coatings rather than joining. In the case of PTA process, Stellite 6 powder was fed into the arc column to obtain the coating. The average grain size of the Stellite 6 powder is 120 ± 10 μm. As there will be significant dilution during PTA process, two layers of coatings, one over the other, were deposited by PTA process.

The primary process parameters involved and used for PTA and FS processes are listed in Tables 2 and 3, respectively.

Table 2. Process parameters used for the PTA process.

Current (A)	100
Weld Speed (mm/s)	1.20
Polarity	DCEN
Oscillation width (mm)	10
Preheating (°C)	300
No. of layers	2
Powder feed rate (g/min)	30
Shielding gas(Ar) flow rate (L/min)	12
Powder feed gas flow rate (L/min)	3.5
Torch work distance (mm)	15
Plasma gas(Ar) flow rate (L/min)	4

Table 3. Process parameters used for the FS process.

Spindle speed (rpm)	800
Axial/Vertical load (kN)	0.9
Substrate transverse feed (mm/s)	1.6
Dwell time (s)	30

2.2. Microstructural Characterization

For microstructural characterization, the samples were sectioned in the transverse direction from the FS, PTA, and consumable rod. Subsequently, the samples were polished by following standard metallographic polishing procedures and were etched by aqua-regia. The microstructures of the prepared samples were observed by optical microscope (OM) (Leica DM200, Wetzlar, Germany) and scanning electron microscopy (FEI Quanta 200, SEM, Eindhoven, The Netherlands) with back scattered electron (BSE) and energy dispersive spectroscopy (EDS) detector. To identify/assess the various phases present in the samples, X-Ray diffractometer (XRD, Panalytical, Almelo, The Netherlands) employing Cu-Kα radiation was used. Transmission electron microscope (TEM, Philips CM12, Eindhoven, The Netherlands) characterization was employed to study microstructure evolution during FS and to identify the second phase particles in the coated samples. For TEM studies, thin samples of 300 μm were sectioned transversely through EDM cutting machine. These thin samples were mechanically grounded to 100 μm and punched in to 3 mm disc. Then, these samples were twin-jet electro polished.

2.3. Mechanical Testing

The micro Vickers hardness was measured across the coating/substrate interface from the top surface of the Stellite 6 coating to the LCS base metal. A (Matsuzawa MMT-7, Akita-shi, Japan) digital microhardness tester was employed to conduct the experiments. The microhardness measurements were conducted at 500 g load applied for 15 s.

2.4. Wear Testing

Dry sliding wear tests were carried out at room temperature using pin-on-disk equipment. Pin of 15 mm length and 6 mm diameter was prepared for all the above-mentioned material process conditions. A 56 mm disk of hardened alloy steel with a hardness of 64 HRC was used as the rotating disk for the machine. The disk was rotated at 650 rpm and load applied was 30 N. Each sample was characterized for sliding distance of 1000 m. These parameters were selected based on commonly used conditions in wear studies for hard-facing coatings and align with the ASTM G99 standard for pin-on-disk wear testing. Further, these parameters were chosen to simulate moderate contact stresses and sliding velocities encountered in real-world tribomechanical applications. Before the wear test, each sample was mechanically ground in SiC paper to an average surface roughness of about 0.8 μm and cleaned by using acetone in an ultrasonic bath for 15 min. The pin and disc after the wear test are shown in Figure 2.

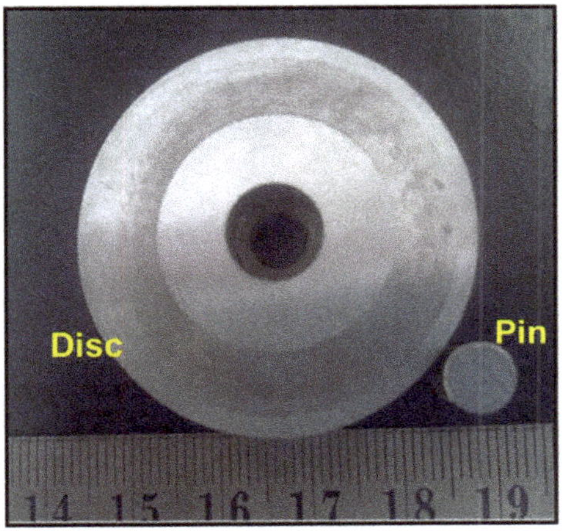

Figure 2. Pin-on-disc wear test set-up.

3. Results and Discussion

3.1. Microstructures

3.1.1. Substrate and Consumable Rod

Figure 3A shows the optical micrograph of the LCS substrate comprising bands of α-ferrite and pearlite phases. Figure 3B indicates the microstructure of the cast Stellite 6 consumable rod, showing a coarse dendritic structure with γ-dendrites and an inter-dendritic carbide network. The results of the EDS spot analysis taken from the dendrite core (spot 1 in Figure 3B) and inter-dendritic regions (spot 2 in Figure 3B) of cast Stellite 6 are given in Table 4. The dendritic regions showed higher Co concentration than the inter-dendritic regions, and the inter-dendritic regions showed higher Cr than the dendrite core.

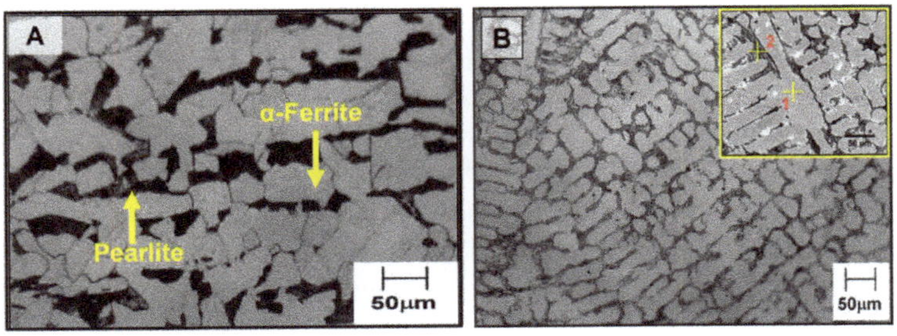

Figure 3. Optical micrographs of (**A**) hot rolled low carbon steel substrate and (**B**) cast Stellite 6 consumable rod. The inset in (**B**) shows the high magnification image with the EDS spot analysis location (1 and 2) indicated.

Table 4. EDS spot analysis results of the Stellite 6 consumable rod. The EDS spot analysis locations (1 and 2) are indicated in the inset of Figure 3B.

Location	Composition (Wt.%)			
	Co	Cr	W	Fe
Location—1 (dendrite Core)	63	24	7	5
Location—2 (inter-dendritic region)	36	50	9	4

3.1.2. Plasma Transferred Arc Coating

The microstructures of the PTA-deposited Stellite 6 coating are shown in Figure 4. PTA provides a solidified dendritic structure analogous to cast Stellite 6; however, the microstructures are much finer in the former compared to the latter. The average grain size in the HAZ of the PTA-coated sample is found to be 6 ± 1 µm. The EDS spot analysis on the dendrite core and inter-dendritic regions exhibits distinct Co and Cr concentrations; however, the difference is notably less than the cast Stellite 6. The EDS spot analysis results are listed in Table 5. The rapid cooling condition during the PTA process compared to the casting would provide less time for the elements to segregate to the inter-dendritic regions, thereby reducing segregation effects. Nevertheless, segregation is still significant in the PTA coatings.

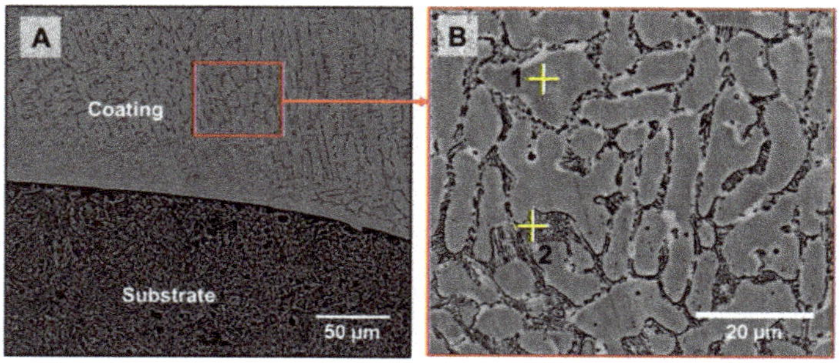

Figure 4. Microstructures of the PTA coating (**A**) at low magnification and (**B**) at higher magnification.

Table 5. EDS spot analysis on the PTA coating. The EDS spot analysis locations (1 and 2) are indicated in Figure 4B.

Location	Co	Cr
Location 1 (dendrite core)	56	22
Location 2 (inter-dendritic region)	45	37

3.1.3. Friction Surfacing

The photograph of the Stellite 6 coated sample prepared through FS is shown in Figure 5. Indeed, the Stellite 6 FS Coating is completely metallurgically bonded with the substrate LCS. The microstructures of the FS-coated sample and its interface are shown in Figure 6. The microstructure of the FS-coated sample does not exhibit any solidified structure, unlike the Stellite 6 consumable rod and PTA coating (Figures 3B and 4B). Further, the interconnected carbide network at the inter-dendritic regions of the consumable rod is fragmented into extremely fine discrete particles, because of the severe plastic deformation of the consumable rod during FS. The average grain size of the FS-coated sample in the HAZ is 1 ± 0.5 µm. The frictional heating and the blending activity prompted a stream of the material in the whole surface of the bar, prompting the fracture of hard second stage/phase particles. The chemical composition of Co and Cr in the FS coating is mentioned in Table 6, which provides the evidence that composition homogeneity was achieved and different locations of Co or Cr were discovered to be homogeneous, demonstrating that FS produces compositional homogenization throughout the coated substrate [20].

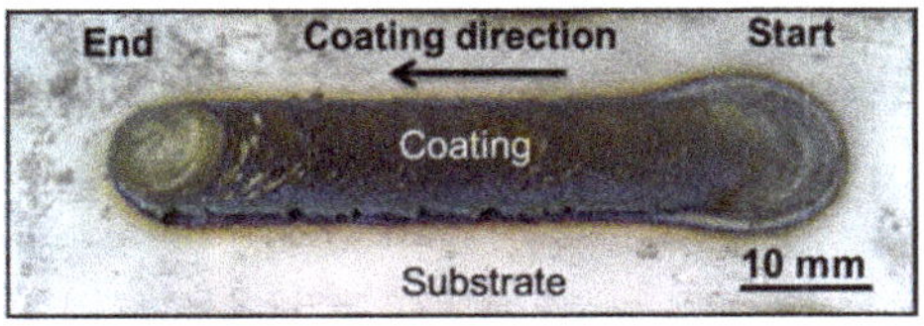

Figure 5. Photograph of friction surfaced Stellite 6 coating.

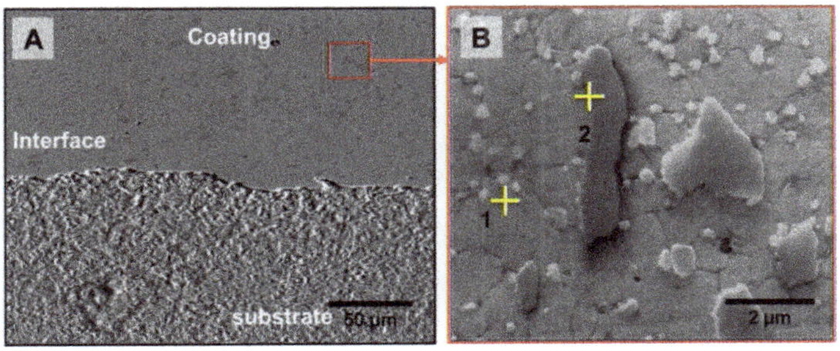

Figure 6. (**A**) SEM microstructures of friction surface-coated samples and (**B**) magnified view of the friction surface coating.

Table 6. EDS spot analysis on PTA coatings and FS coatings at different locations.

Location	PTA Coating (Wt. %)			FS Coating (Wt. %)		
	Co	Cr	Fe	Co	Cr	Fe
4	60	28	3			
3	58	26	5			
2	55	24	12	61	27	2.7
1	54	22	15	61	28	2.6
0	51	20	18	60	27	2.5

3.2. Dilution Studies

To determine the extent of dilution, the samples coated with Stellite 6 employing FS and PTA were subjected to the EDS examination as shown in Figures 7A and 7B respectively. The average coating thicknesses of the FS- and PTA-coated samples are 1 ± 0.5 mm and 3 ± 0.5 mm, respectively, as shown in Figures 7A and 7B. According to the EDS investigation, the amount of Co and Cr decreases as we approach the interface from the top surface of the coating, with a dilution of 15% for Co in PTA and just 1.6% in FS, and a dilution of 28.5% for Cr in PTA coated sample, and a negligible dilution for Cr in FS. Further, there is significant dilution of Fe from the base material. When compared from the extreme end of the coating to the base metals surface (which is closest to the HAZ), there is a case of enrichment of Fe in PTA. The Fe, which was originally 3 (Wt. %) at the extreme end of the coating, increases to 18 (Wt. %) at the base metals surface, which is an increase of six times. As we approach the HAZ, the Fe in the case of FS is diluted by 7.4% from the surface of the coating to the surface of the base metal.

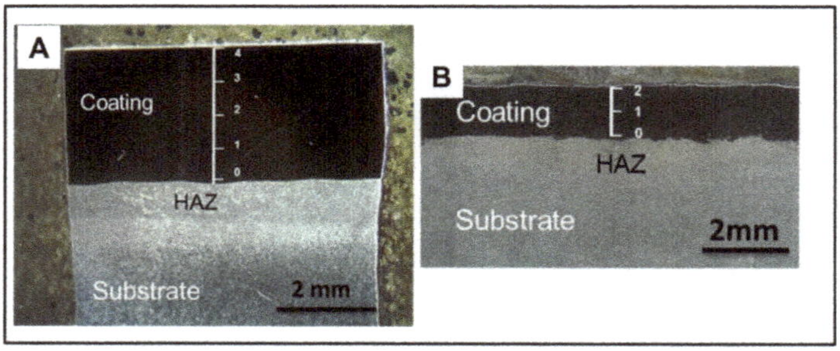

Figure 7. Macrograph images showing location of EDS spot analysis on PTA coatings (**A**) and FS coatings (**B**).

3.3. *Phase Analysis*

The results of X-ray diffraction patterns obtained from FS, PTA, and the Stellite 6 consumable rod are shown in Figure 8. The phases found in the deposited coating are γ-Co (fcc) and carbides C_3Cr_7, $Cr_{23}C_6$, and W_2C. The highest peak intensity is recorded for the γ-Co (fcc) and there is no Co-HCP phase identified in the FS and PTA coatings. Further, while the PTA and the Stellite 6 consumable rod showed a solidification texture (with highest intensity peak at <200> plane, <100> is the easy growth direction for cubic materials), the FS coatings showed no texture effects because of the severe plastic deformation leading to dynamic recrystallization. The occurrence of dynamic recrystallization in the FS coatings can be inferred from the TEM micrographs (Figure 9A). Equiaxed grains have varying low, medium, and high dislocation densities (indicated with 4a, 4b, and 4c, respectively, in Figure 9A) and dislocation sub-cell boundaries (indicated as 1 in Figure 9A) are present. Further, fine discrete carbide particles are seen in FS coatings, whereas a lamellar, interconnected carbide network is observed in the PTA coatings (Figure 9B).

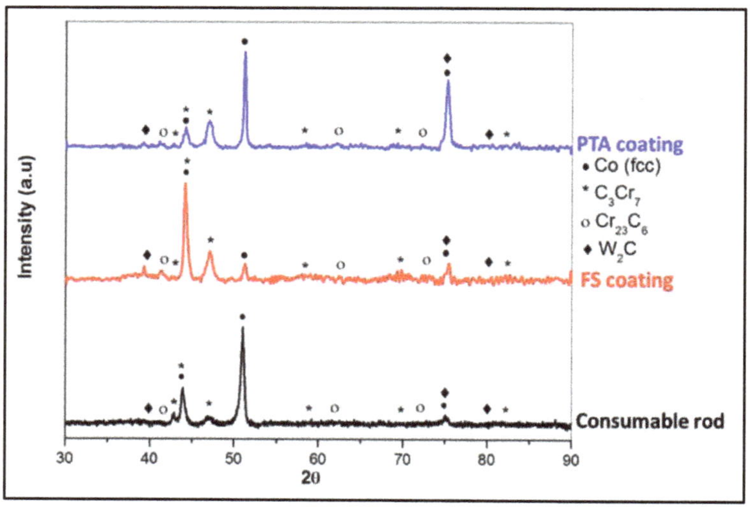

Figure 8. X-Ray diffraction patterns of the consumable rod, friction surfacing, and plasma transfer arc coatings.

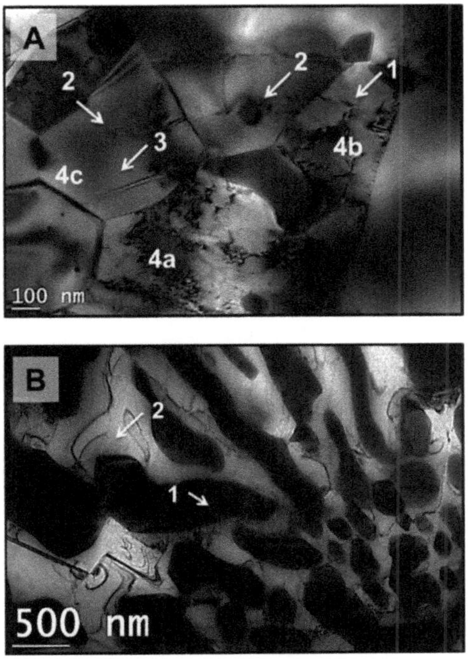

Figure 9. TEM microstructures of coated samples: (**A**) Friction surfacing, (**B**) Plasma Transfer Arc Coating. The annotations in (**A**): 1—dislocation sub-cell formation; 2—Fine carbide particles; 3—fine grain microstructure; and 4a, 4b, 4c—grains with low, medium, and high dislocation densities, respectively. The annotations in (**B**): 1—interconnected carbide network and 2—coarse dendritic structure.

3.4. Mechanical Properties

3.4.1. Microhardness

Figure 10 depicts the hardness profile obtained from the vertical cross-section of the FS and PTA samples. As illustrated, the PTA coatings have a lower hardness of 450 HV compared to the 550 HV of the FS coating, demonstrating that the FS coatings perform better in terms of hardness than the PTA. The increased hardness in the FS coatings can primarily be attributed to the fine grain structure with uniform distribution of fine carbide particles in FS. This can be noted clearly from Figure 11A. Importantly, the hardness increases in the HAZ to a larger width in the PTA-coated samples, ~6 mm, compared to the FS-coated samples, ~1 mm (Figure 11). This indicates that the width of the HAZ during FS will be significantly less compared to the PTA coatings. The difference in heat input between the two processes has mostly contributed to the HAZ being substantially thicker in the case of PTA compared with FS. It is highly desirable to minimize the HAZ as much as possible because it has adverse effects on the base metal strength and toughness [31]. It is worth noting that although with no preheating in FS, the HAZ hardness in the FS is similar to PTA samples. During FS, the peak temperature was reported to be ~0.8 T_m, and the cooling rates were rapid (with no preheating) [28]. These conditions resulted in lower HAZ width in the FS samples compared to PTA coatings (which experienced peak temperatures greater than the melting point of the alloy with high preheating temperatures of about ~300 °C).

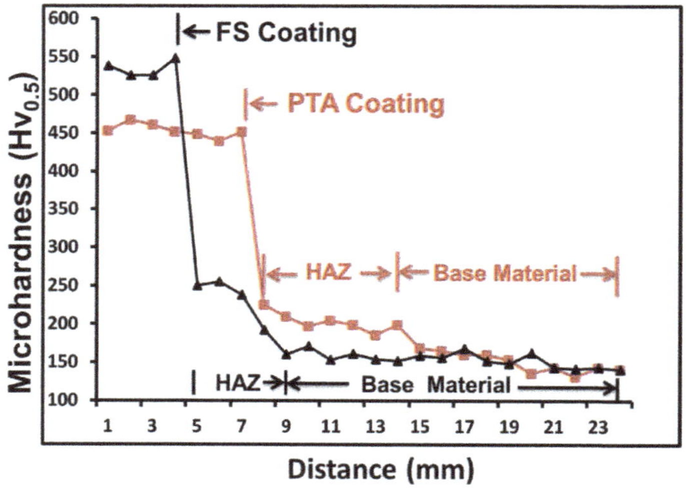

Figure 10. Micro Vickers hardness plots across the coating/substrate interface of the FS- and PTA-coated samples.

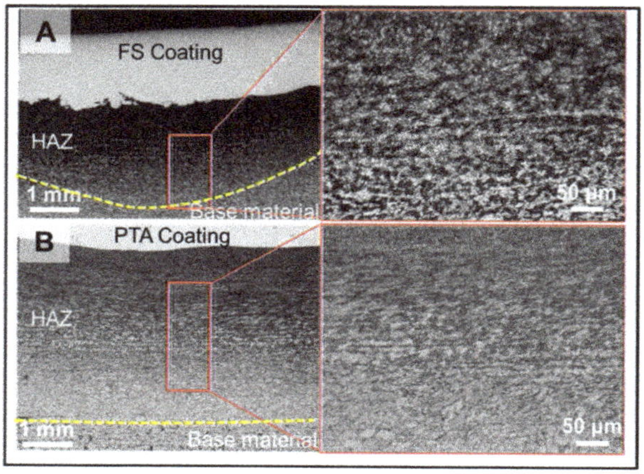

Figure 11. Optical microscope microstructures of the HAZ; (**A**) Friction surfacing and (**B**) PTA coatings.

3.4.2. Wear Characterization

The pin-on-disc wear test results of FS and PTA Stellite 6 coatings are presented in Figure 12. Here, two parameters, volume loss (mm^3) and coefficient of friction (μ), are compared. With the FS coating, the volume loss is 1.205 mm^3, and it is 6.005 mm^3 with PTA. It is evident that FS has significantly less volume loss than PTA, primarily because the FS sample possesses higher hardness comparatively. Additionally, the coefficient of friction is 0.505 and 0.302 for PTA and FS samples, respectively. In other words, the higher the coefficient of friction, the more quickly that metal will wear [6,7,9,11,21–23,32,33]. The FS approach is superior to PTA when comparing the two aforementioned characteristics. As mentioned in [34], the resistance to abrasive and adhesive wear is seen in both FS and PTA coatings due to the presence of hard carbide particles. Additionally, FS coating outperforms PTA since it has more hardness, meaning no further work hardening is possible, which is mainly attributed to the grain refinement from fine grains. It was observed that the wear resistance of FS coating increases as the material resistance to plastic deformation increases. However, more plastic deformation is seen in the case of PTA.

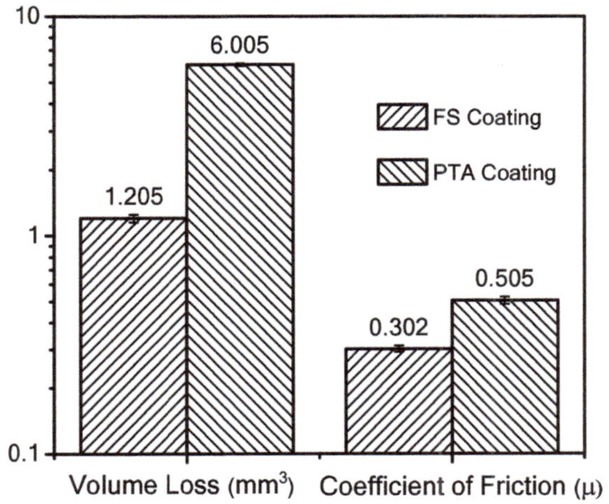

Figure 12. Results of pin-on-disc wear test at room temperature for the FS and PTA coatings.

Figure 13 represents the surface morphology of the PTA sample after the wear test. There are ledge areas and delamination formed as shown in Figure 13A, which are indications of plastic deformation. It should be emphasized that the accumulation of dislocations beneath the wear surface is the primary cause of the wear lamellas. Dislocations begin to coalesce close to the surface during sliding contact as seen in Figure 13B. Small cavities will occur in such regions because the production of voids and cavities in regions with a high density of dislocations reduces the energy [35]. As voids continue to expand as a result of applied stresses, they eventually reach a critical length where they can start to crack. The wear particles separate off the surface in the form of a thin film due to the convergence of these cracks.

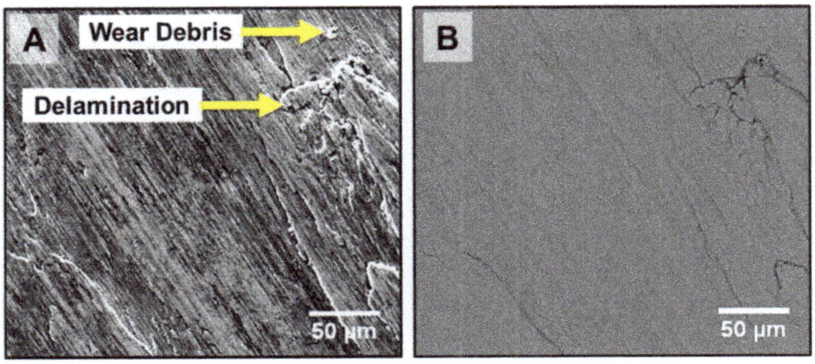

Figure 13. Surface morphology of PTA-coated sample after wear test: (**A**) SEM-SE Image, (**B**) SEM-BSE Image.

Figure 14 shows the surface morphology of the FS sample after the wear test. It can be observed that there is fine distribution of carbide particles throughout the matrix as shown in Figure 14B with abrasive grooves seen in Figure 14A; this formation of high-density coatings led to a drastic increase in obtained hardness values with a reduction in the coefficient of friction value; this is a significant factor that shows that FS has greater wear resistance than PTA. The relationship between the change in wear rate and the change in coating hardness holds true for all coatings. Consequently, increasing the coating hardness is necessary to reduce the wear rate. According to [36], the hardness and wear rate are inversely related. Due to the presence of fine dispersion of carbide particles, equiaxed grains exhibit improved wear properties. Further, an adhesive wear mechanism is observed in the worn surface of the FS-coated samples. It can be seen that, as the hardness of the material decreases, the depth and width of the grooves increases, as does the amount of the abrasive mechanism.

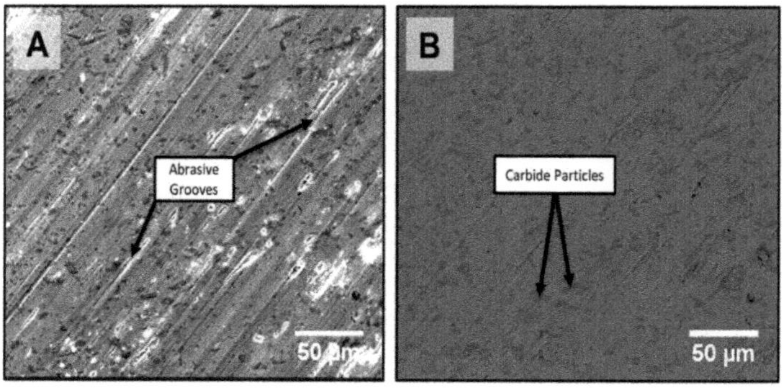

Figure 14. Surface morphology of FS-coated sample after wear test; (**A**) SEM-SE Image, (**B**) SEM-BSE Image.

The FS Stellite 6 coatings in the current study exhibit superior wear resistance compared to PTA coatings and previously reported results. The volume loss of FS coatings (1.205 mm^3) is significantly lower than PTA (6.005 mm^3) and those reported in [13] at a much lesser load (15 N), highlighting the enhanced performance at higher loads. Similarly, the coefficient of friction for FS (0.302) is lower than PTA (0.505) and the values reported by Magarò et al. [37] (0.4–0.65 at low speeds), suggesting better tribological properties. Yoon et al. [38] found that dilution negatively affects Stellite 6 wear resistance, while our FS coatings, free from dilution effects, demonstrate improved hardness and resistance to plastic deformation. These findings indicate that FS coatings offer a more effective wear-resistant solution compared to conventional PTA methods. The worn surface morphologies in the current study indicate abrasive and adhesive wear mechanisms in both FS and PTA coatings. Similar findings were reported by Magaro et al. [38], who observed subsurface cracking and plastic deformation in Stellite 6 coatings under reciprocating dry sliding conditions [38]. However, in the present study, FS coatings exhibited finer carbide dispersion and reduced plastic deformation, leading to improved wear performance.

4. Conclusions

- The FS and PTA coatings exhibited excellent bonding at the substrate/coating interface. The FS coatings developed a wrought microstructure with dynamically recrystallized grains containing fine, uniformly distributed carbides, while the PTA coatings showed a cast microstructure with coarse γ-dendrites and an inter-dendritic carbide network.
- Dilution and chemical uniformity varied between the two processes. PTA coatings exhibited significant dilution and localized compositional variations, whereas FS coatings achieved a more uniform composition across the thickness with minimal dilution.
- The HAZ in FS coatings (~1 mm) was significantly narrower than in PTA coatings (~6 mm), indicating that the solid-state nature of FS processing leads to reduced heat input and more controlled thermal cycles.
- The hardness of FS coatings was superior to that of PTA coatings, which is attributed to the formation of dynamically recrystallized grains with finely dispersed carbide particles.

- Wear resistance testing demonstrated that FS coatings (1.205 mm^3 wear loss) exhibited significantly better wear resistance compared to PTA coatings (6.005 mm^3 wear loss). The improved performance is primarily due to the refined microstructure and higher hardness in FS coatings.

In summary, this study demonstrates that FS is a viable and effective alternative to PTA deposition for Stellite 6 coatings on LCS substrates. The superior microstructural refinement, reduced dilution, and enhanced hardness of FS coatings result in significantly improved wear resistance, making FS a promising process for applications requiring wear-resistant coatings. These findings highlight the potential of FS for industrial applications such as aerospace, automotive, and tooling industries, where high-performance wear coatings are critical. Future studies can explore the optimization of FS process parameters and FS's applicability to other coating materials for broader engineering applications.

Author Contributions: Conceptualization, G.M.K. and R.D.; methodology, M.F.K. and R.D.; validation, G.M.K., M.F.K. and R.D.; formal analysis, G.M.K., M.F.K. and R.D.; investigation, R.D.; resources, M.F.K. and H.A.; data curation, G.M.K. and R.D.; writing—original draft preparation, M.F.K.; writing—review and editing, G.M.K. and H.A.; visualization, H.A.; supervision, G.M.K.; project administration, M.F.K.; funding acquisition, M.F.K. and H.A. All authors have read and agreed to the published version of the manuscript.

Funding: This work was supported by the Deanship of Scientific Research, Vice Presidency for Graduate Studies and Scientific Research, King Faisal University, Saudi Arabia. Grant No. KFU250185.

Institutional Review Board Statement: Not Applicable.

Informed Consent Statement: Not Applicable.

Data Availability Statement: The original contributions presented in this study are included in the article. Further inquiries can be directed to the corresponding authors.

Conflicts of Interest: The authors declare no competing interests.

Abbreviations

The following abbreviations are used in this manuscript:

FS	Friction Surfacing
PTA	Plasma Transferred Arc
LCS	Low Carbon Steel
GTAC	Gas Tungsten Arc Coating
HVOF	High Velocity Oxygen Fuel
SEM	Scanning Electron Microscopy
XRD	X-Ray Diffractometer
EDS	Energy Dispersive Spectroscopy
TEM	Transmission Electron Microscope
EDM	Electrical Discharge Machining
HAZ	Heat Affected Zone

References

1. Singh, R. *Applied Welding Engineering: Processes, Codes, and Standards*, 2nd ed.; Elsevier: Amsterdam, The Netherlands, 2016. [CrossRef]
2. Rao, K.P.; Damodaram, R.; Rafi, H.K.; Ram, G.J.; Reddy, G.M.; Nagalakshmi, R. Friction surfaced Stellite6 coatings. *Mater. Charact.* **2012**, *70*, 111–116. [CrossRef]
3. Corchia, M.; Delogu, P.; Nenci, F.; Belmondo, A.; Corcoruto, S.; Stabielli, W. Microstructural aspects of wear-resistant stellite and colmonoy coatings by laser processing. *Wear* **1987**, *119*, 137–152. [CrossRef]
4. D'Oliveira, A.S.C.M.; Da Silva, P.S.C.P.; Vilar, R.M.C. Microstructural features of consecutive layers of Stellite 6 deposited by laser cladding. *Surf. Coat. Technol.* **2002**, *153*, 203–209. [CrossRef]

5. Sidhu, B.S.; Prakash, S. Erosion-corrosion of plasma as sprayed and laser remelted Stellite-6 coatings in a coal fired boiler. *Wear* **2006**, *260*, 1035–1044. [CrossRef]
6. Vite, M.; Castillo, M.; Hernández, L.H.; Villa, G.; Cruz, I.H.; Stéphane, D. Dry and wet abrasive resistance of Inconel 600 and stellite. *Wear* **2005**, *258*, 70–76. [CrossRef]
7. Shin, J.C.; Doh, J.M.; Yoon, J.K.; Lee, D.Y.; Kim, J.S. Effect of molybdenum on the microstructure and wear resistance of cobalt-base Stellite hardfacing alloys. *Surf. Coat. Technol.* **2003**, *166*, 117–126. [CrossRef]
8. Malayoglu, U.; Neville, A.; Lovelock, H. Assessing the kinetics and mechanisms of corrosion of cast and HIPed Stellite 6 in aqueous saline environments. *Corros. Sci.* **2005**, *47*, 1911–1931. [CrossRef]
9. Lima, C.R.C.; Belém, M.J.X.; Fals, H.D.C.; Rovere, C.A.D. Wear and corrosion performance of Stellite 6® coatings applied by HVOF spraying and GTAW hotwire cladding. *J. Mater. Process. Technol.* **2020**, *284*, 116734. [CrossRef]
10. Sidhu, T.S.; Prakash, S.; Agrawal, R.D. Hot corrosion studies of HVOF NiCrBSi and Stellite-6 coatings on a Ni-based superalloy in an actual industrial environment of a coal fired boiler. *Surf. Coat. Technol.* **2006**, *201*, 1602–1612. [CrossRef]
11. Madadi, F.; Shamanian, M.; Ashrafizadeh, F. Effect of pulse current on microstructure and wear resistance of Stellite6/tungsten carbide claddings produced by tungsten inert gas process. *Surf. Coat. Technol.* **2011**, *205*, 4320–4328. [CrossRef]
12. Singh, P.K.; Mishra, S.B. Studies on solid particle erosion behaviour of D-Gun sprayed WC-Co, Stellite 6 and Stellite 21 coatings on SAE213-T12 boiler steel at 400 °C temperature. *Surf. Coat. Technol.* **2020**, *385*, 125353. [CrossRef]
13. Zhu, Z.; Ouyang, C.; Qiao, Y.; Zhou, X. Wear Characteristic of Stellite 6 Alloy Hardfacing Layer by Plasma Arc Surfacing Processes. *Scanning* **2017**, *2017*, 6097486. [CrossRef] [PubMed]
14. Cinca, N.; Guilemany, J.M. Cold Gas Sprayed Stellite-6 Coatings and their Wear Resistance. *J. Mater. Sci. Eng.* **2013**, *2*, 122. [CrossRef]
15. Wei, L.; Dong, Y.; Zhao, Y.; Xu, Z.; Li, H.; Wu, Y. The Influence of Ti and Ni on the Microstructure, Wear Resistance, and Thermal Fatigue Performance of Laser Cladded Stellite 6 Coatings. *J. Mater. Eng. Perform.* **2024**, *5*. [CrossRef]
16. Félix-Martínez, C.; Salgado-López, J.M.; López-Martínez, A.; García-Salas, L.D.; González-Carmona, J.; Cruz-González, C.E. Microstructure; hardness, and wear resistance at room and high temperature of Stellite-6/WC-6Co coatings deposited by laser cladding process. *Int. J. Adv. Manuf. Technol.* **2024**, *130*, 3355–3371. [CrossRef]
17. Manukonda, S.; Bijjam, R.R. Wear Resistance of Stellite-6/TiC Coating on Stainless Steel 316L Produced by Laser Cladding Process. *Ann. Chim. Sci. Mater.* **2023**, *47*, 75–80. [CrossRef]
18. Moradi, M.; Ashoori, A.; Hasani, A. Additive manufacturing of stellite 6 superalloy by direct laser metal deposition—Part 1: Effects of laser power and focal plane position. *Opt. Laser Technol.* **2020**, *131*, 106328. [CrossRef]
19. Nair, A.; Ramji, V.; Raj, R.D.; Veeramani, R. Laser cladding of Stellite 6 on EN8 steel—A fuzzy modelling approach. *Mater. Today Proc.* **2020**, *39*, 348–353. [CrossRef]
20. Nair, A.; Khan, A. Studies on effect of laser processed stellite 6 material and its electrochemical behavior. *Optik* **2020**, *220*, 165221. [CrossRef]
21. Rajeev, G.P.; Rahul, M.R.; Kamaraj, M.; Bakshi, S.R. Microstructure and high temperature mechanical properties of wire arc additively deposited Stellite 6 alloy. *Materialia* **2020**, *12*, 100724. [CrossRef]
22. Singh, J.; Kumar, S.; Mohapatra, S.K. Erosion tribo-performance of HVOF deposited Stellite-6 and Colmonoy-88 micron layers on SS-316L. *Tribol. Int.* **2020**, *147*, 105262. [CrossRef]
23. Thawari, N.; Gullipalli, C.; Katiyar, J.K.; Gupta, T.V.K. Influence of buffer layer on surface and tribomechanical properties of laser cladded Stellite 6. *Mater. Sci. Eng. B Solid State Mater. Adv. Technol.* **2021**, *263*, 114799. [CrossRef]
24. Wang, G.; Zhang, J.; Shu, R.; Yang, S. High temperature wear resistance and thermal fatigue behavior of Stellite-6/WC coatings produced by laser cladding with Co-coated WC powder. *Int. J. Refract. Met. Hard Mater.* **2019**, *81*, 63–70. [CrossRef]
25. Moradi, M.; Hasani, A.; Beiranvand, Z.M.; Ashoori, A. Additive manufacturing of stellite 6 superalloy by direct laser metal deposition—Part 2: Effects of scanning pattern and laser power reduction in differrent layers. *Opt. Laser Technol.* **2020**, *131*, 106455. [CrossRef]
26. Sidhu, H.S.; Sidhu, B.S.; Prakash, S. Solid particle erosion of HVOF sprayed NiCr and Stellite-6 coatings. *Surf. Coat. Technol.* **2007**, *202*, 232–238. [CrossRef]
27. Sidhu, B.S.; Puri, D.; Prakash, S. Mechanical and metallurgical properties of plasma sprayed and laser remelted Ni-20Cr and Stellite-6 coatings. *J. Mater. Process. Technol.* **2005**, *159*, 347–355. [CrossRef]
28. Rajoria, V.; Nain, G.; Vijayan, S.; Prasad, C.H.; Damodaram, R.; Karthik, G.M.; Khan, F. Materials Today: Proceedings Development of SS 304L composite coatings on mild steel substrate using friction surfacing and wear characterization. *Mater. Today Proc.* **2022**, *62*, 746–754. [CrossRef]
29. Khan, M.D.F.; Rokkala, U. Development of high strength and corrosion resistance Mg-Zn-Dy/HA-Ag composite for temporary implant applications. *Mater. Lett.* **2023**, *347*, 134604. [CrossRef]
30. Sadhu, A.; Karmakar, D.P.; Mypati, O.; Muvvala, G.; Pal, S.K.; Nath, A.K. Performance of additive manufactured Stellite 6 tools in friction stir processing of CuCrZr sheet. *Opt. Laser Technol.* **2020**, *128*, 106241. [CrossRef]

31. Silwal, B.; Li, L.; Deceuster, A.; Griffiths, B. Effect of Postweld Heat Treatment on the Toughness of Heat-Affected Zone for Grade 91 Steel. *Weld. Res.* **2013**, *92*, 80–87.
32. Opris, C.D.; Liu, R.; Yao, M.X.; Wu, X.J. Development of Stellite alloy composites with sintering/HIPing technique for wear-resistant applications. *Mater. Des.* **2007**, *28*, 581–591. [CrossRef]
33. Ramachandran, C.S.; Balasubramanian, V.; Varahamoorthy, R. Comparative evaluation of dry sliding wear behaviour of plasma transferred arc hardfaced surfaces by the pin-on-roller method. *Proc. Inst. Mech. Eng. Part J J. Eng. Tribol.* **2010**, *224*, 91–106. [CrossRef]
34. Cao, H.T.; Dong, X.P.; Pan, Z.; Wu, X.W.; Huang, Q.W.; Pei, Y.T. Surface alloying of high-vanadium high-speed steel on ductile iron using plasma transferred arc technique: Microstructure and wear properties. *Mater. Des.* **2016**, *100*, 223–234. [CrossRef]
35. Mousavi, S.E.; Naghshehkesh, N.; Amirnejad, M.; Shammakhi, H.; Sonboli, A. Wear and Corrosion Properties of Stellite-6 Coating Fabricated by HVOF on Nickel–Aluminium Bronze Substrate. *Met. Mater. Int.* **2021**, *27*, 3269–3281. [CrossRef]
36. Rahmati, Z.; Aval, H.J.; Nourouzi, S.; Jamaati, R. Effect of copper reinforcement on the microstructure, macrotexture, and wear properties of a friction-surfaced Al-Cu-Mg coating. *Surf. Coat. Technol.* **2022**, *438*, 128380. [CrossRef]
37. Magarò, P.; Furgiuele, F.; Maletta, C.; Tului, M.; Wood, R.J.K. Wear Mechanisms of Cold-Sprayed Stellite-6 During Reciprocated Dry Sliding Under Different Sliding Speeds. *J. Therm. Spray Technol.* **2023**, *32*, 2336–2350. [CrossRef]
38. Yoon, B.H.; Lee, C.H.; Kim, H.J. Effect of Dilution on wear performance of plasma transferred Arc deposited layers. *ISIJ Int.* **2017**, *57*, 913–920. [CrossRef]

materials

MDPI

Article

Friction and Wear Performances and Mechanisms of Graphite/Copper Composites Under Electrical Contact in Marine Environments

Nenghui Wang [1,2], Chuanfeng Wang [2], Wenhu Xu [3], Weiping Cheng [2], Haihong Wu [2] and Hongsheng Li [1,*]

1 School of Instrument Science and Engineering, Southeast University, Nanjing 210096, China
2 CSSC Jiujiang Marine Equipment (Group) Co., Ltd., Jiujiang 332005, China
3 School of Advanced Manufacturing, Nanchang University, Nanchang 330031, China
* Correspondence: haiyan_8453@163.com

Abstract: Marine environment-induced apparatus failures have led to substantial losses in marine engineering. Graphite/copper composites, known for their excellent electrical conductivity and wear resistance, are extensively utilized in various electric contact devices. However, research on the current-carrying friction and wear behavior of graphite/copper composites in marine environments is still limited. This study investigates the effects of mating materials, graphite content (30 wt.% and 45 wt.%), and electric voltage on the friction and wear mechanisms of graphite/copper composites in seawater. The results show that under seawater coupled with electricity, no mass loss was observed in the 30 wt.% graphite composites after friction tests against different counterparts. Electric voltage (3 V) affects the composite's damage mechanism, inducing delamination wear, arc erosion and accelerating corrosion. Specifically, the electricity factor promotes oxidation recreations while inhibiting chlorine formation. Notably, when the composite is paired with gold-coated copper, it undergoes electrochemical reactions, leading to the formation of needle-like copper oxide. These oxides alter the surface morphology, elevate the mass of worn composites, and raise the friction coefficient of the tribopair to approximately 0.3, an increase from 0.2.

Keywords: marine engineering; electrical contacts; graphite/copper composite; current-carrying friction and wear; mating materials

1. Introduction

Abundant green energy resources in the blue ocean, including tidal, wind, biomass and wave energy, are vital for the sustainable development of human society [1–3]. Nevertheless, the unique marine environment significantly affects the functionality of diverse marine engineering equipment. High salinity and humidity can cause the formation of a liquid electrolyte film on the surfaces of electrical contacts in marine environment [4,5]. This film participates in the interactions at the electrical contact interface, resulting in wear and corrosion, which affects the service life of materials and equipment [6–9].

Graphite/copper composites integrate the lubricity, low thermal expansion and low density of graphite with the high electrical conductivity, thermal conductivity and ductility of copper, thereby exhibiting excellent wear resistance, arc erosion resistance and anti-fusion welding properties [10]. Lin et al. [11] incorporated copper-plated graphite powder into the copper matrix, improving the C–Cu interface bonding, which enables the composites to retain the physical and mechanical properties of the Cu matrix while achieving a lower friction coefficient and wear rate. Song et al. [12] developed copper-graphite gradient composite materials to address the need for optimizing damage in current-carrying friction materials. Andrej et al. [13] investigated the coefficient of thermal expansion of graphite/copper composites and discovered that the coefficient decreases as the graphite volume fraction increases. Kumar et al. [14] found that graphite addition not only decreased the friction coefficient of copper but also improved material strength. Li et al. [15] ascribed the good tribological performance of graphite/copper composites to the synergistic effect of an oxidized friction layer and a graphite lubricating film. Liu et al. [16] investigated the relationship between graphite content and the evolution of contact resistances, revealing that the graphite/copper composites with high graphite content exhibit good resistance to fretting wear but compromise current transmission efficiency. Yang et al. [17,18] studied the effects of graphite content and raw material powder on the current-carrying friction and wear properties of the composite material. Currently, graphite/copper composites are widely used in various mechanical and electrical components. To enhance the tribological properties of copper-graphite composites in applications such as slip-ring assemblies and locomotive pantograph strips, Grandin et al. [19] examined the friction, wear and contact resistance of the graphite/copper composite when rubbed against copper. Xu et al. [20] combined large-size cluster graphite and copper matrix to prepare pantograph sliders with high impact toughness. Wang et al. [21] prepared a Cu-graphite brush using the powder metallurgy process and identified the dominant wear mechanism of the brush under the current as electrical wear.

However, most studies on graphite/copper composites have concentrated on the preparation and friction performance of the material in dry or wet conditions, with limited investigation into their friction behavior in marine environment and the impact of graphite content and current-carrying factors on friction characteristics. Consequently, a thorough study of the friction and wear of copper/graphite composites in the marine environment is of importance for unlocking the application potential of these materials and extending the service life of electrical contact components.

In this study, two graphite/copper composites with varying graphite contents were prepared using the powder metallurgy method. Subsequently, friction pairs of these composites were paired with three mating materials of varying hardness. Current-carrying friction tests were then performed in a simulated seawater environment. Scanning electron microscopy (SEM) and energy dispersive spectroscopy (EDS) were utilized to analyze the microscopic morphology of the wear samples and explore the wear mechanism. The aim of this study is to provide theoretical insights and experimental data to support the selection and application of these materials in marine engineering.

2. Materials and Methods

2.1. Materials

The addition of a small amount of lead to graphite/copper materials significantly enhances their hardness and improves their friction behavior [22]. Therefore, in this study, some lead was added to the graphite/copper. Copper powder (~400 mesh) was purchased from Beijing Beike New Material Technology Co., Ltd. (Beijing, China), graphite powder (~200 mesh) was provided by Tianjin Gaokexin Material Co., Ltd. (Tianjin, China), and lead powder (~400 mesh) was sourced from Hebei Tebo Metal Material Co., Ltd. (Xingtai, China), all with purities of no less than 99%.

Graphite/copper composites were prepared by the powder metallurgy pressure sintering method. The specific process involved the following steps: First, the raw material powders were weighed. Next, the powders were mixed in a ball mill (MITR YXQM-4L, Changsha, China) with a ball-to-powder ratio of 10:1, using 2 mm copper balls, at a rotational speed of 200 rpm for 24 h. Following mixing, the powders were compressed into rectangular compacts, with a dimension of 4 mm × 4 mm × 25 mm, using a hydraulic press (Riyi FBSY-C05, Wenzhou, China) at a pressure of 300 MPa for 15 s. Finally, the compacts were sintered in a hydrogen furnace (Kejing KSL-1200X, Hefei, China) at 900 °C for 1 h, followed by furnace cooling. Prior studies have shown that incorporating approximately 30~40 wt.% of graphite can improve the current-carrying performance and mechanical strength of graphite/copper composites [16,23]. The chemical element contents of the two graphite/copper composites are detailed in Table 1 below.

Table 1. Chemical element composition of graphite/copper composites.

Material ID	C (wt.%)	Pb (wt.%)	Cu (wt.%)
A1	45	3	52
A2	30	3	67

The counterpart materials were red copper (T2), zirconium bronze (QCr0.8), and gold-plated red copper (Au/T2 with a gold plating layer thickness of 5 μm). Their hardness values were 70, 150, and 200 HV, respectively.

2.2. Methods

The composite hardness was determined by a nanoindenter (Nanotest Vantage, Wrexham, UK). Following fixation and leveling of the sample on the test bench, a load of 200 mN is applied. To enhance the precision of the test results, each sample underwent three repetitions, with all indentations made within the sample matrix.

Friction and wear tests were performed using a multifunctional friction and wear testing machine (Tribolab, Billerica, MA, USA) to investigate the wear behavior of graphite/copper in 3.5% NaCl solution under the condition of current-carrying. The dimensions of the graphite/copper composite pin samples were 4 mm × 4 mm × 25 mm, while the counterpart disks had dimensions of Φ70 mm × 5 mm. Prior to the tests, the samples were ground using sandpaper until they achieved similar surface roughness (Figure 1a,b). The tribopairs were immersed in the NaCl solution, and a constant voltage (as depicted in Table 2) was applied during the tests. A load of 5 N and a rotational speed of 75 rpm were applied. The sliding distance was 2986 m, and the sliding time was 1 h. Each set of friction experiments was repeated three times.

Table 2. Voltage under different operation conditions.

Operation Condition	Voltage (V)
C1	0
C2	3

The wear rate W was calculated using the following formula:

$$W = \frac{m}{s}$$

where W is the wear rate (mg/mm^2), m is the wear mass (mg), and s is the wear area (mm^2).

SEM, EDS (Thermo Scientific/Apero 2S HiVac, Waltham, MA, USA), and laser confocal microscope (LSK, Bruker, Billerica, MA, USA) were applied to test the surface morphology, microstructure, and elemental distribution samples. At an accelerating voltage of 15 KV, the step sizes of SEM and EDS were 0.45 μm and 0.31 μm, respectively.

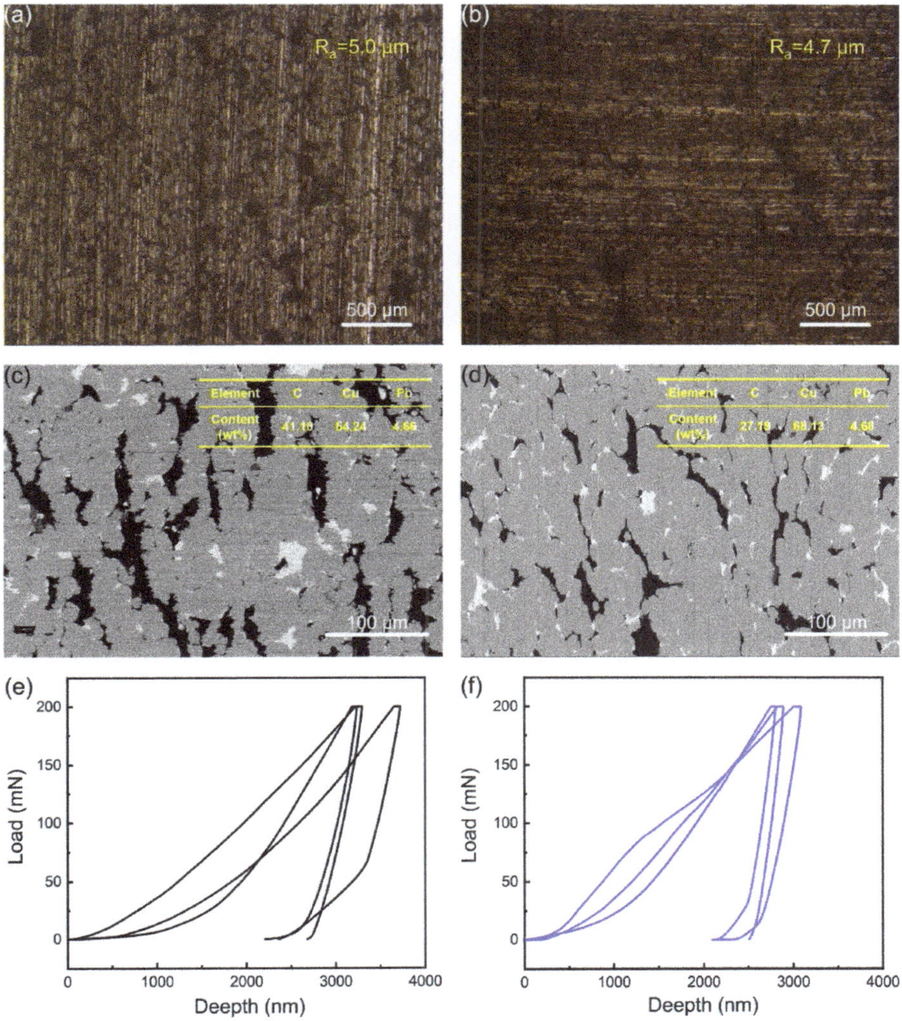

Figure 1. The initial surface topography of sample C1 (**a**) and sample C2 (**b**); The BSE image of sample C1 (**c**) and sample C2 (**d**); The load-depth curves of sample C1 (**e**) and sample C2 (**f**) under loads of 200 mN.

3. Results and Discussion

3.1. Microstructure and Hardness

The distribution of graphite and Pb components within the copper matrix before testing is shown in Figure 1c,d. These images reveal a gray copper matrix, with the graphite phase appearing as black, elongated islands of varying sizes and the Pb phase as white, worm-like structures uniformly distributed. Additionally, the content of the various elements approximated the preset values. Figure 1e,f show the nanoindentation test results. A greater maximum indentation depth under the same load corresponds to lower material hardness. Due to the softness of graphite, its addition to copper reduces the overall hardness [15,24]. Sample A2 exhibited a lower maximum indentation depth compared to sample A1, indicating that it possesses greater hardness. The hardness of sample A1 is 0.81 GPa, and the hardness of sample A2 is 1.05 GPa.

3.2. Friction Coefficient

Figure 2 illustrates the friction coefficients (COF) of these tribopairs of graphite/copper composites and their counterpart materials under C1 and C2 conditions. As shown in Figure 2a–c, the two graphite/copper samples display similar trends in friction coefficient variations in seawater. As sliding time increases, the friction coefficient decreased and stabilized after the initial run-in period, during which the contact interface achieved an optimal state through plastic deformation and wear. The reduction in friction coefficient over time can be ascribed to the formation of a graphite layer and surface oxidation. As sliding time increases, more graphite is stripped from the copper matrix, leading to the gradual formation of a lubricant film. It reduces the likelihood of direct contact between the tribopairs, causing friction to occur between the lubricating film and the opposing surface. This diminishes the adhesion between metals, leading to a reduction in the friction coefficient [15,16]. Furthermore, during friction, the surface of the composites generally forms many oxides, which facilitates the sliding. This mechanism is in alignment with other established findings [15,25]. Due to the combined effects of graphite and oxides, the friction coefficient deceases as sliding time progresses. Figure 2d demonstrates that the average friction coefficient of these tribopairs decreases as the hardness of the counterpart material increases in a seawater environment.

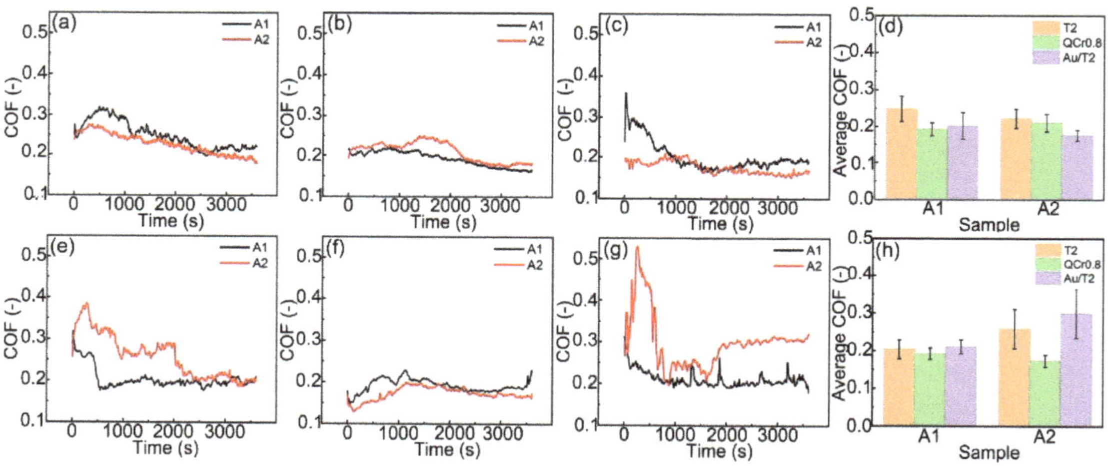

Figure 2. Instantaneous and average friction coefficients of graphite/copper-T2, graphite/copper-QCr0.8, and graphite/copper-Au/T2 (**c**) under C1 (**a–d**) and C2 (**e–h**) conditions.

Furthermore, when a constant voltage was applied to the A1's tribopair, the time required for the contact interface to stabilize was reduced, and the friction coefficient stabilized at approximately 0.2, consistent with the friction coefficient observed in seawater. Under the effect of voltage, the A2 sample exhibited significant fluctuation in the friction coefficient when paired with Au/T2, and its average friction coefficient increases significantly (Figure 2g,h). Additionally, the friction coefficient of the A2–Au/T2 finally stabilized at approximately 0.3, higher than that of other tribopairs under C1/C2 conditions, indicating a change in the lubricating film at the contact interface (Figure 2h).

3.3. Macroscopic Wear Morphology and Wear Rate

Apparently flat surfaces exhibit some roughness due to micro-asperities. Figure 3 exhibits the wear morphology of the graphite/copper composites after friction tests in seawater. Sample A2 exhibited greater hardness, indicating enhanced deformation resistance [26]. The worn surface of sample A1 was less flat than that of sample A2. These micro-convex bodies on the T2 surface repeatedly pressed and slid against the A1 and A2 samples, causing scratches on their surfaces. During this process, cracks also appeared on the friction surface, and they arise to excessive plastic deformation of the surface. As shown in Figure 3b,e, the cracks on these composites are approximately 0.3–0.4 mm in length. Additionally, wear debris accumulated on the sample surfaces, forming red protrusions in the 3D topography images of A1 (Figure 3c). The maximum height of these protrusions was approximately 260.06 μm. Meanwhile, numerous small pits were observed, resulting from the rolling of abrasive particles during three-body wear [27]. These protrusions and pits increased the roughness of the A1 surface, rendering it more uneven compared to A2.

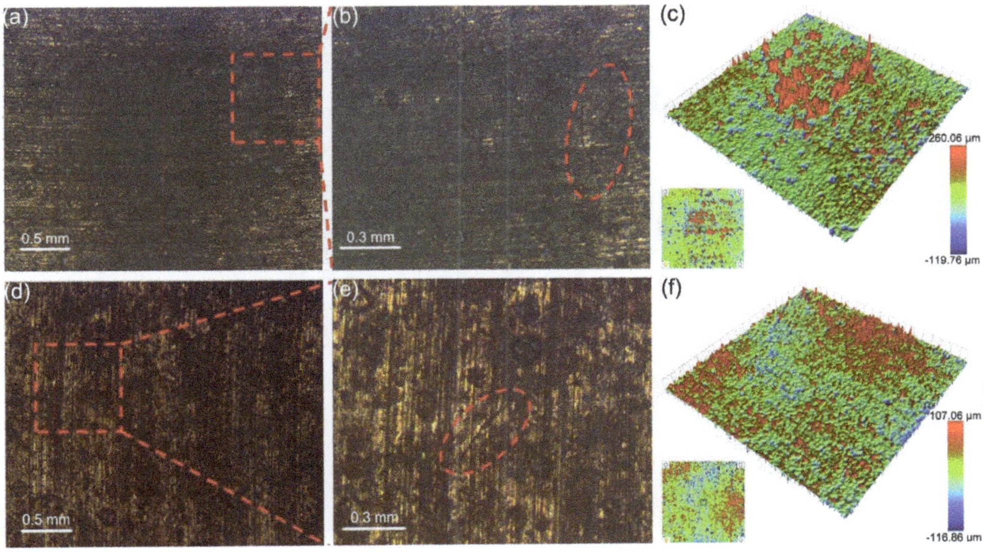

Figure 3. Surface morphology of samples A1 (**a**–**c**) and A2 (**d**–**f**) after friction with T2 under C1 conditions. (**b**) and (**e**) are, respectively, the enlarged image of the area marked with red rectangles in (**a**) and (**d**), with cracks marked by red elliptical dashed lines.

Figure 4 compares the macroscopic wear morphology of A1 and A2 samples after friction with the harder QCr0.8. For the sample A1, the increased presence of protrusions led to higher surface roughness. Because A2 has higher hardness, during the friction process, A2 exhibited greater resistance to deformation against the surface convex of QCr0.8, resulting in wear morphology and surface roughness similar to those observed after friction with T2.

As the hardness of the mating material increases, the wear morphology of the two graphite/copper composites in seawater underwent further changes. Figure 5 demonstrates that the surfaces of A1 and A2 show an increase in protrusions, indicating higher surface roughness after abrasion with the hardest Au/T2. Analysis of the density and height of these protrusions reveals that the surface roughness of A1 is significantly greater than that of A2. Additionally, these small pits observed indicate that there is three-body wear during the friction (Figure 5f).

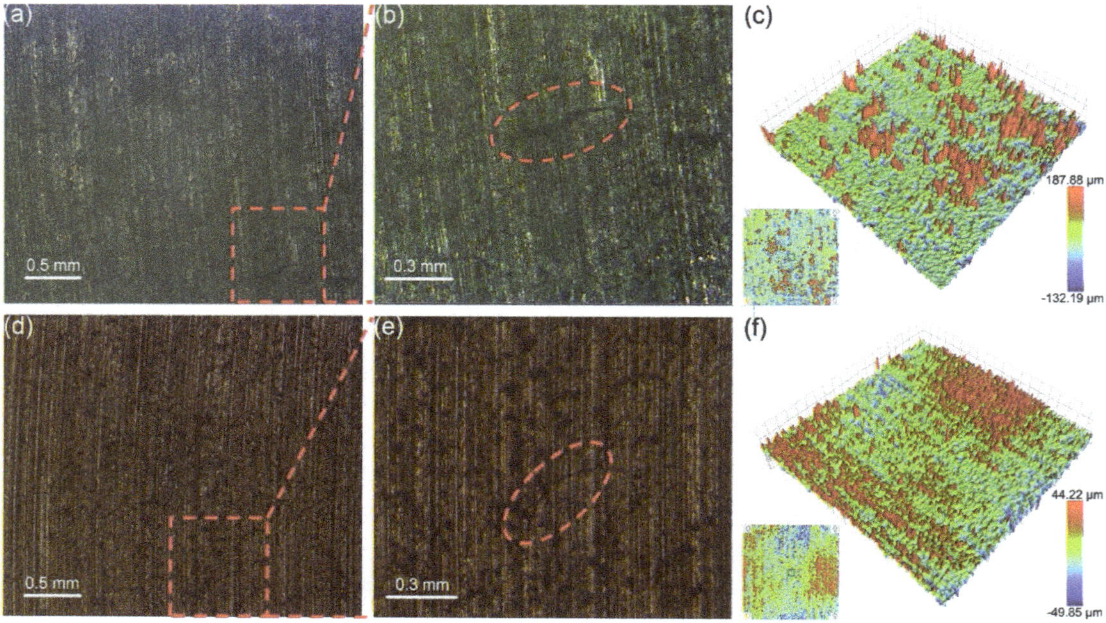

Figure 4. Surface morphology of samples A1 (**a**–**c**) and A2 (**d**–**f**) after friction with QCr0.8 under C1 conditions. (**b**) and (**e**) are, respectively, the enlarged image of the area marked with red rectangles in (**a**) and (**d**), with cracks marked by red elliptical dashed lines.

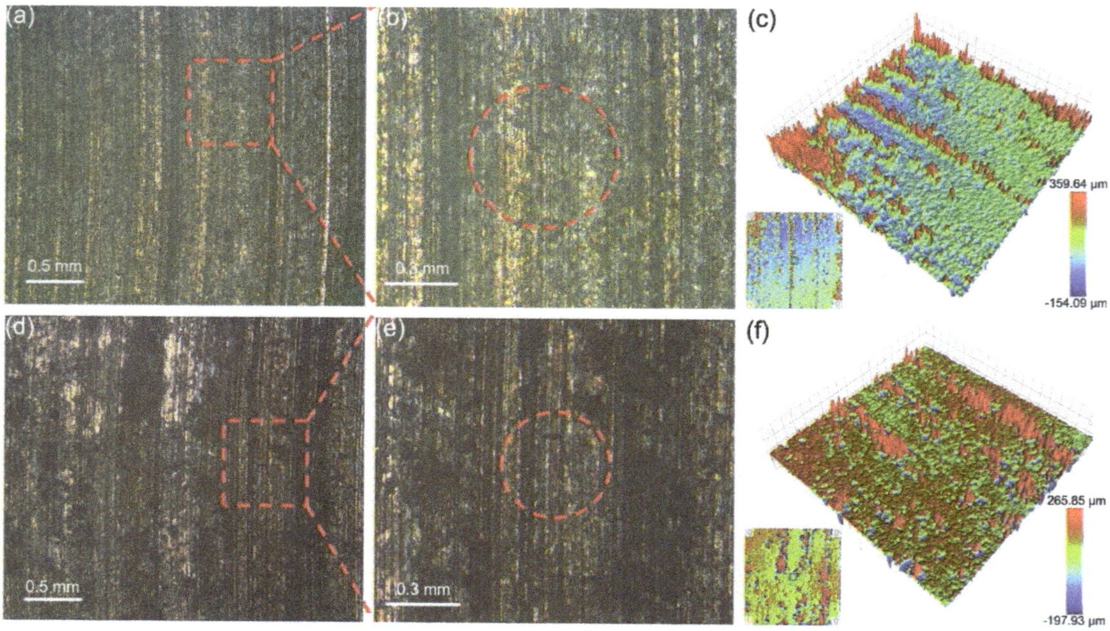

Figure 5. Surface morphology of samples A1 (**a**–**c**) and A2 (**d**–**f**) after friction with Au/T2 under C1 conditions. (**b**) and (**e**) are, respectively, the enlarged image of the area marked with red rectangles in (**a**) and (**d**), with cracks marked by red elliptical dashed lines.

Figure 6 exhibits the macroscopic wear morphologies of A1 and A2 samples after friction with different counterparts under C2 conditions. The wear morphology of A1 and A2 did not change significantly compared to those observed without applied voltage. Cracks were present on the worn surfaces of the composites, with lengths of approximately 0.5 mm (Figure 6). Similarly, protrusions were more easily observed on the worn surfaces of sample A1 compared to sample A2, with majority having heights of 100~300 μm. Most of these protrusions on worn surfaces of sample A1 were 100~200 μm in height. As the hardness of the counterpart material increased, the surface roughness of the graphite/copper samples increased accordingly. Similarly, under identical counterpart material conditions, the surface roughness of A1 was higher than that of A2.

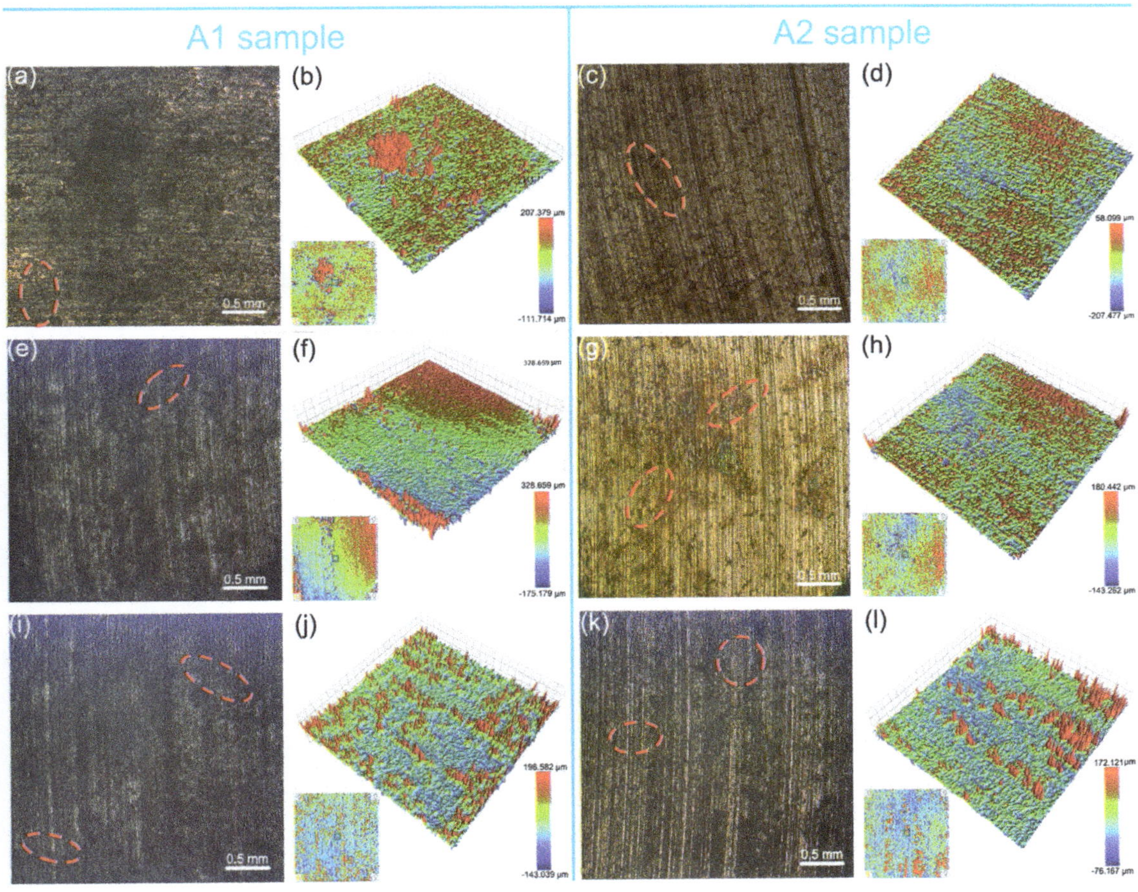

Figure 6. Surface morphology of A1 and A2 samples after friction with T2 (**a–d**), QCr0.8 (**e–h**), and Au/T2 (**i–l**) under C2 conditions, with cracks marked by red elliptical dashed lines.

To further investigate the morphology changes in the graphite/copper composites under C1 and C2 conditions, the surface roughness of the samples was characterized using a laser confocal microscope. As shown in Figure 7a, the surface roughness of the composites increases with the hardness of the counterpart material, which is consistent with the macroscopic morphology observations. Furthermore, under identical conditions and counterpart materials, the surface roughness of A1 was higher than that of A2. As the hardness of the counterpart material increased, the change in roughness of A2 was less than that of

A1. These characterization results demonstrate that as the hardness of the counterpart material increases, the interaction at the contact interface intensifies, leading to greater plastic deformation and the accumulation of wear debris on the sample surfaces, thereby increasing the surface roughness. Reducing the graphite content effectively enhances the resistance of graphite/copper composites to plastic deformation, resulting in more stable roughness changes when interacting with counterparts of varying hardness. Additionally, Figures 2 and 7 illustrate that increased surface roughness may lead to a decrease in the friction coefficient, as the reduced adhesion due to the diminished actual contact area, which is evident in the A1 specimens' friction behavior [28].

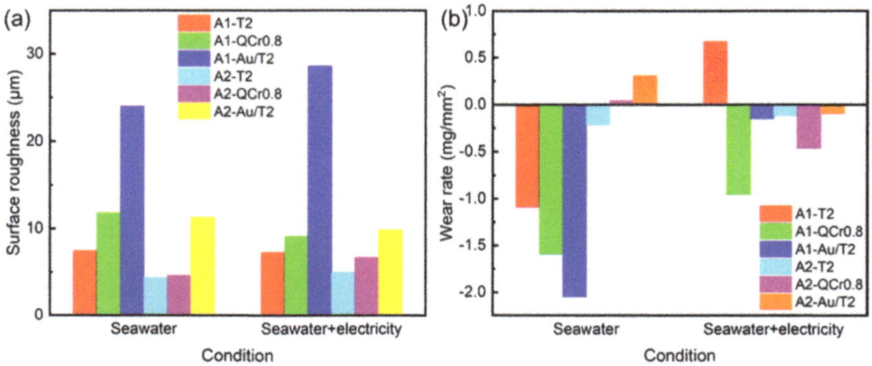

Figure 7. Surface roughness (**a**) and wear rate (**b**) of A1 and A2 samples after friction with T2, QCr0.8, and Au/T2 under C1 and C2 conditions.

The wear rate is a key indicator of a material tribological behavior and performance. Figure 7b shows the wear rates of A1 and A2 samples after friction with different counterpart materials under conditions C1 and C2. Under the C1 conditions, two graphite/copper samples exhibited negative wear rates, indicating an increase in the post-wear mass of samples A1 and A2. This phenomenon was also observed in C2 conditions. The possible reasons for this include chemical reactions between the copper and substances in the seawater, forming corrosive products, or material transfer from the counterpart disk to the sample surface. Due to the absence of mass loss observed in all A2 specimens under C2 conditions, and considering the importance of controlling wear rates to ensure long-term material durability and stable equipment performance, A2 composites are considered more appropriate for electric contact devices in marine environments.

3.4. Microscopic Wear Morphology and Wear Mechanism

To reveal the wear mechanisms of graphite/copper composites under these two conditions, the microscopic wear morphology and surface element distribution of A2 samples were characterized using SEM and EDS.

Figure 8 displays the microscopic wear morphology of the composite under voltage and non-voltage conditions. As depicted in Figure 8a,b, A2 presents various worn morphologies, suggesting different wear mechanisms during friction. Under the C1 conditions, a broken surface was observed, with numerous granular debris evenly distributed across the surface. The wear debris indicates that the A2 composite has experienced abrasive wear. Upon the application of voltage, little debris was found on the worn surface of A2. However, sheet-like stacking was observed on the surface, a characteristic of delamination wear. Concurrently, arc ablation due to electric field breakdown from unstable dynamic contact was observed on the worn surface (Figure 8b).

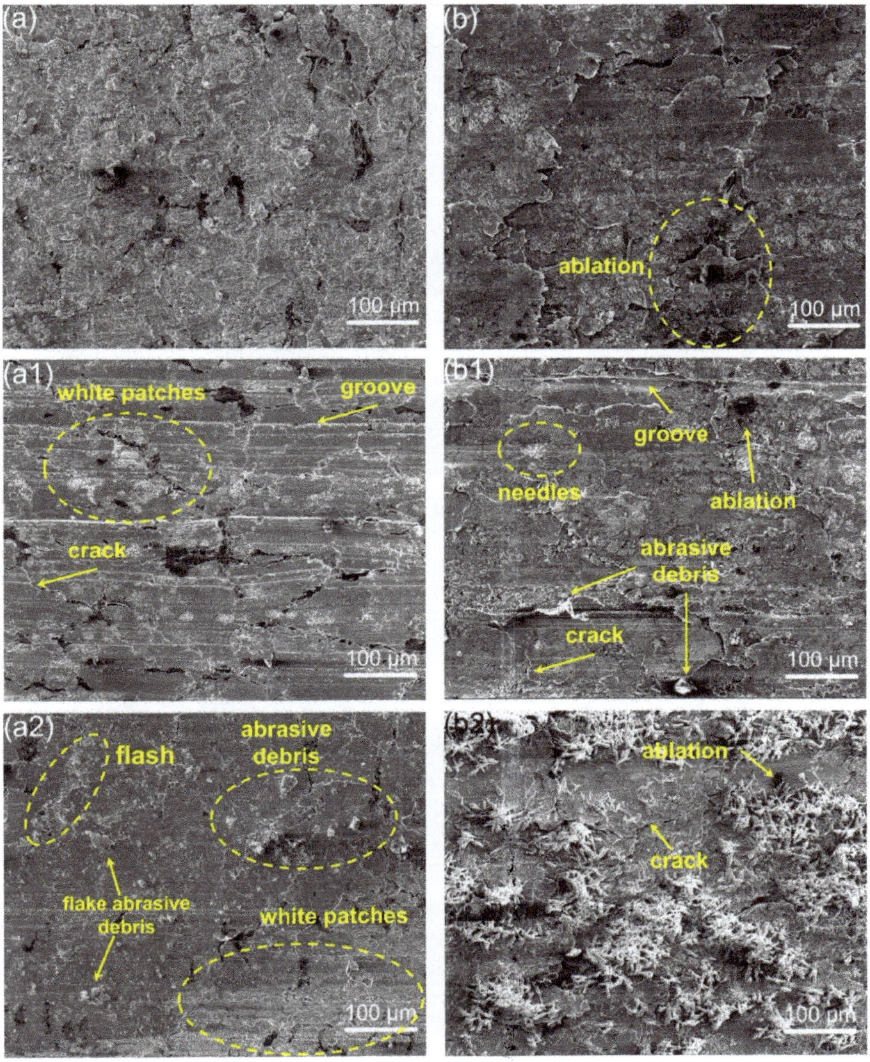

Figure 8. Microscopic wear morphology of A2 samples after friction with T2, QCr0.8, and Au/T2 under C1 (**a–a2**) and C2 (**b–b2**) conditions.

Figure 8(a1,b1) exhibits morphologies of A2 grounded with QCr0.8 under two operational conditions. Under the non-voltage condition, numerous grooves indicate that abrasive wear remains the primary mechanism of surface damage (Figure 8(a2)). At the same time, some white patches were observed. Under the C2 condition, both sheet-like stacking, crack and abrasive debris indicate simultaneous occurrence of delamination and abrasive wear during the friction of A2. Similarly, damage from arc ablation persisted. Furthermore, it is noteworthy that white patches formed on the A2 worn surface under non-voltage condition, while needle-like structures were observed on the worn surfaces under the voltage condition.

Figure 8(a2,b2) illustrates the worn surfaces of the composite grounded with Au/T2 under both C1 and C2 conditions. For the A2 sample under C1 condition, the presence of flash and flake debris indicates adhesive wear. These granular debris suggest the occurrence of abrasive wear as well. The rolling of debris on the surface induces the appearance of the small pits (Figure 5f). Under C2 conditions, the A2 sample exhibited persistent delamination wear, arc ablation and increasing needle-like substances.

To further analyze the wear mechanisms of A2 samples, we performed elemental analysis on the wear surfaces of A2 samples after friction with different hardness counterparts under different conditions using EDS. Figure 9 shows the EDS analysis results of the wear surfaces of A2–T2 tribopairs under C1 and C2 conditions. In a seawater environment, the wear surface of A2 mainly contained C, O, Cl, and Pb elements. The graphite phase was separated from Cu matrix under shear forces, resulting in a significant reduction in the relative content of C. As shown in Figure 9(a2–a4), O, Cu, and Cl elements appear to be more concentrated in some regions due to the accumulation of wear debris on the sample surface. The presence of O and Cl elements indicates that the copper matrix reacts chemically with oxygen and chloride ions in the seawater. This confirms that the wear process of A2 involves not only abrasive wear but also corrosion wear. After applying voltage, besides the appearance of Na elements due to the evaporation of residual seawater, the oxygen content on the A2 wear surface significantly increased, while Cl disappeared. This suggests that under C2 conditions, the corrosion type is oxidative corrosion (Figure 9b). Additionally, the ratio of $W_{(O+Cl)}/W_{(Cu+Pb)}$ for A2 wear samples under C1 conditions was 0.13, while under C2 conditions, it was $W_{(O)}/W_{(Cu+Pb)} = 0.28$, indicating that the electricity factor accelerates the corrosion of A2 samples. The uniform distribution of C across the entire wear surface of A2 under C2 conditions indicates the formation of a graphite lubricating film during friction.

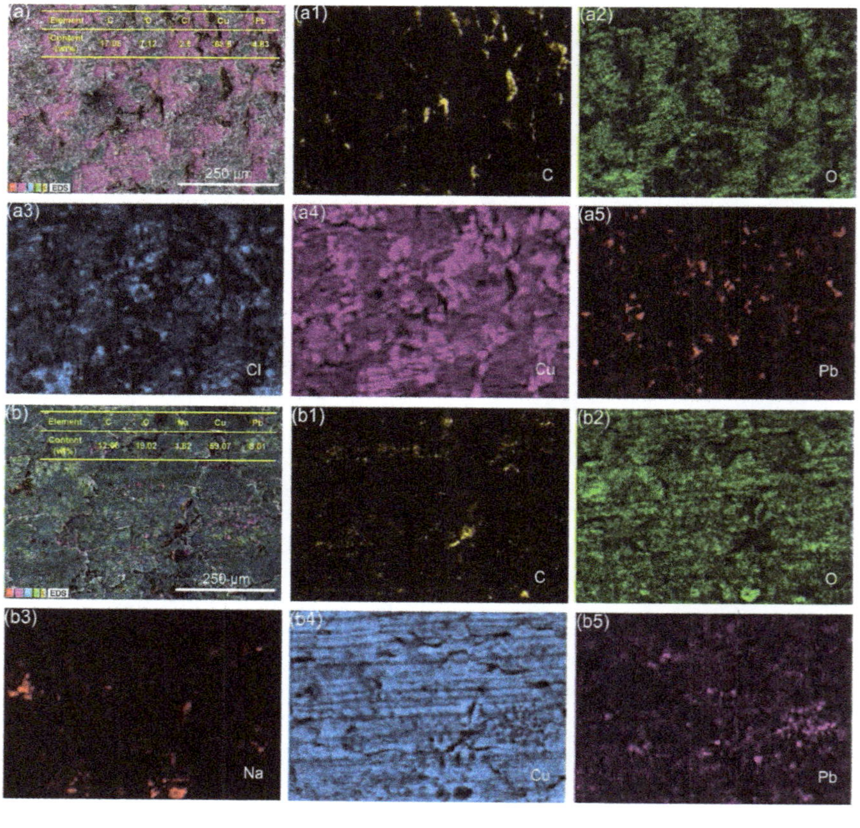

Figure 9. EDS spectra of A2 samples after friction with T2 under C1 (**a–a5**) and C2 (**b–b5**) conditions.

Figure 10 compares the elemental distribution on the wear surfaces of A2–QCr0.8 tribopairs under C1 and C2 conditions. After friction with the harder QCr0.8, A2 samples also underwent corrosion reactions, with $W_{(O+Cl)}/W_{(Cu+Pb)} = 0.13$. For the A2–QCr0.8 tribopair under C2 conditions, the oxide content on the A2 wear surface increased, while chlorides nearly disappeared, with $W_{(O+Cl)}/W_{(Cu+Pb)} = W_{(O)}/W_{(Cu+Pb)} = 0.22$. The observed phenomenon was consistent with the elemental characterization results of the A2–T2 tribopair under C2 conditions, indicating that although the electric factor accelerates the corrosion of graphite/copper composites, it has opposite effects on the formation of oxygen-containing and chlorine-containing corrosive products. Additionally, as shown in Figure 10(b1), a graphite lubricating layer exists at the contact interface of the A2–QCr0.8 tribopair under C2 conditions.

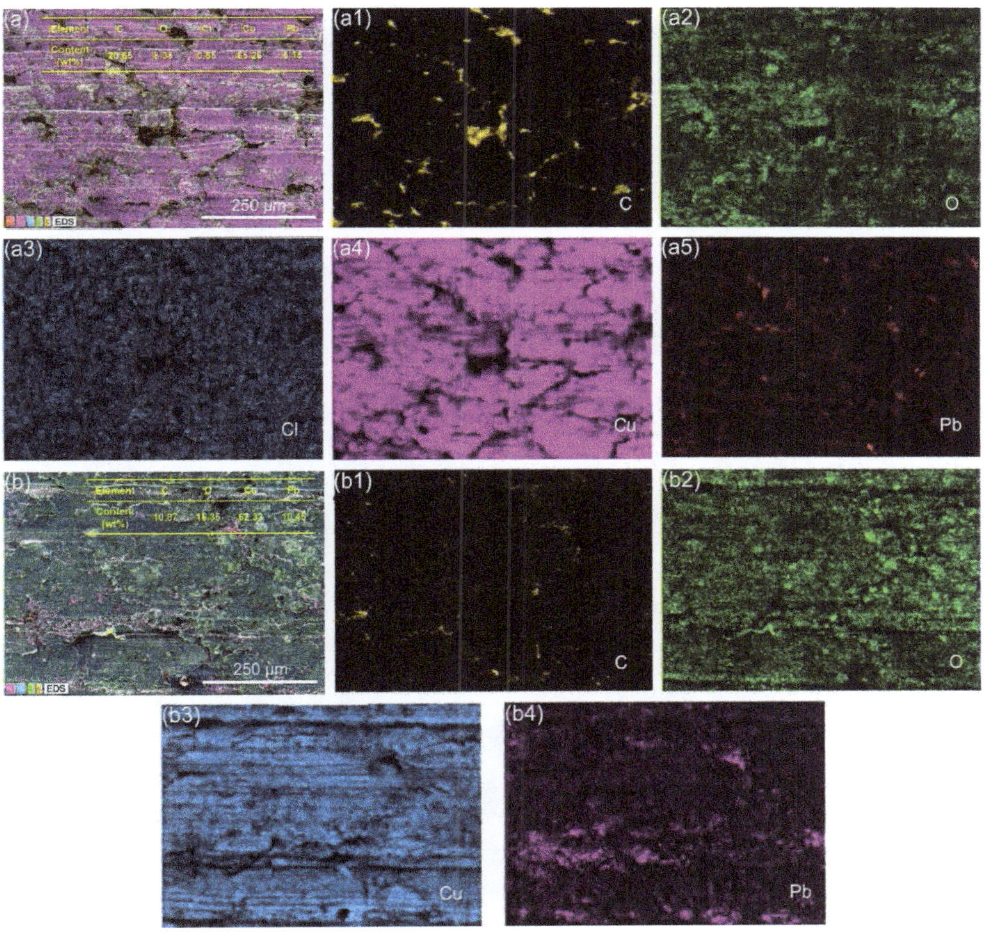

Figure 10. EDS spectra of A2 samples after friction with QCr0.8 under C1 (**a**–**a5**) and C2 (**b**–**b4**) conditions.

For the A2–Au/T2 tribopair, the wear surface of A2 samples under C1 conditions showed a similar distribution of O, Cl, Cu, and Pb elements to that observed in the A2–T2 and A2–QCr0.8 tribopairs, with $W_{(O+Cl)}/W_{(Cu+Pb)} = 0.15$. Due to adhesive wear, some Au from the Au/T2 counterpart transferred to the A2 sample surface (Figure 11(a5)). Similarly, as show as Figure 11(b4), Au elements are also observed on the worn surface of A2 under C2 conditions. Furthermore, the weight ratio $W_{(O+Cl)}/W_{(Cu+Pb)} = W_{(O)}/W_{(Cu+Pb)} = 0.50$, significantly higher than that of A2 samples in friction with other counterparts under C2 conditions, indicating that the corrosion is further exacerbated. This is because gold has a higher electrode potential than copper in the electrochemical series, making gold more likely to act as the cathode and copper more likely to lose electrons. Additionally, as shown in Figure 11, these unknown substances are oxides/chlorides of copper. Given the minimal distribution of Cl in the region where the unknown substances were located, it is reasonable to infer that these substances were mainly copper oxides. XPS analysis of Cu element reveals that the white patches consist of a mixture of CuO and Cu_2O, with Cu_2O as the predominant substance (Figure S1). Similarly, the needle-like objects are also a mixture of CuO and Cu_2O, predominantly CuO.

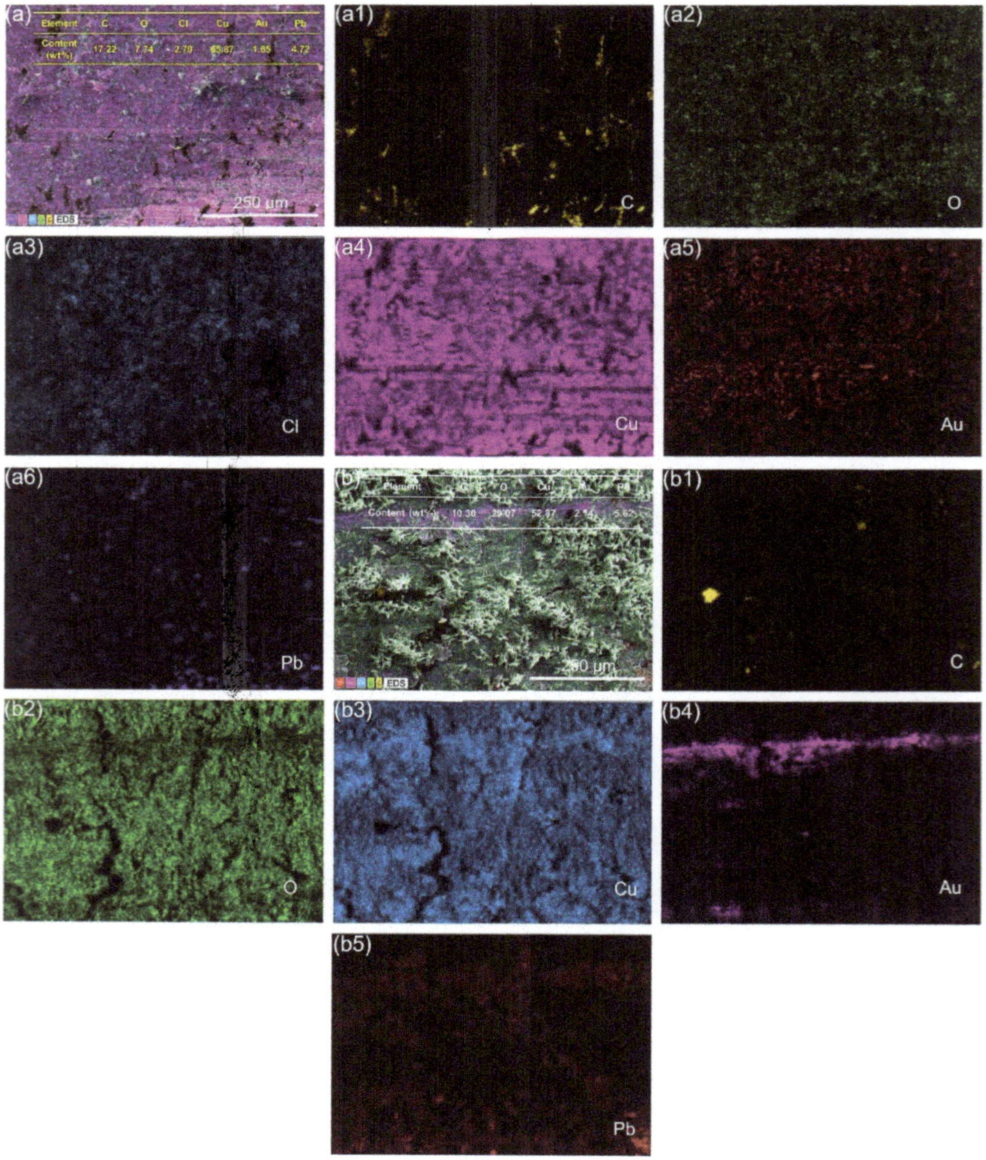

Figure 11. EDS spectra of A2 samples after friction with Au/T2 under C1 (**a**–**a6**) and C2 (**b**–**b5**) conditions.

4. Application Recommendation

This study provides critical insights into optimizing graphite/copper composites for marine electrical systems, addressing urgent challenges in reliability and durability. The findings hold transformative potential for marine engineering applications in extending the service life of critical components, such as subsea connectors, slip ring assemblies and dynamic cable systems, etc. In addition, the study provides an instruction for material pairing optimization. The electrochemical reaction mechanism identified in gold-coated copper pairings provides actionable guidelines for preventing unintended oxide formation in high-reliability marine electronic. This work establishes a framework for designing next-generation marine electrical components, with particular relevance to blue energy infrastructure (wave, tidal and wind power converters), deep-sea exploration equipment and smart port electrification systems. By bridging the knowledge gap in current-carrying tribocorrosion mechanisms, it provides essential data for international standards development in marine material selection.

5. Conclusions

In this study, we prepared graphite/copper composites with powder metallurgy to investigate the effects of graphite content, mating materials and electrical voltage on the friction behavior of these composites in a marine environment. The research provides valuable experimental data and theoretical insights into the wear behavior of graphite/copper composites, contributing to their potential application in marine engineering. Based on this research, we draw the following conclusions:

(1) Two graphite/copper composites, differing in graphite content, displayed similar friction coefficients (ranging from 0.2 to 0.3) during the final stage of the friction process in both C1 and C2 conditions.
(2) With the increase in counterpart material hardness, the interaction at the contact interface intensified, resulting in more plastic deformation on the surfaces of the graphite/copper samples, which in turn elevated the surface roughness. Meanwhile, graphite/copper composites of less graphite exhibited lower surface roughness due to higher hardness.
(3) Under C1 and C2 conditions, both graphite/copper samples exhibited a negative wear rate, suggesting that the corrosive products generated during friction tests lead to an increase in the mass of the worn samples instead of decrease.
(4) Under C1 conditions, the composite primarily underwent abrasive wear, while adhesive wear intensifies when grounded with harder Au/T2. Under C2 conditions, delamination wear and arc ablation became predominant, and minor amount of abrasive wear was also present. Similarly, adhesive wear arose when the composite was grounded with Au/T2. Furthermore, corrosive wear, particularly oxidative wear, was observed on the worn surfaces of all specimens.
(5) In a seawater context, the electricity factor not only induced electrical damage to the graphite/copper surface but also aggravated corrosion, leading to an increased oxygen concentration on the wear surface while suppressing the formation of chlorine-containing corrosive substances.

Supplementary Materials: The following supporting information can be downloaded at: https://www.mdpi.com/article/10.3390/ma18071516/s1, Figure S1: XPS results of the worn surfaces of A2 after grounded with Au/T2 under C1 (a) and C2 (b) conditions.

Author Contributions: N.W., conceptualization, writing—original draft, data curation; C.W., investigation, formal analysis; W.X., writing—review and editing, investigation; W.C., formal analysis;

H.W., project administration, supervision; H.L., supervision, project administration, writing—review and editing. All authors have read and agreed to the published version of the manuscript.

Funding: This work was supported by the Major Science and Technology Projects of Jiangxi Province (20223AAE01002) and the Gan-Po Talent Support Program of Jiangxi Province (20243BCE51035, 20243BCE51183).

Institutional Review Board Statement: Not applicable.

Informed Consent Statement: Not applicable.

Data Availability Statement: The original contributions presented in this study are included in the article/Supplementary Materials. Further inquiries can be directed to the corresponding author.

Conflicts of Interest: Author Nenghui Wang, Chuanfeng Wang, Weiping Cheng and Haihong Wu were employed by the China Ship Jiujiang Marine Equipment (Group) Co., Ltd. The remaining authors declare that the research was conducted in the absence of any commercial or financial relationships that could be construed as a potential conflict of interest.

References

1. Miller, K.A.; Thompson, K.F.; Johnston, P.; Santillo, D. An overview of seabed mining including the current state of development, environmental impacts, and knowledge gaps. *Front. Mar. Sci.* **2018**, *4*, 418. [CrossRef]
2. Pendleton, L.; Evans, K.; Visbeck, M. We need a global movement to transform ocean science for a better world. *Proc. Natl. Acad. Sci. USA* **2020**, *117*, 9652–9655. [CrossRef] [PubMed]
3. Constant, C.; Kotarbinski, M.; Stefek, J.; Green, R.; DeGeorge, E.; Baring-Gould, I. Accelerating ocean-based renewable energy educational opportunities to achieve a clean energy future. *Prog. Energy* **2021**, *3*, 042002. [CrossRef]
4. Zheng, Z.; Xie, J.; Liang, P.; Liu, J. Study on measurement technology and influencing factors of atmospheric salt fog concentration at the seaside—A case study of sanya, China. *Asia-Pac. J. Atmos. Sci.* **2023**, *59*, 311–326. [CrossRef]
5. Cole, I.S.; Paterson, D.A. Possible effects of climate change on atmospheric corrosion in Australia. *Corros. Eng. Sci. Technol.* **2013**, *45*, 19–26. [CrossRef]
6. Wood, R.J.K. Marine wear and tribocorrosion. *Wear* **2017**, *376–377*, 893–910. [CrossRef]
7. Chen, Z.; He, X.; Xiao, C.; Kim, S. Effect of humidity on friction and wear—A critical review. *Lubricants* **2018**, *6*, 74. [CrossRef]
8. Drach, A.; Tsukrov, I.; DeCew, J.; Aufrecht, J.; Grohbauer, A.; Hofmann, U. Field studies of corrosion behaviour of copper alloys in natural seawater. *Corros. Sci.* **2013**, *76*, 453–464. [CrossRef]
9. Lv, W.Y.; Zheng, K.Q.; Wang, Y.B.; Liu, Z.W. Influence of electric current on the corrosion behavior of Cu for the electric contact in the isolation switch. *Mater. Corros.* **2019**, *70*, 1900–1906. [CrossRef]
10. Rohatgi, P.K.; Ray, S.; Liu, Y. Tribological properties of metal matrix graphite particle composites. *Int. Mater. Rev.* **1992**, *37*, 129–149. [CrossRef]
11. Bai, L.; Ge, Y.; Zhu, L.; Chen, Y.; Yi, M. Preparation and properties of copper-plated expanded graphite/copper composites. *Tribol. Int.* **2021**, *161*, 107094.
12. Song, Y.; Yang, Z.; Jiao, J.; Li, W.; Zhao, M.; Shangguan, B.; Zhang, Y. Research on damage optimization of copper-carbon gradient composite materials for current-carrying friction. *Tribol. Int.* **2024**, *195*, 109635. [CrossRef]
13. Opalek, A.; Emmer, S.; Cicka, R.; Beronska, N.; Oslanec, P., Jr.; Kovacik, J. Structure and thermal expansion of Cu-90 vol. % graphite composites. *Materials* **2021**, *14*, 7089. [CrossRef] [PubMed]
14. Kumar, M.; Ankit; Majhi, M.R.; Chaturvedi, R.K. Wear and friction behaviour of Cu-Gr-B_4C composites and correlating to the topographical characteristics. *Surf. Topogr. Metrol. Prop.* **2024**, *12*, 015012.
15. Li, H.; Liu, Y.; Zheng, B.; Wang, S.; Yi, Y.; Zhang, Y.; Li, W. On the tribological behaviors of Cu matrix composites with different Cu-coated graphite content. *J. Mater. Res. Technol.* **2023**, *25*, 83–94. [CrossRef]
16. Liu, X.L.; Cai, Z.B.; Xiao, Q.; Shen, M.X.; Yang, W.B.; Chen, D.Y. Fretting wear behavior of brass/copper-graphite composites as a contactor material under electrical contact. *Int. J. Mech. Sci.* **2020**, *184*, 105703.
17. Yang, Z.; Ge, Y.; Xu, Z.; Bao, S.; Zhang, Y.; Zhang, J. Effect of carbon content on friction and wear properties of copper matrix composites at high speed current-carrying. *Materials* **2019**, *12*, 2881. [CrossRef]
18. Yang, Z.; Ge, Y.; Zhang, X.; Bao, S.; Zhang, Y.; Wang, Y. Effect of particle size on current-carrying friction and wear properties of copper-graphite composites by spark plasma sintering. *Materials* **2019**, *12*, 2825. [CrossRef]
19. Grandin, M.; Wiklund, U. Wear phenomena and tribofilm formation of copper/copper-graphite sliding electrical contact materials. *Wear* **2018**, *398*, 227–235.

20. Xu, E.; Huang, J.; Li, Y.; Zhu, Z.; Cheng, M.; Li, D.; Zhong, H.; Liu, J.; Jiang, Y. Graphite cluster/copper-based powder metallurgy composite for pantograph slider with well-behaved mechanical and wear performance. *Powder Technol.* **2019**, *344*, 551–560. [CrossRef]
21. Wang, C.; Xiao, J.K.; Xiao, S.X.; Xu, G.M.; Chen, J.; Zhang, C. Effect of electrical current on the tribological property of Cu-graphite brush. *Tribol. Lett.* **2024**, *72*, 29.
22. Gan, Z.Y.; Liu, Y.; Lü, X.Q.; Tan, Y.N.; Zou, J.P. Lubrication mechanism of copper-based solid self-lubricating materials by lead. *Mater. Sci. Eng. Powder Metall.* **2018**, *23*, 347–353.
23. Li, Y.S.; Zhang, K.; Zhang, G.H.; Xu, C.F.; Hu, Z.; Wang, Y.F. Heat resistance and friction properties of copper-based friction materials with different graphite content. *Powder Metall. Ind.* **2023**, *33*, 68–73.
24. Kumar, D.; Bharti, A.; Azam, S.M.; Kumar, N.; Tripathi, H. Investigations of mechanical properties of copper matrix hybrid composite. In *Advances in Mechanical Engineering, International Conference on Recent Innovations and Developments in Mechanical Engineering (ICRIDME), Shillong, India, 8–10 November 2018*; Springer: Singapore; Natl. Inst. Technol. Meghalaya: Shillong, India, 2020; pp. 671–676.
25. Hu, J.; Li, H.; Li, J.; Huang, J.; Kong, J.; Zhu, H.; Xiong, D. Structure, mechanical and tribological properties of TaCx composite films with different graphite powers. *J. Alloys Compd.* **2020**, *832*, 153769.
26. Escherova, J.; Krbata, M.; Kohutiar, M.; Barenyi, I.; Chochlikova, H.; Eckert, M.; Jus, M.; Majersky, J.; Janik, R.; Dubcova, P. The influence of Q & T heat treatment on the change of tribological properties of powder tool steels ASP2017, ASP2055 and their comparison with steel X153CrMoV12. *Materials* **2024**, *17*, 974. [CrossRef]
27. Kohutiar, M.; Krbata, M.; Escherova, J.; Eckert, M.; Mikus, P.; Jus, M.; Polasek, M.; Janik, R.; Dubec, A. The influence of the geometry of movement during the friction process on the change in the tribological properties of 30CrNiMo8 steel in contact with a G40 steel ball. *Materials* **2023**, *17*, 127. [CrossRef]
28. Shen, G.; Zhang, J.; Culliton, D.; Melentiev, R.; Fang, F. Tribological study on the surface modification of metal-on-polymer bioimplants. *Front. Mech. Eng.* **2022**, *17*, 26.

MDPI AG
Grosspeteranlage 5
4052 Basel
Switzerland
Tel.: +41 61 683 77 34

Materials Editorial Office
E-mail: materials@mdpi.com
www.mdpi.com/journal/materials

www.ingramcontent.com/pod-product-compliance
Lightning Source LLC
Chambersburg PA
CBHW061608210926
51850CB00009B/2257
9783725852970